# Electromechanical and Electronic Controls for HVAC/R

*Billy C. Langley*

Prentice Hall
*Upper Saddle River, New Jersey     Columbus, Ohio*

**Library of Congress Cataloging-in-Publication Data**
Langley, Billy C., 1931–
    Electromechanical and electronic controls for HVAC/R/Billy C.
Langley.
        p.    cm.
    Includes index.
    ISBN 0-13-907569-0
    1. Refrigeration and refrigerating machinery—Automatic control.
2. Air conditioning—Control.   3. Heating—Control.   I. Title.
TP492.7.L325   1999
621.5'6—dc21                                                    98-5682
                                                                CIP


Cover art: Billy C. Langley
Editor: Ed Francis
Production Editor: Stephen C. Robb
Design Coordinator: Karrie M. Converse
Text Designer: Carlisle Publishers Services
Cover Designer: Rod Harris
Production Manager: Patricia A. Tonneman
Production Supervision: Carlisle Publishers Services
Marketing Manager: Danny Hoyt

This book was set in Univers and Times Roman by Carlisle Communications, Ltd. and was printed
and bound by Quebecor Printing/Book Press. The cover was printed by Phoenix Color Corp.

 © 1999 by Prentice-Hall, Inc.
Simon & Schuster/A Viacom Company
Upper Saddle River, New Jersey 07458

Earlier editions, entitled *Electric Controls for Refrigeration and Air-Conditioning,* ©1988, 1974 by
Prentice-Hall, Inc.

Printed in the United States of America

10  9  8  7  6  5  4  3  2  1

ISBN: 0-13-907569-0

Prentice-Hall International (UK) Limited, *London*
Prentice-Hall of Australia Pty. Limited, *Sydney*
Prentice-Hall Canada, Inc., *Toronto*
Prentice-Hall Hispanoamericana, S.A., *Mexico*
Prentice-Hall of India Private Limited, *New Delhi*
Prentice-Hall of Japan, Inc., *Tokyo*
Simon & Schuster Asia Pte. Ltd., *Singapore*
Editora Prentice-Hall do Brasil, Ltda., *Rio de Janeiro*

# Preface

This book evolved from an earlier Prentice Hall text that I wrote, *Electric Controls for Refrigeration and Air-Conditioning*. The most obvious change—the new title, *Electromechanical and Electronic Controls for HVAC/R*—more precisely reflects the content of the book. Even more important is the book's new format, specifically, the use of units rather than chapters. Each unit discusses *only one* type of control, so each subject is presented in a concise, readable, and easily understood manner.

This book is dedicated to the operation, troubleshooting, and repair of both electric and electronic controls. It introduces a method of checking and troubleshooting a control when a problem occurs. The electrical and electronic theory has been omitted so that more types of controls can be presented.

The operation of each of the controls presented is covered to ensure that the different uses of each control are easily understood. Troubleshooting problems are presented at the end of each unit so the reader can immediately put the knowledge gained to practical use. Questions are also included to test the reader's comprehension. The language in the text closely relates to everyday conversation but still keeps the jargon that is used in the refrigeration and air-conditioning field.

Every attempt has been made to cover the latest materials and devices used in the field today; however, it is impossible to include all the controls in *any* book because so many manufacturers are developing new controls almost daily, especially in the electronics area.

Students who learn the control applications, uses, and troubleshooting will always be in demand. This book will certainly impart knowledge that will give readers a boost on the journey into the refrigeration and air-conditioning industry. Technicians should also be aware that it takes constant study and review to keep *current* with HVAC/R controls and control systems as they are changed or improved. Remember, the more you study and learn, the more valuable you will be to yourself and to your employer.

# Acknowledgments

The author gratefully acknowledges the following reviewers for their insightful suggestions: W. Del Winston, Northland Career Center; Norm Christopherson, San Jose City College; Bennie Barnes, Live Oaks Vocational Technical Center; Craig Barnett, Los Angeles Trade Technical College.

*Billy C. Langley*

# Contents

**Section 1: Types of Control Systems   1**
    Unit 1: Electric Control Systems   1
    Unit 2: Self-Contained Control Systems   4
    Unit 3: Control System Components   6
    Unit 4: A Simple Control System   11
    Unit 5: Control Modes   13

**Section 2: Basic Control Theory   19**
    Unit 6: Controlled System Characteristics and Elements   19
    Unit 7: Fundamentals of Electric Control Circuits   24
    Unit 8: Classification of Electric Control Circuits   28
    Unit 9: ON-OFF Two-Position (Low-Voltage) Controls   34
    Unit 10: Two-Position, Spring Return Control Applications   38
    Unit 11: Floating Controls   45
    Unit 12: Unidirection Control Applications   51
    Unit 13: Line-Voltage Two-Position Control Applications   61
    Unit 14: Proportioning (Modulating) Control Applications   64

**Section 3: Electromechanical Controls   81**
    Unit 15: Transformers   81
    Unit 16: Contactors and Starters   87
    Unit 17: Electromagnetic Relays   100
    Unit 18: Thermal Relays   107
    Unit 19: Motor Starting Devices   112
    Unit 20: Compressor Motor Overloads   123
    Unit 21: Solenoid Valves   128
    Unit 22: Reversing (Four-Way) Valves   133
    Unit 23: Pressure Controls   140

Unit 24: Oil-Failure Controls    154
Unit 25: Temperature Controls    157
Unit 26: Humidistats    184
Unit 27: Airstats and Enthalpy Controllers    186
Unit 28: Gas Valves    188
Unit 29: Pilot Safety Devices    196
Unit 30: Fan Controls    201
Unit 31: Limit Controls    205
Unit 32: Combination Fan and Limit Controls    209
Unit 33: Oil Burner Controls    212
Unit 34: Hydronic Heating Controls    218
Unit 35: Hydronic Cooling Controls    231
Unit 36: Discharge Pressure Controls (Cooling Towers)    235
Unit 37: Air-Cooled Condenser Discharge Pressure Controls    240
Unit 38: Modulating Motors    247
Unit 39: Step Controllers    253
Unit 40: Timers and Time Clocks    256
Unit 41: Commercial Defrost Systems    271
Unit 42: Domestic Refrigeration Defrost Controls    279
Unit 43: Heat Pump Defrost Controls    281
Unit 44: Lockout Relay    288

## Section 4: Electronic (Solid-State) Controls    291

Unit 45: Electronic Thermostats    291
Unit 46: Solid-State Fan Controls (Condensers)    293
Unit 47: Solid-State Gas Burner Ignition Controls    296
Unit 48: Solid-State Oil Burner Controls    315
Unit 49: Discharge Gas Temperature Protector    320
Unit 50: Delay-on-Make Adjustable Solid-State Timers    323
Unit 51: Delay-on-Break Adjustable Solid-State Timers    327
Unit 52: Off-Delay/On-Delay Adjustable Timers    330
Unit 53: Bypass Timers    333
Unit 54: Brownout Protectors and Low-Voltage Monitors    336
Unit 55: Line Monitors    340
Unit 56: Plug-In Time-Delay Relays    342
Unit 57: Current-Sensing Relays    346
Unit 58: Solid-State Motor Starting Devices    347
Unit 59: Solid-State Heat Pump Defrost Controls    353
Unit 60: Solid-State Motor Protectors    355
Unit 61: Solid-State Heat Pump Controls    358
Unit 62: Solid-State Humidistats    360
Unit 63: Solid-State Crankcase Heaters    362
Unit 64: Economizer Control Package    364
Unit 65: Integrated Furnace Control Modules    365

**Section 5: Six-Step Troubleshooting Procedure   367**

**Laboratory Workbook   379**
    Laboratory 1: Control System Components   379
    Laboratory 2: Fundamentals of Electric Control Circuits   381
    Laboratory 3: Starters and Contactors   383
    Laboratory 4: Electromagnetic Relays   385
    Laboratory 5: Thermal Relays   387
    Laboratory 6: Potential Starting Relays   389
    Laboratory 7: Sizing Potential Starting Relays   392
    Laboratory 8: Pressure Controls   394
    Laboratory 9: Temperature Controls   396
    Laboratory 10: Fan Controls   399
    Laboratory 11: Limit Controls   401
    Laboratory 12: Discharge Gas Temperature Protector   403
    Laboratory 13: Delay-on-Make Adjustable Solid-State Timers   405
    Laboratory 14: Current-Sensing Relay   408
    Laboratory 15: Solid-State Crankcase Heaters   410

**Glossary   413**
**Index   433**

# 1 Types of Control Systems

## Introduction

Control systems are usually rated according to their source of power. The controls used to make up a control system are either electromechanical, self-contained, or electronic. Each is discussed in the following sections.

## Unit 1: ELECTRIC CONTROL SYSTEMS

### Introduction

Obviously, electricity is the power source for electric control systems. These devices are wired directly to the power source, which supplies either line-voltage or low-voltage current, as follows:

- *Line-voltage controls.* Line-voltage controls can be used on either 120- or 240-volt supply, with the control circuit itself powered either by 120 or 240 VAC.
- *Low-voltage controls.* Low-voltage controls can operate only on 24 VAC. A step-down transformer is used to reduce 120 or 240 VAC line current to the appropriate lower voltage.

Line voltage in either control system can be 120-VAC single-phase, 240-VAC single-phase, or 240-VAC three-phase. Single-phase electricity is commonly used to power the transformer in low-voltage systems.

A line-voltage thermostat signals a heating or cooling system when to run—when either heating or cooling is wanted—and when to stop running when the room thermostat is satisfied. See Figure 1–1.

1

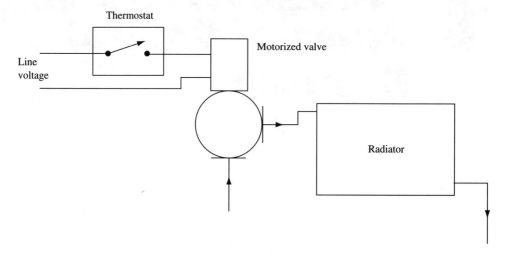

**Figure 1–1**
Line-voltage two-position control.

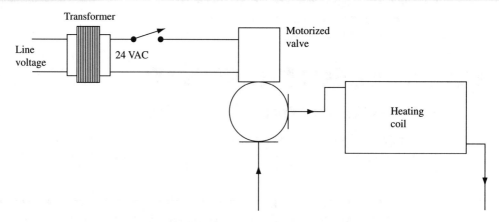

**Figure 1–2**
Low-voltage two-position control.

A simple use of a low-voltage electric control would be a thermostat controlling the low voltage supplied by a transformer. See Figure 1–2.

Twenty-four VAC systems are popular for several reasons. Low voltage is safer for the user and the building; smaller controls have less mass and react faster to temperature and pressure changes; and the controls are less expensive than line-voltage controls. With exception of the transformer, this system operates the same as the line-voltage system previously described.

# Summary

- Control systems are electromechanical, electronic, and self-contained.
- The controls in an electric control system are connected directly to the power source.
- The purpose of the transformer is to reduce the 120- or 240-voltage supply to 24 volts.

# Service Calls

*Service Call 1.*    A customer complains that his air-conditioning system is not working. A check of the system reveals that it uses single-phase 240 VAC electricity and a 24 VAC control circuit. The technician tests the control circuit voltage at the transformer secondary terminals. The technician finds none. The technician then tests the primary terminals for voltage and finds none. A check of the circuit breakers finds them tripped. When the technician resets the breakers, the system starts running. The technician then checks the voltage at the compressor terminals while the compressor is running and finds it to be 210 volts. A check of the voltage at the line side of the contactor shows 210 volts. The voltage at the circuit breaker box is also at 210 volts. The unit is drawing too much current compared with the nameplate rating. The technician notifies the power company about the low voltage at the job site. After the power company repairs the problem, the voltage at the compressor terminals while running is 240 volts. The amperage draw is now within the nameplate rating and the technician is now satisfied that the system is repaired.

*Service Call 2.*    A customer complains that her residential air-conditioning system is not working. A check of the system reveals that it is a 240 VAC single-phase system using a 24 VAC control system. Nothing is running. The technician tests the voltage at the transformer secondary terminals and finds none. Voltage at the transformer primary terminals tests at 240 VAC. The transformer is bad. The technician replaces it with the proper size and then places the unit back in operation. The amperage draw in the secondary control circuit is now within the transformer VA rating. The technician is satisfied that the system is operating properly.

# Student Troubleshooting Problem

A customer complains that his residential air-conditioning system is not cooling. A check of the system reveals that nothing is running. The customer says that the transformer was replaced only a few days earlier. The unit is using 240 VAC, single-phase electricity. The secondary terminals of the transformer show no voltage when checked. The primary terminals show voltage when checked. The transformer is bad. The present transformer is rated for 18 VA. What is the likely cause of the problem?

# Questions

1. Name the three general types of control systems.
2. How are control systems rated?
3. Why are transformers used in air-conditioning systems?

# Unit 2: SELF-CONTAINED CONTROL SYSTEMS

## Introduction

Self-contained controls use an enclosed sensing element as their power source. The final control mechanism, such as a valve, and the sensing element are combined in a single unit. Self-contained controls include a bellows and a bulb connected by a length of capillary tubing that contains a charge of some type of liquid or vapor whose volume changes as the room temperature changes. See Figure 2–1.

A rise in the temperature of the sensing element will cause the fluid inside to swell with enough force to open the valve. A drop in temperature of the sensing element will cause the liquid or vapor to shrink and allow the valve to close. This may be a snap-acting system. As the pressure inside the sensing element changes, the valve or final controls element will either open or close as the temperature demands. See Figure 2–2. The purpose of the valve is to control some type of fluid, such as steam or hot water, flowing through the heating coil or gas entering the combustion chamber of a gas-fired furnace. It is directly operated by the changes in pressure inside the bellows as the temperature of the surrounding medium (room air) changes.

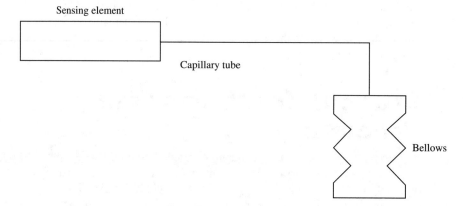

**Figure 2–1**
Self-contained element.

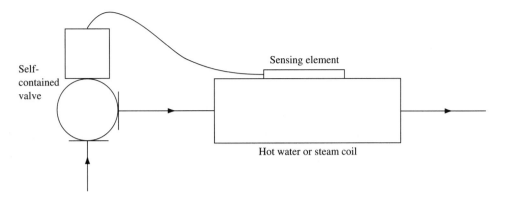

**Figure 2–2**
Self-contained operating system.

# Summary

- Self-contained controls use an enclosed sensing element as their source of power.
- Either a liquid or a vapor inside the sensing element causes the control to function.
- When the temperature surrounding the sensing element changes, the pressure inside changes to operate the control.

# Service Calls

A customer complains that his hot water heating system is not heating the building. A check of the system reveals that the space temperature is too cool. The system is controlled by a self-contained water-regulating valve. The technician checks the hot water coil and finds it cooler than normal. A check of the hot water valve shows that it is not fully open as it should be under these conditions, and the sensing element (bulb) is located correctly in the return air stream. See Figure 2–3. The technician then places the bulb in cold water to see if it will open further, but it will not, so he replaces the valve. The system is now placed back in operation. The coil is now full of hot water and the system is heating the building. The technician feels satisfied that the system is repaired.

# Student Troubleshooting Problem

A customer complains that the heating system in an office building is getting too warm. A check of the system reveals that the hot water flow is controlled by a self-contained valve, and a full flow of hot water is directed into the heating coil. What could be causing the problem?

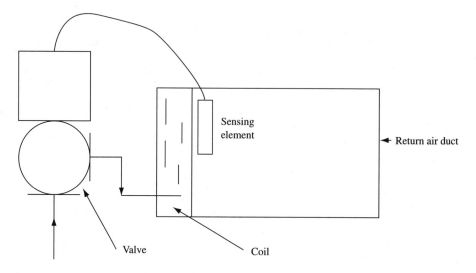

**Figure 2–3**
Placement of sensing element of a self-contained control system.

## Questions

1. How do self-contained controls operate?
2. Of what does a self-contained control usually consist?

# Unit 3: CONTROL SYSTEM COMPONENTS

## Introduction

A control system can be shown by a simple block diagram. See Figure 3–1. Every block in this diagram represents an essential component of the control system.

**The controls process.**    The control system enables the equipment to operate as it was designed. When the equipment and the control system are working together correctly the space will be kept at the wanted conditions. One example is when the control system is keeping the temperature in a hospital operating room steady by controlling the workings of a steam boiler.

**Disturbance sensing element.**    Anything that will affect the operation of the control circuit is known as an external influence. This external influence is the *disturbance.* A disturbance may be caused by changing the thermostat or other control,

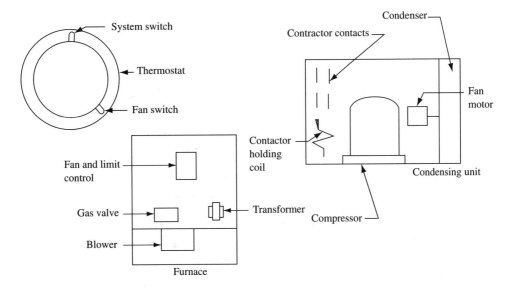

**Figure 3–1**
Simple air conditioning with control elements labeled.

temperature, or pressure setting, which is known as the *set point*. The following variables will affect system operation: the supply, such as an air-conditioning or refrigeration system that has cooled the space to the wanted temperature; space demand, such as an air-conditioning or refrigeration system that has just started and the space has not yet reached the wanted temperature; when the demand on the system is high; the environment, both inside and outside the space; when the air outside the space is blowing and hot, the space will need more from the unit than it would on a warm, calm day; and the opening of a door that would let outside air blow on the thermostat and cause the heating or cooling unit to come back on sooner than normal.

Any disturbance (change) present will cause a change in the controlled variable. The controlled variable is the quantity or the condition being controlled, such as the air temperature inside a conditioned space. The controlled variable is therefore a quality of the controlled medium. The controlled medium is the substance being controlled. For example, when controlling the temperature of a space, the controlled variable is the air temperature and the controlled medium is the air itself. See Table 3–1.

For a disturbance (change) to be corrected, it must first be detected and measured, which is the job of the control system. The control system must know how much change is needed in the controlled variable (air temperature) to correct the disturbance (change). Detection and measuring are the responsibility of the disturbance sensing element, a thermostat, or pressure sensing control. Because it is not possible for a human to know the pressure inside a pipe by looking at the outside of it or to know the temperature inside a combustion chamber by looking at the color of the flame, we must rely on disturbance sensing elements. Many types of transducers (devices that change one type of energy to

**TABLE 3–1**  EXAMPLES OF CONTROLLED MEDIUM
AND CONTROLLED VARIABLE

| Controlled medium | Controlled variable | |
|---|---|---|
| Air | Temperature | °F |
|  | Flow | CFM |
|  | Humidity | % |
| Water | Temperature | °F |
|  | Flow | GPM |
|  | Pressure | PSIG |

another) are used to determine when and what type of disturbance is present and send this information to the control system. For example, a temperature sensor will sense the surrounding temperature and send this signal to the control system. Some of the more common types of disturbance sensing elements that give measurement information are pressure sensing (refrigerant pressure controls), temperature sensing (a thermostat), and humidity sensing (humidistats).

**Controller.**   The controller, such as an air-conditioning thermostat, is the control that receives the measured information (temperature) from the disturbance sensing element (thermostat bimetal) and interprets it to decide what is needed to keep the desired space conditions. The disturbance sensing element will often be a part of the controller mechanism. In a thermostat this would be the bimetal that moves the control contacts. The temperature lever placed at the wanted temperature creates the set point of a thermostat. When the space conditions match the set point, the conditioning process is at work. However, when space conditions are outside the set point differential, the controller signals the equipment to begin operating to bring the space conditions to the desired temperature.

In some cases the set point may be manually set, such as the temperature setting of a thermostat, and will stay at this setting until it is manually changed. In some situations the set point may be changed automatically, such as in a boiler system where the steam needs are sent to the fuel input. The combustion airflow is then changed in response to the steam needs. This constantly changing condition needs continuous and automatic control.

**Final control element.**   The final control element (the refrigerant flow control device) changes the value of the manipulated variable (the refrigerant) in response to a signal from the thermostat or other controller. The manipulated variable is the refrigerant flowing through the evaporator and thus causing a change in air temperature. For example, refrigerant flows through a coil to cool the air that cools the space; the refrigerant is the control agent, and the amount of refrigerant flowing through the coil is the manipulated variable. The flow of refrigerant is changed in response to any of the following changes: the room temperature, the controlled variable, the air, or the controlled medium in the room or space.

The final control element is the device that causes the manipulated variable (the refrigerant flow) and the control agent (the refrigerant) to make a correction in system operation. In this case, the control system is an error-sensitive, self-correcting system, usually called a closed-loop feedback system.

# Summary

- The control process is probably the most important requirement of any control system.
- The process can be a steady temperature such as in a hospital operating room by controlling a steam boiler.
- Any disturbance present will cause a change in the controlled variable.
- For a disturbance to be corrected, it must first be detected and measured.
- Common types of disturbance sensing elements that provide measurement information are pressure sensing, temperature sensing, and humidity sensing.
- The controller is the device that receives the measured information from the disturbance sensors and interprets it to decide how the process is going.
- A final control element is a device that changes the value of the manipulated variable in response to the error signal from the controller.
- The final control element is the device that causes the manipulated variable and the control agent to make a corrective action on a process operation.
- The control system is an error-sensitive, self-correcting system, usually called a closed-loop feedback system.

# Service Calls

*Service Call 1.*    A customer complains that her new system has not properly heated or cooled since it was installed. A check of the system reveals that it is a 240 VAC, single-phase system. Everything is running. The technician checks the system pressures, voltage to the unit, the amperage draw of the unit, the refrigerant charge, and the air temperature differential across the evaporator. All are within the normal operating range. The system is operating correctly. The technician then checks the thermostat and finds it located where it cannot sense the average return air temperature. The technician relocates the thermostat where it will sense the average return air temperature. The system is now cooling the space. The technician feels satisfied that the system is repaired.

*Service Call 2.*    A customer complains that a chilled water system is not cooling properly. See Figure 3–2. A check of the system reveals that the chiller is frozen. After de-icing the chiller, the technician checks the system to ensure that it is fully charged with refrigerant which would show that the chiller tubes are not broken. The technician checks further to find the cause of the frozen chiller. The sensing element for the chilled water thermostat is out of its bulb well and not sensing the water temperature at the chiller outlet. The

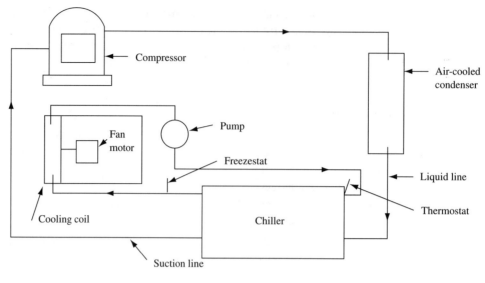

**Figure 3–2**
Simple chilled water system.

technician then places the bulb back in the well and seals it with the proper sealant. The compressor now cycles at the correct water temperature. The system is now operating properly. The technician feels satisfied that the system is repaired.

# Student Troubleshooting Problem

A customer complains that an older residential air-conditioning system is not cooling properly. A check of the system reveals that the suction line is sweating too much and it is also cold to the touch. The system has an air-cooled condenser and a thermostatic expansion valve. The inside return air temperature is 85°F. The outdoor temperature is 100°F. The system is charged with HCFC-22. The suction pressure is 320 psig. What could be the problem?

# Questions

1. To what is the conditioning process always subjected?
2. What factor cannot be changed by the system?
3. What is the condition being controlled known as?
4. What is a characteristic of a controlled medium?
5. What must happen before a disturbance is corrected by the system?

6. What are the different sensors known as?
7. In a mercury bulb type thermostat, what would sense a disturbance?
8. When the surrounding conditions and the controller set point are not the same, what occurs?
9. In a steam heating system, what is the control agent?
10. What is a closed-loop feedback control system?

# Unit 4:  A SIMPLE CONTROL SYSTEM
# Introduction

The parts of a control system may be shown with a single-stage residential heating or cooling system. See Figure 4–1. Refer to the block wiring diagram in Figure 3–1 to help in comparing the two illustrations.

The purpose of the control system is to keep a specific temperature inside the space. The space temperature is the controlled variable. The air in the space is the controlled medium. A disturbance (change) occurs when the space temperature rises above or falls below the thermostat set point. The thermostat bimetal senses this change in the air (the disturbance sensing element). The change in temperature is then sent to the cooling or heating equipment. The temperature level is set on the thermostat, which changes the set point. The thermostat, or the controller, then sends a signal to the compressor contactor, or the gas ignition system, which is the final control element. The volume of refrigerant or gas flow, which is the manipulated variable, is thus changed by the action of the flow control device, or the gas valve, to make the needed change in the refrigerant or gas flow, which is the controlled variable. In this particular system, either the refrigerant or the gas is the control agent and the controlled medium.

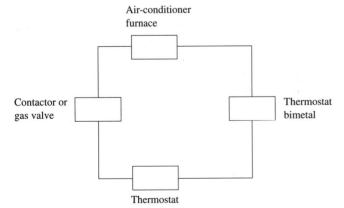

**Figure 4–1**
Single-stage residential cooling system.

# Summary

- In an air-conditioning system the purpose of a control system is to keep a definite temperature inside the space.
- The temperature level is the controlled medium.
- A disturbance occurs when the space air temperature changes to vary the amount of refrigerant, or gas, flowing into the equipment.
- The thermostat bimetal is the disturbance sensing element.
- Moving the temperature lever is an example of changing the set point of the controller.
- The volume of refrigerant, or gas, flowing is the manipulated variable.
- The refrigerant, or gas, itself is both the control agent and the controlled medium.

# Service Call

A customer complains that his refrigeration system keeps going on and off and is not cooling the fixture. A check of the system reveals that the head pressure is 350 psig for an air-cooled HCFC-22 system. It is cycling on the head pressure control. The technician checks the condenser coil and finds it full of dirt and debris. The coil is not cooling and condensing the refrigerant as it should, so the technician cleans the dirt and debris from the condenser coil. He then starts the unit and allows it to run until the head pressure is steady. The head pressure is now 260 psig, which is within the needed range for proper operation for HCFC-22. The system is now operating and cooling down the fixture, and the technician feels satisfied that the system is repaired.

# Student Troubleshooting Problem

A customer complains that a residential air-conditioning system will get too cool before turning off and too warm before coming back on. A check of the system reveals that it is an air-cooled unit operating on 240 VAC and is charged with HCFC-22. The return air temperature is 80°F. The outdoor temperature is 95°F. The suction pressure is 70 psig. The discharge pressure is 280 psig. The technician places a recording thermometer in the return air of the system and leaves it there until the next day for a recheck. The recorder shows that the temperature goes from 80°F to 70°F and back to 80°F. The thermometer shows that this is a steady cycle. What could be the problem?

# Questions

1. What is the purpose of a control system?
2. In Figure 4–1, what is the water level called in a control system?
3. What is present when a position of the output signal changes in a control system?

4. What detects a change in the controlled variable in a control system?

5. How is the set point of a controller changed?

# Unit 5: CONTROL MODES
# Introduction

A control mode is the method used by a control system to make the wanted changes in response to a disturbance. The control mode causes the final control element to respond to the signal from the disturbance sensing element. When the thermostat is properly matched to the wanted process, the overall performance of the control system will be satisfactory. The four basic control types are ON-OFF (two position), multiposition (multistage), floating, and proportioning (modulating).

**ON-OFF (two-position) control.**    Two-position controls are used only for ON-OFF operation of the equipment. The control will either make or break the control circuit, as there are no in-between positions available. When enough disturbance (change) happens in the controlled variable (the space temperature) from the thermostat set point, the final control element will move either to the full open or full closed position, depending on whether heating or cooling is being used. The amount of time it takes the control to move between the ON and OFF positions will change in response to the needs of the space. See Figure 5–1.

When the space temperature rises to the ON temperature, the thermostat sends a signal to the contactor to start the cooling unit. The compressor will remain on until the space temperature drops to the thermostat temperature OFF setting. The thermostat then signals the contactor to stop cooling. The system will remain off until the space temperature again rises to the thermostat ON setting.

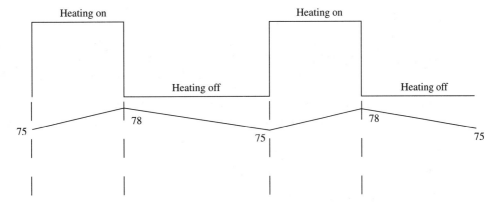

**Figure 5–1**
Temperature changes in an ON-OFF (two-position) mode of system control.

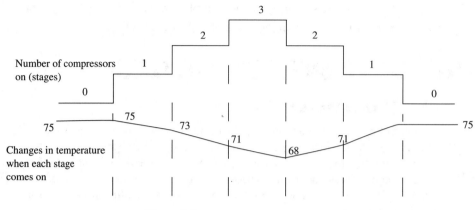

**Figure 5–2**
Multiposition mode of system control.

The ON-OFF type control system is the simplest. It also has some definite disadvantages. It will allow the space temperature to vary over a temperature differential rather than settling down to an almost steady temperature. This differential is sometimes quite wide. If it is too narrow, the controller will be worn out by the constant switching on and off. The differential is the space between the ON and OFF positions of the controller (thermostat).

**Multiposition (multistage) control.**    The multiposition control system is an extension of the ON-OFF system. It has two or more stages. In installations where the range between the ON and OFF positions is too wide to give the wanted operation, multiple stages with smaller differentials are used. The range is the operating limits of the control. Each of these independent stages will also have the ON-OFF operation, but many positions are available by simply adding more switches, thus resulting in a steplike operation. The more stages used, the smoother the temperature will be. As the load increases, more steps turn on. As the load decreases, more steps turn off. In most cases, the multiposition control will provide from two to ten operating stages. See Figure 5–2.

In Figure 5–2, when the thermostat is satisfied, no stages (compressors) are on. As the space temperature rises to the first-stage ON setting, stage 1 (compressor 1) will start. If the space temperature continues to rise, stage 2 (compressor 2) will start. This will continue until either all the stages are on or enough cooling occurs that no more stages (compressors) are needed. As the space temperature starts dropping, the last stage to come on will be the first to go off. Each stage will stop a compressor as long as the space temperature is dropping; however, if the space temperature starts rising again, the action of the controller will be reversed to bring on more stages (compressors). This process will continue until the space temperature drops to the OFF setting of the room thermostat.

**Floating control.**    The floating control is a completely different type of control from the two-position or the multiposition control because the final control element can

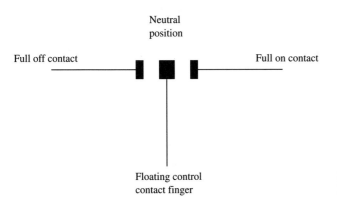

**Figure 5–3**
Neutral position of floating controller.

assume any position between its ON and OFF positions. The control (thermostat) has only two positions with a neutral zone in between. See Figure 5–3.

When the controller is in the neutral zone, the final control element (staging motor) will not move from its position. When the temperature change is large enough, the thermostat will move to one of its positions. This action causes the final control element to move in one of the two directions in response to the position of the thermostat contacts. The final control element makes this change at a constant speed. Therefore, this mode is commonly called a single-speed floating control. As the controller moves into the neutral zone, the final control element will stop. It will remain in this position until the thermostat again moves close to one of its two sets of contacts. This is called a floating mode because the final control element stops when the thermostat is floating between its two sets of contacts.

**Proportioning (modulating) control.**    In proportioning control systems, as in the floating type, the final control element can take any position between the two extreme positions of the thermostat. These systems are different from the floating control, because they have no neutral zone. Therefore, a temperature change, no matter how small, will cause the thermostat to move. Every movement of the thermostat is accompanied by an exact amount of movement for the final control element (water or steam valve, compressor unloader, compressor contactor, etc.). The movements are made as often as the temperature changes. The proportioning control causes a relationship between the disturbance (temperature change), the controller (thermostat) action, and the position of the final control element. See Figure 5–4.

**Complex variations of the proportioning control.**    When there is a large change in the load, a difference may occur between the temperature of the controlled variable and the thermostat set point. A constant change of the set point is needed to keep the controlled variable at the same temperature through the complete load range of the equipment.

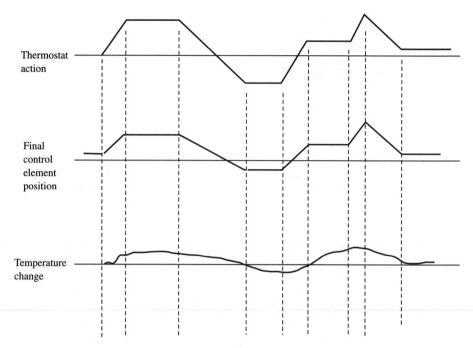

**Figure 5–4**
Proportioning (modulating) mode of system control.

# Summary

- A two-position control system provides either ON or OFF operation.
- A control mode is the method used by a control system to make the wanted changes in response to a disturbance. It is generally, therefore, a function of the controller itself.
- The four basic control types are ON-OFF (two position), multiposition (multistage), floating, and proportioning (modulating).
- The ON-OFF control system is either fully on or fully off.
- Time between the ON and OFF position will vary according to the needs of the load.
- The multiposition control system has two or more stages.
- Each of these stages will have an ON and OFF position, but many functions are available by simply adding more switches.
- The floating control system can take any position between fully ON and fully OFF.
- As the thermostat moves into its neutral zone, the final control element will stop at that position. This is called the floating mode because the final control element can take any position between the fully ON and fully OFF positions of the thermostat.
- In the proportioning control system, the final control element can assume any position between the fully ON and the fully OFF positions of the thermostat.

- Proportioning type controllers have no neutral zone.
- Proportioning type control causes a linear relationship between the temperature difference, the amount of thermostat action, and the position of the final control element.

# Service Calls

*Service Call 1.*   A customer complains that his residential air-conditioning system is either too cool or too warm. A check of the system finds it to be operating properly. The system is controlled by an ON-OFF thermostat. Using an accurate thermometer the technician determines that the thermostat has a differential of about 12°F (too large for the customer's comfort), so the technician replaces the thermostat. After the temperature of the thermostat has settled down, the system operates satisfactorily. The technician feels satisfied that the system is repaired.

*Service Call 2.*   A customer complains that the air-conditioning unit in the offices of a commercial building is not cooling. A check of the system reveals that the unit is controlled by a floating type control system. The offices are warmer than desired. The technician checks the refrigeration system and finds it operating properly. A check of the control system discovers that the damper motor is not working. The technician checks the resistance of the motor potentiometer and finds it has infinite resistance from the center connection to one of its other two terminals. He then replaces the motor. The technician then places the system back in operation, checks the system, and sees that the controls are working correctly. The technician feels satisfied that the system is repaired.

# Student Troubleshooting Problem

A customer complains that a restaurant air-conditioning system will not get cool enough. A check of the system finds that the unit is controlled by a floating type motor with stages. Only about half of the stages are turned on. The thermostat is turned up, but the motor still does not move. The technician checks the voltage at the motor terminals and finds it to be 24 VAC. What could be the problem?

# Questions

1. What is the name for response to a change in a control system?
2. What determines the overall performance of the control system?
3. Name the four basic control types.
4. Which type of control system has no in-between positions?
5. Which type of control system allows a change in the space temperature?
6. To which control system are switches added to provide the needed operation?
7. How is smoother operation gained in the multiposition control system?

8. In what type of control system can the final control element take any position between its two extremes?

9. Why is the floating control system known as the single-speed control system?

10. Why is the proportioning control system different from the floating control system?

11. In the proportioning control system, why does the damper motor move each time the thermostat sensing element moves?

12. Why is a constant changing of the set point required when an offset control is used?

13. What type of control system will stop only when the error signal becomes zero?

14. In what type of control system is the error signal from the thermostat changed in proportion to the rate of the disturbance?

# 2 Basic Control Theory

## Introduction

Automatic control systems vary from simple domestic refrigerator temperature controls to the accurate control of complicated industrial processes. Therefore, automatic control systems can be used anywhere that a changing condition needs to be controlled. The changing condition may be temperature, pressure, humidity, or the flow rate of a substance such as air motion or water flow through a pipe.

The most important consideration in controlling these conditions is how the control operates, the controlled variable, and the controlling devices.

## Unit 6: CONTROLLED SYSTEM CHARACTERISTICS AND ELEMENTS

### Introduction

For an automatic control system to operate correctly, some variable must be controlled. Control of this first variable is usually accomplished through an automatic control system designed to control the second variable, which is known as the *manipulated variable,* such as the refrigerant in a refrigeration system. The manipulated variable is what causes the needed changes in the controlled variable, such as the space temperature.

**Controlled systems.** A controlled system is made up of all the equipment in which the controlled variable exists, except for the automatic control equipment. This would be the air-conditioning, heating, or refrigeration system and the ductwork and other air-handling equipment.

**Controlled variable.**    The controlled variable is that which is measured and controlled. See Figure 6–1. Thus, the controlled variable is the controlled medium. For example, when the air temperature is controlled, the controlled condition is the temperature and the air is the controlled medium. When considering an air-conditioning system, the temperature of the space is most important, whereas the air is simply the medium used to change the space temperature.

**Manipulated variable.**    A manipulated variable is that quantity of the condition being controlled by the automatic control system which causes the wanted change in the controlled variable. The manipulated variable is part of the control agent. As an example, consider a heat exchanger in a gas-fired furnace used to heat a space. A room thermostat can measure the temperature (the controlled variable) of the room air (the controlled medium) and operate the gas valve that regulates the gas flow (the manipulated variable) into the combustion chamber. The heat from the heat exchanger is used to heat the room air. See Figure 6–2.

**Control equipment.**    Following are several terms with which the technician must become familiar before our discussion about automatic control systems.

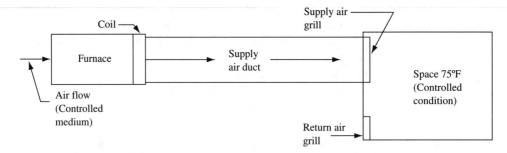

**Figure 6–1**
Controlled variable.

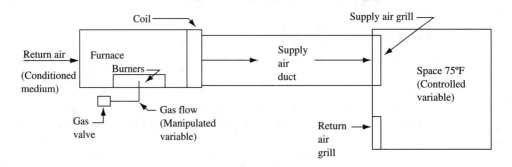

**Figure 6–2**
Manipulated variable.

*Set Point.*    The set point is the position on the controller at which the controller indicator is set; or, the temperature at which the thermostat is set.

*Control Point.*    The control point is the point of the controlled variable (the space air temperature) that the controller (thermostat) attempts to keep; or, the average temperature of the space air in an air-conditioning system.

*Desired Value.*    The desired value is the value of the controlled variable (space temperature) that is wanted or desired to maintain. This is the actual space temperature.

*Deviation.*    The deviation is the difference between the set point (the thermostat setting) and the value of the controlled variable (air temperature) at any instant in time; for example, the difference in temperature between the thermostat's actual setting and the actual space temperature.

*Corrective Action.*    The corrective action causes a change in the manipulated variable (airflow and temperature) and is started by a deviation (the difference between the thermostat setting and the air temperature). When the space temperature deviates from the thermostat temperature setting, it causes the equipment to run and bring the space temperature to the wanted value.

*Differential Gap.*    The differential gap is a term applied to two-position controllers (thermostats). It is the smallest range through which the controlled variable (space temperature) must pass while moving the final control element from fully open to fully closed. It is also called the *differential,* which is the difference between the ON and OFF points (temperature) of the controller.

*Proportional Band.*    The proportional band is the point of a positional controller (thermostat) through which the controlled variable (space temperature) must pass when moving the final control element through its full operating range. It may also be called the throttling range and modulating range.

*Cycling.*    Cycling, also known as hunting, is a repetitive change in the controlled variable (space temperature) from one point in the conditioning process to another. Cycling may include both the normal ON and OFF operation of a system, such as the compressor short cycles. The compressor cycling on the thermostat is normal operation.

*Offset.*    The offset is a kept variation between the point of the controlled variable (the space temperature) that matches the set point and the control point. An example is the temperature difference between the temperature setting of the thermostat and the actual space temperature.

*Lag.*    Lag is the delay in the effect of a changed condition at one point in the control system to another condition to which it may be related. For example, lag is the time passed between when the thermostat called for conditioning and when the conditioning was actually sensed by the thermostat. See Figure 6–3.

*Overshoot.*    Overshoot is the rise in temperature above the thermostat setting. An example is the amount of heat added to the space after the thermostat setting has been reached. See Figure 6–3.

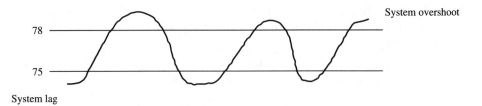

**Figure 6–3**
System lag and overshoot.

***Primary Element.***    The primary element is that part of the controller (thermostat) that uses any energy gained from the controlled medium first to cause a condition that represents the point of the controlled variable. For example, the bimetal element in a thermostat is a primary element. As the space temperature changes, the bimetal changes the position of the switch to either start or stop the equipment it is controlling.

***Final Control Element.***    The final control element is that part of the controller that directly acts to cause a change in the manipulated variable. Examples include the compressor contactor used in the control circuit of a refrigeration system or the flow regulating valve, or motor, in a hydronic heating and/or cooling system.

***In-Contacts and Out-Contacts.***    The in-contacts in a control system are those that are held closed when the relay coil is energized. The out-contacts are those that are closed when the relay coil is deenergized. In-contacts are NO when the relay coil is deenergized, whereas out-contacts are NC when the relay coil is energized.

***Balancing Relay.***    A balancing relay has a pivoted armature that swings between two electromagnetic coils. See Figure 6–4. In operation, when the current flow is stronger in one circuit leg than in the other, the coil with the stronger current flow will have a stronger magnetic field and will pull the armature toward it. Balancing relays usually have two sets of contacts used to make the circuit when the armature reaches a certain position in its swing.

**Automatic control systems and basic functions.**    To be fully automatic, a control system performs the following six functions.

1. Measures any changes in one or more of the controlled conditions or variables. Performed by the sensing and measuring element of the proper controller, such as the bimetal in a thermostat.
2. Converts those changes into energy that can be used by the final control element. Performed by the thermostat in air-conditioning systems.
3. Transmits energy from where it was converted to the corrective device. Performed by the wiring used to connect the components of the control circuit, or the linkages used for moving mechanical components.

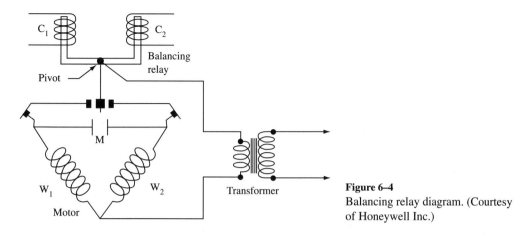

**Figure 6–4**
Balancing relay diagram. (Courtesy of Honeywell Inc.)

4. Uses energy to place the final control element and to cause the proper corrective change in the controlled condition. Performed by the controlled device, such as a valve or motor.

5. Detects the completed corrective change. Performed by the sensing and measuring elements of the particular controller.

6. Ends the call for corrective action and prevents overcorrection of the condition. Performed by the controller, connecting means, and the actuator or the controlled device.

To accomplish this work, there must be a connection between the required number of controllers (thermostats). Usually, the type of energy best suited for the type of control problem being considered will determine the kind of equipment needed for the installation. The most common type of energy available for control systems is electricity; therefore, discussion is limited to electromechanical and electronic control systems.

# Summary

- For an automatic control system to function as it was designed, some variable must be controlled.
- The second variable is known as the manipulated variable.
- A controlled system is made up of all the parts in which the controlled variable exists, except for the automatic control equipment.
- The variable being controlled is that which is measured and controlled.
- A manipulated variable is that quantity of the condition being regulated by the automatic control system which causes the wanted change in the controlled variable.
- For an automatic control system to function as designed, a connection between the required number of controllers is needed.

# Service Call

A property owner complains that the temperature in his building is quite irregular. A check of the system reveals that the space temperature is above the thermostat setting. The technician checks the thermostat and finds it out of calibration; however, it cannot be calibrated, so the technician replaces it. After the thermostat settles to room temperature, it is within the normal operating range. It also controls the equipment at the thermostat setting. The technician is satisfied that the system is repaired.

# Student Troubleshooting Problem

Office workers complain that the air-conditioning system in their building is always too warm. A check of the system reveals that the space temperature is 9°F higher than the thermostat set point, and the pressures, voltage, and amperage are within the manufacturer's specifications. What could be the problem?

# Questions

1. What must be present for a control system to work?
2. Of what does a controlled system consist?
3. When heating water, what is the controlled condition?
4. To what does the manipulated variable belong?
5. What is the set temperature of the thermostat known as?
6. What is the control deviation?
7. To what type of control system is the term *differential gap* applied?
8. What is the bimetal in a room thermostat known as?

# Unit 7: FUNDAMENTALS OF ELECTRIC CONTROL CIRCUITS

## Introduction

It is common practice to use electricity to carry the controller's (thermostat's) measurement of some change in the controlled variable (space temperature) to the controlled parts of the system. Electricity is also used to change that measurement into useful work at the final control element. Therefore, electricity has the following advantages.

1. Electricity rapidly amplifies (increases) the weak signal received from the sensing element in the controller (thermostat), making it possible to properly control systems that would usually be difficult to control.

2. Electricity allows the system to be controlled from some remote point. Thus, the controller can be at some distance from the final control element.

3. The installation of electric wiring is usually quite simple.

4. The signal received from the controller element (thermostat bimetal) can be directly used to produce one or several different sequences in the electrical output. Therefore, one signal can be used for several different functions.

5. Electricity is available wherever power lines may be installed.

**Definitions.**    The following definitions apply to electric control systems.

*Line Voltage.*    Line voltage is the term that applies to the normal electrical service supplying voltage between 120 and 240 VAC, single phase to residential areas and 480 VAC, three phase to commercial and industrial buildings. The line voltage may do the work directly, such as in an electric heating unit, or it may be connected to the primary winding of a transformer to reduce the line voltage to low voltage, 24 VAC, for the control circuit. In air-conditioning and refrigeration work, low voltage ranges between 24 and 30 VAC. In some electronic control systems DC voltage as low as 12 volts may be used.

*Low Voltage.*    Low voltage, in automatic control jargon, refers to electrical devices that use 25 volts or less. Most of the control manufacturers use the 24 VAC type of circuit in their systems. The low voltage is used in control circuits because it is safer for both the user and the building, the installation is less costly, and the controls are more responsive to changes.

*Potentiometer.*    A potentiometer is a variable resistor made from many turns of high-resistance wire wrapped around a core and has connections for linking to the system. It also has a movable sliding contact as a voltage divider. See Figure 7–1. The center connection is to the sliding contact, or movable finger or wiper blade, that moves over the entire length of the coil and makes a complete circuit wherever it touches the coil. The blade can move along the coil either manually or automatically. The less resistant side of the coil will allow more voltage to flow in that part of the circuit. The higher voltage is used to energize another control system component.

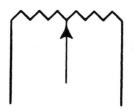 

**Figure 7–1**
Typical potentiometer symbols.
(Courtesy of Honeywell Inc.)

*Balancing Relay.*   A balancing relay has a pivoted armature that swings between two electromagnetic coils. See Figure 7–2. As the amount of electric current flowing through the two coils changes, the magnetism of the coils also changes. The armature will move toward the stronger of the two magnetic fields. The armature relay is usually equipped with two contacts adjusted so that certain circuits are completed at certain positions of the armature.

*In-Contacts and Out-Contacts.*   Relay contacts making an electric circuit when the armature is pulled into the relay coil when it is energized are called in-contacts (these are NO contacts). Those contacts that complete a circuit when the relay coil is deenergized are called out-contacts (these are NC contacts). The relay contacts will be shown in the out position unless they are otherwise labeled. See Figure 7–3.

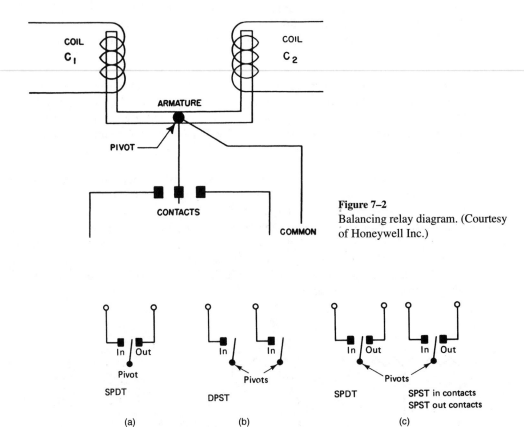

**Figure 7–2**
Balancing relay diagram. (Courtesy of Honeywell Inc.)

**Figure 7–3**
Relay contact. (Courtesy of Honeywell Inc.)

# Summary

- Electricity is used to translate a measurement of some type into useful work at the final control element.
- Electricity rapidly increases the weak signal received from the sensing element in the controller, thus making it possible to properly control systems that would otherwise be difficult to control.
- Electricity allows the system to be controlled from some remote point.
- The installation of electric control wiring is usually quite simple.
- The signal received from the controller element (thermostat bimetal) can be directly applied to cause one or several different combinations or sequences in the electrical output.
- Electricity is available wherever power lines may be installed.
- Line voltage may be connected to the transformer primary to reduce the line voltage to low voltage for the control circuit.
- The center connection of a potentiometer is a movable finger, or wiper blade, that moves over the entire length of the coil and makes a complete circuit wherever it touches the coil.

# Service Call

A customer complains that the air-conditioning unit in her large building is not cooling. A check of the system reveals that the compressor is turned off on the manual reset high-pressure control. The condenser is water cooled with a modulating motor connected to a three-way valve. The motor is controlled with a pressure switch equipped with a potentiometer. The technician resets the control and the compressor starts running. The discharge pressure is at 300 psig with HCFC-22, which is extremely high. The technician checks the valve and finds it to be mostly open to the bypass port, and checks the valve motor and finds it in good condition. The pressure switch settings seem to be out of adjustment, so the tech attempts to adjust the switch. The pressure control is bad. The tech replaces the control and puts the unit back in operation. The head pressure drops to about 215 psig. A check of the unit shows it to be operating satisfactorily, and the technician is satisfied that the system is repaired.

# Student Troubleshooting Problem

A customer complains that his commercial air-conditioning system is not cooling properly. A check of the system reveals that it has a water-cooled condenser and is charged with HCFC-22. The water flow is controlled by a three-way valve and a modulating motor connected to a high-pressure control. The space temperature is at 80°F. The suction pressure is 55 psig. The discharge pressure is at 170 psig. The evaporator is iced over, and the water going through the condenser has a 6°F temperature rise. What could be the problem?

# Questions

1. In a control system, where is the electricity changed into useful work?
2. In an electric control system, why can the controller be at some distance from the final control element?
3. How can line voltage be used for low-voltage control systems?
4. At what value is low voltage applied to the control circuit?
5. Of what is a potentiometer made?
6. What is a balancing relay?

# Unit 8: CLASSIFICATION OF ELECTRIC CONTROL CIRCUITS

# Introduction

Modern control systems have seven basic electric control circuits, each having its own characteristics and jobs. These different types of control systems are identified as two position, single speed floating, and proportional. Table 8–1 describes the different control circuits.

**TABLE 8–1**  CONTROL CIRCUIT CLASSIFICATION

| Control mode | Signal circuit | Controller |
|---|---|---|
| Two position | Three wire, low voltage | Two sets of contacts make in sequence to start; break in reverse sequence to stop |
| Two position | Three wire, low voltage | Makes one circuit to start, breaks it and makes a second circuit to stop |
| Two position | Two wire, line voltage | Makes a circuit when the switch is closed, breaks it when the switch is open |
| Two position | Three wire, line voltage | Makes one circuit to start, breaks it and makes a second circuit to stop |
| Single speed floating | Three wire | Line- or low-voltage floating control |
| Two position | Two wire, low voltage | Makes a circuit when the switch is closed, breaks it when the switch is open |
| Proportional | Three wire, low voltage | Modulating action |

Each control is built to meet specific needs in the basic circuit for which it was designed. Generally, controls of a specific type would be used in a control circuit with the same designated type. For example, an ON-OFF line-voltage control would be used in an ON-OFF line-voltage circuit. However, many types of applications for limit control are a different type from the control circuit in which they are installed. Many of these variations will be shown later in this unit.

The above mentioned information discussed the basic control circuits. In operation, these basic controls are sometimes expanded to give additional features such as the following:

1. High-limit protection
2. Low-limit protection
3. Compensated control (mechanical or electrical)
4. Positive cycling sequence

The following information will introduce other common uses of these controls. When using additional controls to meet the functions just listed, remember that the basic control circuit is not changed.

To present more understandable information, the simplest control circuits will be discussed first.

**ON-OFF two-position (line-voltage) applications.** The ON-OFF line-voltage control circuit uses line voltage switched directly by the single-pole single-throw (SPST) action of the ON-OFF line-voltage controller. This type circuit is operated by a two-position type controller (thermostat) that requires only two wires. These circuits are useful in controlling fans, electric motors, and other standard line-voltage equipment in addition to the ON-OFF control motors and relays that are designed for use in the ON-OFF control circuit to provide fail-safe operation. This type control must be used with the proper controlled equipment.

In operation, the equipment being controlled is energized when the controller switch closes and is deenergized when the switch is opened. The principle of this arrangement is quite simple.

In most control circuits, the ON-OFF type controller (thermostat) makes and breaks the air-conditioning, heating, and refrigeration equipment load directly. It is possible, however, for the equipment load to be greater than the current rating of the controller. When used this way, a relay is installed between the load and the controller.

A relay must also be used when the load needs more than one switching action. During these times, the ON-OFF type controller (thermostat) energizes and deenergizes the relay coil, which in turn opens and closes the needed number of contacts in the controlled circuit.

ON-OFF type controllers (thermostats) are designed to operate on line voltage and therefore should be installed and wired accordingly.

**ON-OFF two-position (line-voltage) equipment.** Following are the different types of ON-OFF control equipment and their descriptions.

*Controllers.*     The following list identifies the different types of ON-OFF controllers.

1. Room thermostats
2. Insertion thermostats
3. Pressure controllers
4. Humidity controllers

These types of controllers may have either open contacts or mercury switches to make and break, and electric circuits. Usually, the ON-OFF controllers use the snap-acting type switch.

*Relays.*     ON-OFF type relays use a line-voltage coil that operates an armature that may carry one or more sets of in- (NC) or out- (NO) contacts that have the capacity to handle the circuit current load.

*Motor Units.*     An ON-OFF motor unit may be used, through a mechanical linkage, to operate the wanted valves or dampers. These motors are electrically driven through their complete rotation stroke. The stroke may be either 60 degrees or 160 degrees, depending on the equipment design needs. This rotation occurs when the controller contacts are closed. When the end of the power stroke (end of rotation) is reached, the motor stops and is held in this position by a holding coil if the controller keeps the circuit closed. When the controller breaks the circuit or when a power failure occurs, either an internal or an external return spring returns the motor to the deenergized position. This makes the ON-OFF type controller, when used on motor applications, fail-safe.

*Solenoid Valves.*     ON-OFF solenoid valves are designed to go closed when deenergized, which provides a fail-safe system.

**ON-OFF (line-voltage) control operations.**     A simple ON-OFF control circuit consists of an ON-OFF controller and an ON-OFF type motor. See Figure 8–1. In an ON-OFF control circuit, the internal circuit of the controlled device (valve, motor, etc.) is not considered, except when making the electrical connections needed for proper system operation.

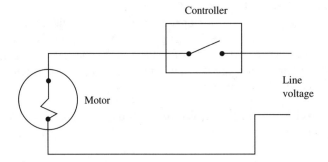

Controller

Line
voltage

Motor

**Figure 8–1**
Two-position, two-wire, line-voltage
control circuit. (Courtesy of Honeywell
Inc.)

Having the controller contacts open stops the current to the valve motor, relay coil, and other controlled devices. See Figure 8–2.

**ON-OFF two-position (line-voltage) control combinations.**    The ON-OFF control circuits are quite simple, as are the possible control combinations that follow:

*Unit Heater Control.*    Control of a unit heater is necessary to keep the fan from running and blowing cold air into the space when the burner is not turned on. See Figure 8–2, which is a wiring diagram of a typical ON-OFF type line-voltage unit heater. Though it may look the same as the diagram shown in Figure 8–3, this diagram does not have the same purpose.

*High-Limit Control.*    High-limit controls are of two types. One makes or breaks a single circuit on a rise or fall in temperature. The other makes a circuit while simultaneously

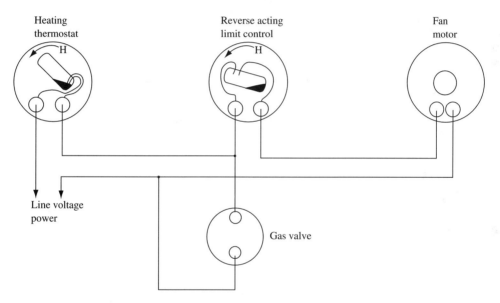

**Figure 8–2**
Two-position, two-wire, line-voltage unit heater control diagram.

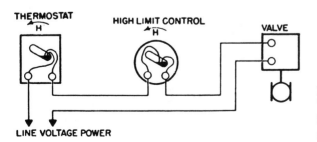

**Figure 8–3**
Two-position, two-wire, line-voltage high-limit control diagram. (Courtesy of Honeywell Inc.)

breaking another, or reverse acting on a temperature rise. This type of ON-OFF high-limit control is wired in series with the thermostat and gas valve, or steam valve, so that it will make an electrical circuit to the fan motor while breaking the electrical circuit to the gas, or steam, valve when the temperature inside the heater rises to a dangerous level. When the temperature inside the unit heater drops to the cut-in setting of the high-limit control, the circuit to the fan motor will be broken and the circuit to the gas, or steam, valve is made. See Figure 8–2. In this circuit, both the thermostat and the high-limit control must have snap-acting contacts to keep the unit from short cycling.

The ON-OFF high-limit control is usually a reverse-acting pressure control installed with its sensing element in the steam supply line to the unit heater. Because it is connected in series with the thermostat and the fan motor, it will not allow the motor to run until the steam pressure rises to the pressure setting of the high-limit control.

***Low-Limit Control.***     A low-limit control is used to stop equipment operation when the pressure or temperature drops to the control cut-out point. The low-limit controls used in this type circuit are connected in parallel with the primary controller (thermostat). See Figure 8–4. In Figure 8–4 the low-limit control can make the electrical circuit through the steam valve motor even though the primary controller has opened the circuit. It may also be used to control other circuit components when wanted.

***Two-Stage Control.***     The ON-OFF two-stage control is primarily used in air-conditioning units to control two stages of the refrigeration unit in sequence. See Figure 8–5.

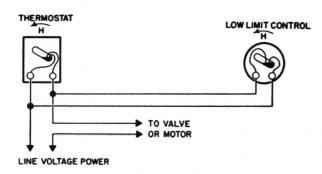

**Figure 8–4**
Two-position, two-wire, line-voltage low-limit control system diagram. (Courtesy of Honeywell Inc.)

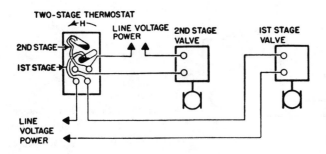

**Figure 8–5**
Two-position, two-wire, line-voltage heating control. (Courtesy of Honeywell Inc.)

When the thermostat bimetal senses a rise in temperature, the lower mercury bulb tips to the left to make the circuit to the first-stage cooling control valve. If the temperature continues to rise, the upper mercury bulb tips to the left to make the circuit to the second-stage cooling control valve.

# Summary

- The seven basic electric control circuits have unique characteristics and jobs.
- Generally, controls of a specific series would be used in a control circuit of the same type.
- The ON-OFF control circuit uses line voltage switched directly by the single-pole, single-throw (SPST) switching action of the ON-OFF controller (thermostat).
- In an ON-OFF control circuit, the equipment being controlled is energized when the controller switch closes and is deenergized when the switch opens.
- ON-OFF type controllers are designed to operate on line voltage and therefore should be installed and wired accordingly.
- In an ON-OFF control circuit, when the end of the motor power stroke is reached, the motor stops and is held in this position by the holding coil because the controller (thermostat) keeps the circuit closed. When the holding coil is deenergized, the return spring returns the motor to the deenergized position.

# Service Call

A customer complains that his steam heating system is overheating his building. A check of the system reveals that the thermostat is satisfied, but warm air is still being blown into the conditioned space. The technician checks the control circuit and finds that the motor is holding the steam valve in the full open position. The tech removes the power from the motor terminals, but the valve remains fully open. The tech then checks the return spring. It is not returning the motor to the deenergized position. The tech therefore replaces the motor unit and returns the system to operation. The technician sets the thermostat temperature above the space temperature and the motor unit returns the steam valve to its open position. The tech then sets the thermostat below the space temperature and the valve closes. With the thermostat at its original setting, the technician is satisfied that the system is repaired.

# Student Troubleshooting Problem

A customer complains that the air-conditioning system in her large office building is getting too cool. A check of the system reveals that the space is cooled with a chilled water unit. The water to the evaporator coil is controlled by a three-way valve connected to a spring-loaded motorized valve. The technician discovers the valve completely open to the coil. The space temperature is 65°F. When the thermostat is turned up, the motor still does not move. The technician checks the voltage to the motor terminals and finds it at 24 VAC. The voltage at the thermostat cooling and common terminals is at 24 VAC. The tech removes the power from the motor terminals and the motor closes the valve to the coil and opens the bypass. What could be the problem?

# Questions

1. How are control components generally chosen?
2. What type of voltage do ON-OFF control systems use?
3. How many wires are needed in an ON-OFF control circuit?
4. What must be done for an ON-OFF control circuit to provide fail-safe operation?
5. What type of control mode does the ON-OFF system provide?
6. What is the purpose of a motor unit in an ON-OFF system?
7. How is an ON-OFF motor unit returned to its deenergized position?
8. In an ON-OFF control system, how is the low-limit control wired?

# Unit 9: ON-OFF TWO-POSITION (LOW-VOLTAGE) CONTROLS
# Introduction

The ON-OFF low-voltage controls are the low-voltage equivalent to the ON-OFF line-voltage controls just described. They are suited for use on systems that need a low-voltage two-position control for a two-wire circuit. They prevent damage to the controls and must be used only on low-voltage (24 VAC) circuits. They are single-pole single-throw (SPST) type circuits.

**ON-OFF (low-voltage) control applications.** The ON-OFF low-voltage control circuits have the following advantages over the ON-OFF line-voltage controls.

1. The electrical contacts of an ON-OFF low-voltage controller are designed to carry a small amount of current; therefore, they and the controlled devices can be made of a smaller mass than the line voltage. Therefore, the controller has less lag and can be made to respond to a narrower differential controller.
2. The needed low-voltage wire, when not installed in a conduit or an armor cable, has a lower cost than the high-voltage wiring. The most popular ON-OFF low-voltage actuators are used on solenoid valves, water valves, and damper motors designed to operate in these circuits.
3. Low-voltage controls are less expensive than line-voltage controls for the same purpose.
4. Low-voltage controls are much safer than line-voltage circuits.

**ON-OFF (low-voltage) control equipment.**    Following is a list of the ON-OFF low-voltage equipment and its description.

1. Room thermostats
2. Insertion thermostats

ON-OFF low-voltage controllers are not designed to directly switch line-voltage loads. A relay must be installed between the controller and any line-voltage load. ON-OFF line-voltage controllers can, however, be used with ON-OFF low-voltage actuators. An actuator is that part of a regulating valve that changes fluid, thermal, or electrical energy into mechanical motion to open or close valves, dampers, and so on.

ON-OFF low voltage must have snap-acting type contacts or a mercury switch. Slow-moving type contacts should not be used with these controllers.

*Relays and Motors.*    The relays and motors used in ON-OFF low-voltage control circuits are designed for use on low-voltage (24 VAC) electricity only.

*Valves.*    Generally, valves used in ON-OFF low-voltage control circuits are designed for use on low-voltage systems only.

**ON-OFF (low-voltage) control operation.**    The diagrams shown in Figure 9–1 (a) and (b) are two simple ON-OFF low-voltage control circuits. Note that they are the same diagrams as those for the ON-OFF line-voltage circuits shown previously, except that a transformer is used in these diagrams to provide the 24 VAC for the control circuit.

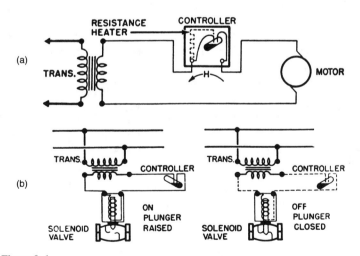

**Figure 9–1**
(a) Two-position, two-wire, low-voltage motor application. (b) Two-position, two-wire, low-voltage valve application. (Courtesy of Honeywell Inc.)

*Heat Anticipation.*    Most ON-OFF low-voltage thermostats use a resistance heater. See Figure 9–1(a). This resistor is wired in electrical series with the control circuit load so that all current flowing in this control circuit must pass through the heat anticipator. The anticipator produces heat equal to the current passing through it. This heat is released inside the thermostat and causes it to cycle the equipment earlier than it would without its use. This heating action tends to flatten both the override spots normally caused by a temperature measurement lag by the thermostat and the flywheel effect of the heat stored in the building walls, ceilings, and furnishings.

**ON-OFF (low-voltage) control combinations.**    The possible ON-OFF low-voltage control combinations are two position, used for cycling the equipment on demand from the controller. However, the following exceptions apply.

1. ON-OFF low-voltage control circuits require a transformer to provide the 24 VAC.
2. ON-OFF low-voltage equipment can be controlled by either an ON-OFF line-voltage controller or an ON-OFF low-voltage controller, but the ON-OFF line-voltage equipment must be controlled by an ON-OFF line-voltage controller designed for use on these types of circuits.

*Solenoid Valve Control.*    The ON-OFF low-voltage solenoid valve and the high-limit control are shown in Figure 9–2. In this diagram a two-wire controller and the high-limit control operate an ON-OFF low-voltage valve. Either a low voltage or an ON-OFF low-voltage controller and high-limit type control can be used, because the high-voltage control can handle the low voltage in the circuit. The solenoid valve uses low voltage that requires a transformer. The high-limit control interrupts the electrical circuit on a rise in the temperature.

This combination can also be used to control other types of ON-OFF low-voltage equipment, such as a gas heating unit, an electric heating unit, or a refrigeration compressor contactor.

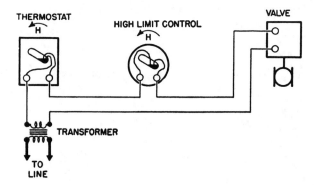

**Figure 9–2**
High-limit control of two-position, two-wire, line-voltage solenoid valve. (Courtesy of Honeywell Inc.)

*Low-Limit Control.* The low-limit control is used to sense a low pressure or temperature and to thus stop the equipment if either one is sensed. A pressure control is used to sense pressure and a temperature control is used to sense temperature. Because they are used in a low-voltage (24 VAC) circuit, a transformer and a low-voltage actuator are needed.

# Summary

- ON-OFF low-voltage controls are the low-voltage equivalent to the ON-OFF line-voltage controls.
- ON-OFF low-voltage controllers are not designed to directly switch line-voltage loads.
- ON-OFF low-voltage thermostats include a resistance heater (heat anticipator).
- ON-OFF low-voltage equipment can be controlled by either an ON-OFF line-voltage controller or an ON-OFF low-voltage controller. The ON-OFF line-voltage equipment, however, must be controlled by an ON-OFF line-voltage controller.

# Service Call

A customer complains that the temperature swing in his residential heating unit is too wide. The technician checks the heating unit and finds it to be operating properly, and checks the thermostat and finds that it turns the equipment on and off at the thermostat settings. Current flow in the control circuit is at 0.4 amps. It is equipped with an adjustable heat anticipator. The heat anticipator is set at 0.9 amps. The heat anticipator is not giving off enough heat, so the thermostat lets it have too much lag and overshoot. The technician sets the heat anticipator at 0.4 amps and is satisfied that the system is repaired.

# Student Troubleshooting Problem

A customer complains that a residential heating unit is not heating. A check of the system reveals that nothing is working. The voltage at the transformer secondary terminals is currently at 24 VAC, but there is no voltage at the gas valve. The tech checks the voltage from each of the gas valve terminals to the transformer common terminal and finds 24 VAC at each gas valve terminal. The tech then checks the voltage across the R and W terminals and finds 24 VAC. What could be the problem?

# Questions

1. What are the advantages of an ON-OFF low-voltage control circuit?
2. In an ON-OFF low-voltage control circuit, what must be done when the controller is to switch line-voltage components?

3. What type of contacts must an ON-OFF low-voltage controller have?

4. What is the purpose of the heat anticipator used in a thermostat?

# Unit 10: TWO-POSITION, SPRING RETURN CONTROL APPLICATIONS

# Introduction

Two-position, spring return circuits are designed for use in the following two position controls:

1. Valves

2. Relays

Two-position, spring return equipment is fail-safe in operation because it is designed to return to its deenergized position in case of power failure.

A basic two-position, spring return control circuit is made up of a controller connected to a relay, a motorized valve, or a solenoid valve. Should limit controls be needed for a particular application, they can be added to the basic control circuit quite easily.

**Two-position, spring return control equipment.**    The following equipment is used in these types of control systems.

*Controllers.*    The different two-position, spring return type controllers are:

1. Room thermostats

2. Insertion thermostats

3. Pressure controls

The primary element used with two-position, spring return controllers operates with two bimetal blades. Each of these blades has its own set of contacts. The bimetal blades engage with their own specific set of contacts in a sequence as the controlled condition (room temperature) reaches two previously chosen points. First, a starting circuit is completed, and then a holding circuit is completed. The blades disengage their contacts in the reverse sequence. As the controlled condition (room temperature) is satisfied, first the holding circuit is broken, and then the starting circuit is broken.

*Relays.*    Two-position relays are operated by the electromagnetic principle. The relay is made up of the following parts.

1. Low-voltage starting circuit

2. Low-voltage holding circuit

3. Built-in transformer

4. Line-voltage switching circuit

Built-in transformers supply the 24 VAC electric current to both the starting and the holding circuits. The current is supplied at a continuous rate on demand by the control circuit unless a manual shutoff is included in the circuit.

When both blades in the controller are touching their contacts, the relay starting circuit is energized. This circuit causes the relay armature to be pulled in against the relay coil. At the same time, both sets of contacts close. One set closes the holding circuit and the other set energizes the load circuit.

*Valves.*    Normal operation of a two-position valve is identical to that of the two-position relay except that the valve coil operates a plunger or a motor armature rather than a set of line-voltage contacts. The low-voltage contacts that close the holding circuit are also used in the valve assembly. Load contacts used in the relay are not used in the valve assembly because the valve itself is acting as the final control element.

**Two-position, spring return control operation.**    A complete, basic two-position control circuit uses the following:

1. Room thermostats

2. Relay

3. Damper motors

In Figure 10–1, the thermostat is satisfied and both sets of movable contacts are positioned away from their matching stationary contacts. The bimetal blades are arranged so that the flexible bimetal will contact the white (W) terminal and the rigid bimetal will contact the blue (B) terminal.

Both relay contacts (1 and 2) are open. See Figure 10–1. For simplicity, the contacts are shown separately in the diagram; however, in actual practice they are mechanically connected so that they will open and close together.

Figure 10–1 also shows only the line-voltage portion of the circuit. Figures 10–2 through 10–7 show the low-voltage circuit as it is used in both relays and valves.

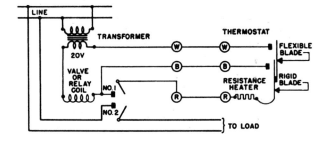

**Figure 10–1**
Complete basic two-position, three-wire, low-voltage control circuit.
(Courtesy of Honeywell Inc.)

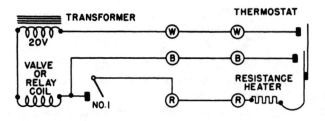

**Figure 10–2**
Thermostat is satisfied. (Courtesy of Honeywell Inc.)

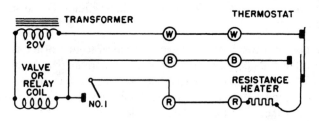

**Figure 10–3**
Slight drop in room temperature. (Courtesy of Honeywell Inc.)

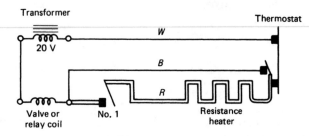

**Figure 10–4**
A further drop in room temperature. (Courtesy of Honeywell Inc.)

When the thermostat temperature selector is moved above the set point, the following occurs (refer to Figure 10–2):

1. Neither bimetal blade is touching its matching contact.
2. All the control circuits are open. No current is flowing.
3. The relay, valve coil, or damper motor is deenergized (whichever part is used). Contacts 1 and 2 are open (2 is not shown).

As the room temperature continues to drop, the following happens (see Figure 10–3):

1. The bimetal blade bends toward the stationary contacts and the flexible blade contacts the W contact.
2. All the other circuits remain open.

When the room temperature continues to drop, the following happens (see Figure 10–4) (the following line markings are used in the next three diagrams):

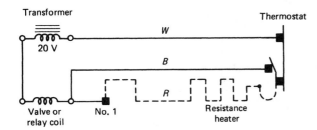

**Figure 10–5**
Holding circuit established. (Courtesy of Honeywell Inc.)

1. A solid black line shows that the maximum current is flowing through that part of the control circuit.
2. A broken black line shows that a tiny amount of current is flowing through that part of the control circuit.
3. A double black line shows that part of the control circuit is not working.

The following describes the operation of the thermostat during this condition.

1. The thermostat bimetal bends a little further, and the rigid contact blade touches terminal B.
2. The control starting circuit is now made.
3. The valve, coil, or relay is now energized.

Following is a description of the operation when the holding circuit is established. See Figure 10–5. The energized coil pulls in the armature. One second later contact 1 is closed and the holding circuit is completed. Because of the heater element resistance (heat anticipator) in the holding circuit, most of the current continues to flow through the starting circuit. The current flow, shown by the broken black line, is strong enough to raise the temperature of the heater element only slightly. Contacts 2 and 1 are closed at the same time. See Figure 10–1.

When the thermostat senses a slight rise in room temperature, the following happens. See Figure 10–6.

1. The rigid thermostat bimetal blade breaks contact with the B terminal, breaking the starting circuit.
2. The full amount of current continues to flow through the resistance heater in the holding circuit. The bimetal is heated.
3. The relay coil continues to keep contacts 1 and 2 closed.

As the room temperature continues to rise, a condition exists as shown in Figure 10–7. The rise in room temperature is aided by both the increase in temperature of the resistance heater and the following:

1. The flexible blade breaks contact with the W contact.
2. The relay is deenergized and drops out.

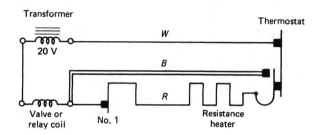

**Figure 10–6**
A slight rise in room temperature.
(Courtesy of Honeywell Inc.)

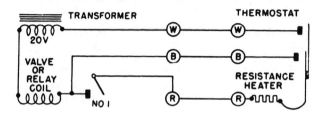

**Figure 10–7**
A further rise in temperature. (Courtesy of Honeywell Inc.)

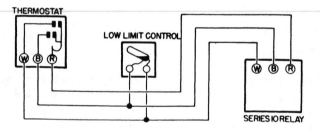

**Figure 10–8**
Two-position, two-wire, line-voltage limit control in a proportioning circuit. (Courtesy of Honeywell Inc.)

3. Both the holding and load circuits are broken.

4. The heater element is also deenergized.

**Two-position, spring return control combinations.** In two-position, spring return control circuits, limit control may be done in several different ways. In some applications, the limit control may consist of a mercury switch type controller that handles line voltage, whereas in other applications it may be a two-position open-contact controller switching the low voltage. See Figure 10–8. Figure 10–8 is an example of how an ON-OFF line-voltage, mercury switch low-limit control can be connected into a two-position, spring return control circuit.

*High- and Low-Limit Two-Position, Spring Return Control Circuits.* Two-position valves and relays can be operated by more than one two-wire controller. The controllers, however, must have mercury switches or snap-acting contacts. A slow switching action is not used in this type of control circuit.

Figure 10–9 shows the wiring connections of a room thermostat, a low-limit control, and a high-limit control.  Note that these controls are the two-wire type and are used to operate a two-position relay.

***Two Controllers and One Relay.***    A common use of this combination is two two-position controllers and one two-position relay. See Figure 10–10. Only one controller has control over the relay at any given time. A double-pole double-throw (DPDT) manual switch is needed to give the proper changeover, because two wires must be broken for a two-position controller to be completely inoperative. A relay or time clock using a DPDT switching action can be used rather than the manual switch as shown.

*Warning:* Do not attempt to control more than one two-position device at any one time from a single two-position controller. The circuit between the two controlled devices will keep the relays from dropping out.

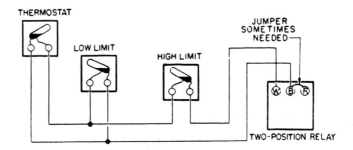

**Figure 10–9**
Two-position, two-wire, line-voltage thermostat and high- and low-limit controls controlling a two-position relay. (Courtesy of Honeywell Inc.)

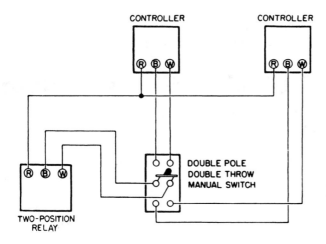

**Figure 10–10**
Two two-position, three-wire, low-voltage controllers controlling one two-position, three-wire, low-voltage relay. (Courtesy of Honeywell Inc.)

# Summary

- Two-position, spring return circuits are used for two-position operation of valves and relays.

- Two-position, spring return equipment is designed to return to its deenergized position when a power failure occurs, thus making them fail-safe in operation.

- A basic two-position, spring return control circuit is made up of a controller connected to a relay and a motorized valve or a solenoid valve.

- The primary controls used with two-position, spring return controllers operate with two bimetal blades, each having its own set of contacts that close as the controlled condition (space temperature) reaches two previously chosen points (temperatures). First a starting circuit is completed, and then a holding circuit is completed.

- When the starting circuit is completed, the relay armature is pulled in against the relay coil and closes both sets of contacts. One set energizes the holding circuit, and the other set energizes the load circuit.

- Complete, basic two-position, spring return control circuits include a room thermostat and a relay.

- When two two-position, spring return controllers are used to control one relay, only one controller has control over the relay at any given time.

# Service Call

A customer complains that the air-conditioning unit in her large building is not cooling properly. The system is a hydronic heating and cooling system. The technician checks the thermostat and finds it in good working condition, and finds the water valve on the coil open to the bypass port. The tech then checks the voltage at the motor and finds that it will energize or deenergize the valve motor with a change in the thermostat setting, but the valve does not change positions. The technician checks the motor and finds no continuity through the windings. The tech then replaces the motor and puts the system back into operation. The motor then starts changing the position of the water valve. The technician is satisfied that the system is repaired.

# Student Troubleshooting Problem

A customer complains that the air-conditioning system in his office building is too hot in some places. A check of the system reveals that it is heated with water and cooled by a direct expansion system to give zone control. The zone control is done by face and bypass dampers. The space temperature is 82°F. The dampers are positioned more to the heating position than the cooling position. The voltage to the damper motor is at 24 VAC. The technician checks the motor windings and finds they have continuity through them, and moves the thermostat temperature lever below the space temperature, but the motor does not move. The tech then sets the thermostat temperature lever above the space temperature, but the motor still does not move. What could be the problem?

# Questions

1. Why are two-position, spring return control circuits fail-safe?
2. What type of refrigerant control can be used with two-position, spring return control circuits?
3. What is the predetermined sequence that occurs when the two-position, spring return controller contacts close?
4. What will happen if more than one two-position, spring return control is operated from a single controller?

# Unit 11: FLOATING CONTROLS
# Introduction

The action of floating control circuits is different from those previously discussed, because they operate with a floating motion rather than the two-position operation. The floating type control is more commonly used with the following:

1. Motorized valves used on tank-level control systems
2. Motorized dampers used for static pressure regulation
3. Specialized pressure and temperature control systems

Floating type control circuits can be used with either line voltage or low voltage, which is determined by the type of equipment used. The basic temperature pattern is similar to that of the two-position control circuit, except that the motor is reversible and the limit switches are substituted for maintaining switches.

In operation, the floating type control circuit has no fixed number of final control element positions. The controlled device—either the valve or the damper—can take any position between fully open and fully closed as long as the controlled variable (space temperature) stays within the temperatures inside the neutral zone of the controller (thermostat). Also, if the controlled variable (space temperature) should drift outside the neutral zone of the controller, the final control element (valve or damper) will move to correct the air temperature until the space temperature moves back into the neutral zone of the controller or until the final control element reaches its maximum position on the controller.

This type control system is best used when the process lag is quite short and in applications when it is possible to use a controller having a tiny amount of lag.

Other common uses for motorized dampers are in duct static pressure regulation in air distribution systems, in liquid level control, and in controlling the suction pressure of several refrigeration compressors connected to a single evaporator.

Note that floating type controls are not generally used to operate dampers in an air-conditioning system because the temperature lag is far too great. The dampers can be controlled with other types of control systems with better results.

The floating control system is a single-speed type circuit. The actuator moves toward its new position at a single speed. This speed must match the natural cycle of the controlled system (air-conditioning or refrigeration). If the actuator moves too slowly, the control would not be able to keep up with sudden temperature or pressure changes. If the actuator moves too fast, two-position control would follow. In most applications, the actuator should move at just the right speed to keep pace with the most rapid changes in the load that can possibly happen.

For example, consider a damper installed in a forced air heating system using a motorized damper. The purpose of the motorized damper is to regulate supply air going into the fan intake. Damper control consists of using a static pressure regulator to measure the air pressure in the duct system. The static pressure controller senses a change in the duct static pressure. This change causes the damper to reposition itself and thus cause the static pressure to return to the value within the neutral zone of the controller.

In this type of situation, when a two-position type control system is used, the damper would be either fully closed or fully open in response to static pressure inside the duct. The floating type control system has no limit to the number of damper positions; therefore, the damper may be positioned at any point in its travel to allow enough air to pass for the fan to keep the wanted duct static pressure. If the air passing through the damper drops below what is needed, the static pressure controller will open the damper so the correct amount of air can pass. In this mode, the damper motor will continue to move the damper toward the open position. It will continue to open until enough air passes to keep the static pressure within the set limits of the controller. At this point, if the static pressure is held constant, the damper motor will move the damper to reduce the airflow until the pressure remains within the wanted limits.

In this way the damper will move toward the open and closed positions, because the duct static pressure is either above or below the set limits of the static pressure controller.

This circuit has no holding or maintaining switches. Therefore, the damper does not need to move to its limit. It can stop at any position to keep the duct air static pressure within the limits of the static pressure controller. The damper thus floats between the maximum limits of its travel path as the static pressure controller keeps the needed static pressure within the wanted limits. In this system, an accurate pressure can be held with a minimum amount of change.

**Floating control equipment.**    The floating type control circuits are designed for single-pole double-throw (SPDT) contacts to allow this type of switching action.

The controller must have a slow movement to allow the switch blade to float between the two contacts, and slow-acting controllers are also used. Snap-acting controllers (thermostats) cannot be used in these circuits.

The line-voltage floating type controller can be used with either line-voltage or low-voltage motors. Line-voltage motors are made up of the following:

1. Reversible capacitor motor unit in both the line- and low-voltage units
2. Limit switches to limit the rotation of the crank arm
3. Gear train to transfer the power from the motor armature to the drive shaft

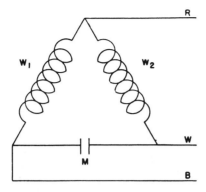

**Figure 11–1**
Internal circuit of the two-position,
three-wire, line-voltage motor diagram.
(Courtesy of Honeywell Inc.)

Figure 11–1 is a wiring diagram showing the coils $W_1$ and $W_2$, the capacitor M, and the circuit connections. The common, or red, terminal is connected to one side of the power line. The motor can run when the other side of the power line is connected to either the W terminal or the B terminal on the motor. The direction of this rotation depends on which terminal the power line is connected. Therefore, the direction of motor rotation can be changed by switching the action of the controller as it directs the electric power to either the W terminal or the B terminal. This change in rotation happens when a corrective action is needed at the final control element.

When the controller is positioned so that the W terminal is energized, the electric current will flow directly through the $W_1$ motor winding. At this time current is also flowing through the $W_2$ terminal, but it must first flow through the capacitor M. The electric current flowing through $W_2$ has a phase shift in relation to that flowing through $W_1$, and the motor turns in the corresponding direction. As the controller switch moves to energize the B terminal, the phase of the electric current is shifted in the other direction. The motor direction is reversed. The motor remains stationary when no current is directed to either the W or the B terminal.

Limit switches are mounted in the motor to break the W and B circuits and stop the motor when it reaches its full travel limit. They also limit the rotation of the motor crank arm to 160 degrees.

**Floating control operation.**    A complete floating control circuit is made up of the following (see Figure 11–2):

1. Static pressure controller
2. Floating control motor

The following is the action on a drop in duct static pressure. See Figure 11–3.

1. The controller blades contact the B terminal, making terminals R to B.
2. The B terminal of the motor is energized.
3. The motor rotates in a direction to cause the final control element to make the corrective change.

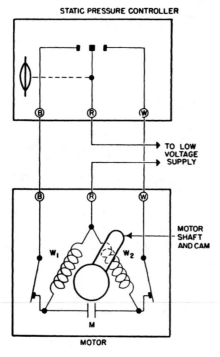

**Figure 11–2**
Diagram of a complete single-speed
floating three-wire control circuit.
(Courtesy of Honeywell Inc.)

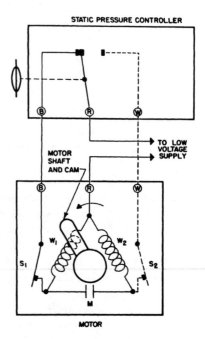

**Figure 11–3**
Action on a drop in pressure.
(Courtesy of Honeywell Inc.)

The following is the action when the static pressure drop cannot be immediately corrected by the floating movement of the controller. See Figure 11–4.

1. The motor continues to drive the final control element (dampers) toward the open position until it reaches its limit of travel.
2. The limit switch $S_1$ is then opened by the motor cam switch. The circuit is deenergized.
3. The motor stops turning.

The following is the action on a rise in duct static pressure. When there is a rise in pressure, the controller blade contacts the W terminal. At this time the motor rotates to close the dampers (final control element) until the rise in pressure has been corrected or until it reaches the limit of its travel path and limit switch $S_2$ is opened.

Any time the controller blade is floating between the W and B terminals the motor is deenergized. The motor circuit is deenergized until the pressure causes changes between

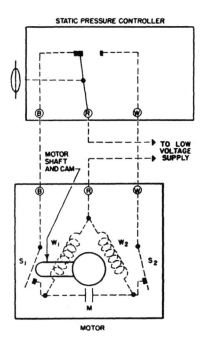

**Figure 11–4**
Action on a continued pressure drop.
(Courtesy of Honeywell Inc.)

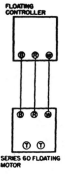

**Figure 11–5**
Single-speed, three-wire floating motor
controller combination. (Courtesy of
Honeywell Inc.)

the upper and lower pressure limits of the neutral zone of the static pressure controller, then
the circuit is again energized.

**Floating control combinations.**   The floating control circuit combinations are
not usually complicated because no limit controls are needed. See Figure 11–5.

# Summary

- Floating controls provide a floating action rather than two-position operation.
- Floating type control circuits are adaptable for either line voltage or low voltage.

- The floating type control has no fixed number of final control element positions.
- Floating type controls are not generally used on dampers in air-conditioning systems because the temperature lag is too great.
- The floating system is a single-speed type circuit, and thus the actuator moves toward its new position at a single speed.
- The floating type circuit has no holding or maintaining switches; therefore, the control element does not need to move to either of its extreme positions before rotating in the opposite direction.
- Direction of motor rotation can be changed by switching the action of the controller.

# Service Call

A property owner complains that the air-conditioning system in his office building gets too cold. A check of the system reveals that it is a hydronic heating and cooling system using a modulating motor to operate the dampers. The dampers are in the full to cooling position. The motor is not returning the dampers to the full bypass position when the thermostat is satisfied. The technician finds that the motor has bad contacts to one of its windings and replaces the motor. The tech then returns the unit to operation and sets the thermostat temperature lever above the space temperature. The dampers move to the full bypass position. When the thermostat temperature lever is set below the space temperature, the dampers move toward the full open to the cooling coil position. The technician sets the thermostat lever to the wanted temperature and is satisfied that the system is repaired.

# Student Troubleshooting Problem

A customer complains that the new air-conditioning unit in her office building gets too warm. A check of the system reveals that the space is heated with hot water and cooled with chilled water. The space temperature is 85°F. The outdoor temperature is 55°F. The voltage at the terminals of both the thermostat and the motor is at 24 VAC. The valve motor is positioned so that the hot water valve is fully open to the coil. The technician checks the resistance of the motor windings and finds it to be within the manufacturer's specifications. What could be the problem?

# Questions

1. In a floating control circuit, how many positions can the controlled device take?
2. Where are floating type controls best used?
3. What type of circuit is the floating control?
4. Why does the floating type control circuit not need to move to either of its extreme positions before rotation?
5. What causes a floating type motor to stop when it reaches its full travel limit?

# Unit 12: UNIDIRECTION CONTROL APPLICATIONS

## Introduction

The unidirection (one-direction) control circuit is designed for use on low-voltage, two-position control circuits made up of the following:

1. Motorized valves
2. Motorized dampers
3. Relays

Unidirection control circuits use one holding and two starting circuits. The motors in these systems will operate in only one direction. They make one half turn each time either of the starting circuits and the holding circuit are energized. The holding circuit is made at the beginning and broken at the end of each half revolution of the cam and switch arrangement located at the motor armature.

A basic unidirection control circuit may be built by combining a unidirection controller with a unidirection motor and adding the necessary limit controls where they are needed.

**Unidirection control equipment.**    The unidirection control equipment is made up of the following parts.

*Controllers.*    Following are the controllers used in unidirectional control circuits.

1. Room thermostats
2. Insertion thermostats
3. Humidity controllers
4. Pressure controllers

The controllers designed for use in unidirectional control circuits use a single-pole double-throw (SPDT) switch. The action may be caused by either an open contact or a mercury switch.

The switch may be snap acting so that the blade will jump from one contact to another, or it may be slow acting so that the blade will float between the two stationary contacts. Regardless of the type used, one circuit is completed on a rise and the other is completed on a fall in the controlled variable (space temperature).

*Relays.*    A relay can be operated by a unidirectional controller. In operation, a universal relay is connected to a unidirectional controller so that it will be energized and deenergized at the correct time during the cycle of the motor.

*Motor Units.*    The unidirectional motor unit can operate either valves or dampers through mechanical linkages. Their construction includes the following:

1. Unidirectional electric motor designed to operate on low voltage
2. Gear trains that provide the turning power at the end of the motor shaft
3. Maintaining switch for the holding circuit

Because the unidirectional motor operates in only one direction, the mechanical linkage between it and the damper or valve must be connected so that the force exerted by it is in one direction during half of the revolution and the other direction during the other half.

To have the two-position operation, the motor and its related circuit are designed so that once the motor is started it cannot stop until it has completed a full, half revolution.

**Unidirectional control operation.**　　A complete unidirectional control circuit is made up of the following (see Figure 12–1):

1. Open contact thermostat
2. Unidirectional motor
3. Step-down transformer

The thermostat has a bimetal blade with a movable contact that moves to touch the B contact with a drop in space temperature. It also causes the movable contact to touch the W contact with a rise in space temperature. See Figure 12–1. Here the thermostat bimetal

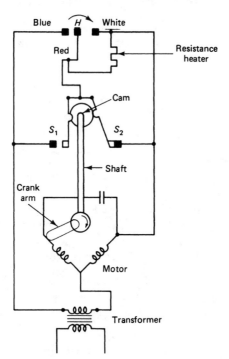

**Figure 12–1**
A complete basic two-position, three-wire, low-voltage control circuit.
(Courtesy of Honeywell Inc.)

blade is shown in the satisfied position (no heating or cooling needed). Also shown are the maintaining switch, the motor, the gear train, and the crank arm in symbolic representation.

Figures 12–2 through 12–6 show the progressive completion of each circuit. In these examples, the cam rotates with the motor armature. Also in these diagrams, the following lines represent the action that is taking place.

1. A solid black line indicates that a maximum current is flowing in that part of the circuit.
2. A broken black line indicates that a tiny amount of current is flowing in that part of the circuit.
3. A double black line indicates that no current is flowing in that part of the circuit.

On a drop in temperature, the following occurs (see Figure 12–2):

1. The thermostat bimetal blade causes the movable contact to touch contact B.
2. The starting circuit is made.
3. The motor starts turning in the clockwise direction.

The holding circuit is made. See Figure 12–3.

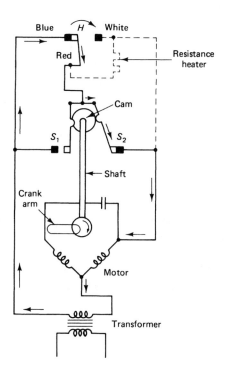

**Figure 12–2**
Action on a drop in temperature. (Courtesy of Honeywell Inc.)

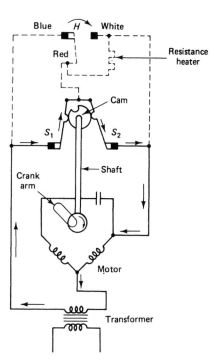

**Figure 12–3**
The holding circuit is established. (Courtesy of Honeywell Inc.)

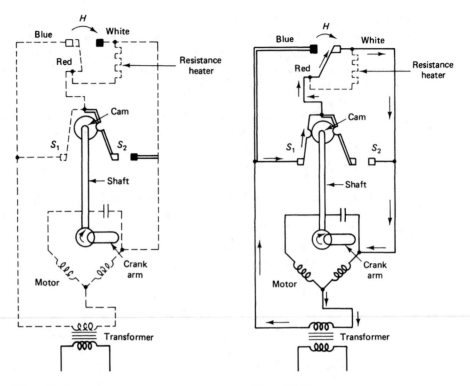

**Figure 12–4**
The holding circuit is broken. (Courtesy
of Honeywell Inc.)

**Figure 12–5**
On a temperature rise. (Courtesy
of Honeywell Inc.)

As the motor and cam rotate, the following occurs:

1. The blade on the left of the maintaining switch makes contact with $S_2$. The holding circuit is made. The holding circuit in unidirectional controls is separate from
   the starting circuit. Once the holding circuit is made, it will direct electric current
   to the motor regardless of thermostat action.
2. Should the thermostat continue to hold the B contact closed, only a small amount
   of current will flow through it. Because the holding circuit has less resistance than
   the starting circuit, most of the current will, at this time, pass through it.

The holding circuit is broken. After the motor shaft has rotated through 180 degrees,
the following occurs (see Figure 12–4):

1. The cam breaks contact $S_2$.
2. All the circuits are open.
3. The controller motor stops turning.

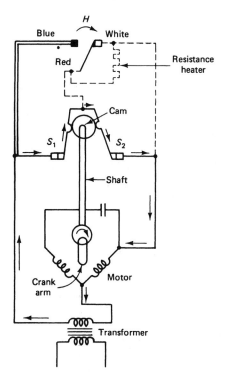

**Figure 12–6**
The holding circuit is reestablished.
(Courtesy of Honeywell Inc.)

4. A small amount of current flows through the heater resistance (heat anticipator). This small amount of current is not enough to cause the motor to turn.

A temperature rise in the controlled space and heat from the resistance heater cause the following to occur (see Figure 12–5):

1. The thermostat bimetal blade and movable contact move to the right and touch the W contact.
2. The starting circuit is made.
3. The motor starts turning in the clockwise direction.

The additional heat from the heat anticipator causes the thermostat to close the R to W contacts more quickly than normal. This type of action tends to smooth out the over-shooting swings that occur in the two-position type control systems.

When the holding circuit is reestablished, the following occurs (see Figure 12–6):

1. The blade and the movable contact on the right of the maintaining switch make contact with $S_2$.
2. The motor continues turning until it has turned one half revolution.

After the motor has made one half revolution, the holding circuit is broken at contact $S_1$, stopping the motor. The cycle is now complete.

**Unidirectional control combinations.**    Several combinations of unidirectional controllers and limit controls are available. See Figures 12–7 and 12–8.

The circuit in Figure 12–7 is completed between the B terminals of the limit control at all times except during a high-limit (high temperature, high pressure, etc.) condition. The electrical connections between the thermostat and the motor generally are straight through, just as if the high-limit control were not in the circuit. This way, the thermostat has direct control over the motor except when a high-limit condition exists.

When a high-limit condition exists, the B circuit is broken. The R to W terminals are made in the limit control. The R to W connection causes the motor to move to the closed position, even when the thermostat is calling for heat.

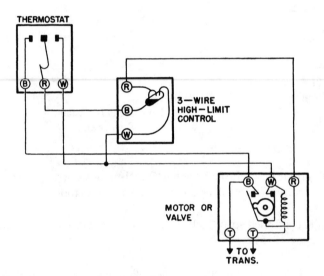

**Figure 12–7**
A typical two-position, three-wire, low-voltage high-limit control circuit. (Courtesy of Honeywell Inc.)

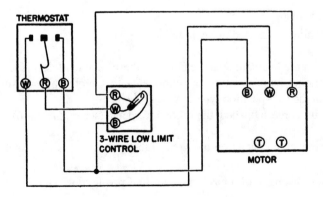

**Figure 12–8**
A two-position, three-wire, low-voltage circuit diagram with a three-wire mercury switch low limit. (Courtesy of Honeywell Inc.)

This particular combination of controls is generally used in hand-fired heating applications when the thermostat and the motor work in combination to open and close a draft damper and a check damper. The high-limit (temperature) control is physically located in the furnace bonnet or boiler outlet. It is used to close the draft damper when the bonnet or water temperature goes above a chosen high-limit temperature setting.

*Low-Limit Control.*    A unidirectional three-wire mercury switch is used as the low-limit control. See Figure 12–8. The only time this control is used is during low boiler water temperature; otherwise, this control does not function in the thermostat or motor circuit. The terminals of the thermostat are connected color to color to the motor terminals.

When a low-limit condition occurs, the mercury bulb will dump to the opposite position. When the bulb dumps, it opens the R circuit to the thermostat and connects the R to B terminals at the motor. This causes the motor to stop regardless of thermostat demands.

A common unidirectional circuit is made up of a relay with a two-position switching action. A two-position manual switch is in control of a unidirectional motor. See Figure 12–9. This combination is commonly used to control a damper on a forced air ventilating system.

In this circuit, the relay coil is connected to the motor side of the fan starting switch so that the common and the in-contacts (NC) are closed when the fan is operating. When in this position, the relay puts the manual switch in control of the damper motor. Should the fan shut down and the operator not close the damper with the manual switch, the relay will cause it to close automatically when it drops out. This action keeps air from entering the ventilating duct when the fan is not running.

*Controlling Two Motors with One Controller.*    This type of unidirectional control circuit is generally made up of two unidirectional control motors being controlled by one controller. See Figure 12–10. The control motors cannot be connected in parallel to the same controller because both motors would operate as long as the holding circuit of either motor is closed. For this type of circuit to operate correctly, it needs perfect synchronization of both motors so that both holding circuits would be broken at the same instant.

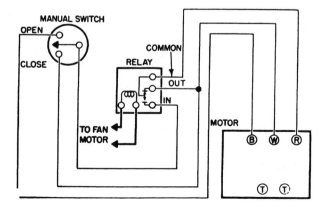

**Figure 12–9**
A two-position, three-wire, low-voltage manual control circuit system.
(Courtesy of Honeywell Inc.)

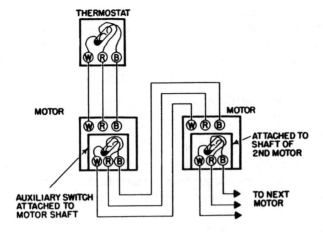

**Figure 12–10**
A diagram for controlling two motors from one two-position, three-wire, low-voltage controller. (Courtesy of Honeywell Inc.)

It is possible, however, to control two motors as shown in Figure 12–10. The first motor is controlled by the controller directly, and the second motor is controlled by an auxiliary switch attached to the shaft of the first motor. Numerous motors can be connected this way, and all can operate from one controller.

Because of the electrical connections, a slight lag will occur between each of the following motors. Note that the practical limit to the number of motors that may be controlled in this way is usually only five or six.

***Unidirectional Controller and Universal Relay.*** The universal relay and the unidirectional controller are connected in series to form this type of control circuit. See Figure 12–11. Study the diagram and note the following occurrences.

1. A drop in space temperature at the thermostat will cause the thermostat to close W to B at the relay.
2. A rise in space temperature at the thermostat will cause it to close X to B at the relay.

As an example, refer to Figures 12–12 and 12–13. In these diagrams the load circuit and the thermostat have been omitted for simplification. Note that the load circuit is closed any time the relay is pulled inward.

When the space temperature decreases at the thermostat, the following occurs:

1. As the space temperature falls, contacts W to B are made by the thermostat, closing the relay circuit.
2. At this time the relay coil is energized and it pulls in the armature, closing both the holding and the load circuits.
3. This connection between W and B may now be broken by the thermostat, and the holding circuit will actually hold in the relay.

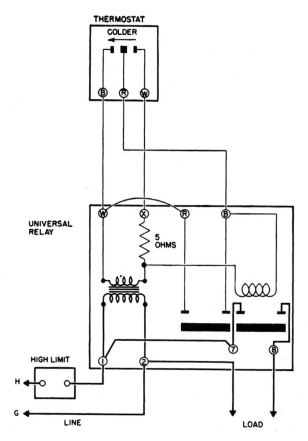

**Figure 12–11**
Two-position, three-wire, low-voltage controller, universal relay, and a high-limit control in series. (Courtesy of Honeywell Inc.)

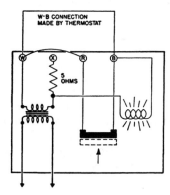

**Figure 12–12**
On a decrease in temperature. (Courtesy of Honeywell Inc.)

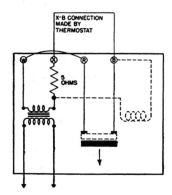

**Figure 12–13**
On an increase in temperature. (Courtesy of Honeywell Inc.)

When the temperature at the thermostat rises, the following occurs:

1. When the space temperature has risen enough, the thermostat makes X to B.
2. This completes an almost direct short circuit in the path shown by the solid lines in Figure 12–13. The 5-$\Omega$ resistor limits the current flowing through this circuit.
3. A limited power type transformer is used as the power source. A critical power drop therefore occurs at the relay coil. (Limited power transformers lower their voltage on an increase in the load.)
4. The relay then drops out, breaking both the holding and the load circuits.

Any time a high-limit control is needed, it is recommended that a unidirectional limit control be used and wired into the circuit shown in Figure 12–11.

# Summary

- The unidirectionl control circuit is designed for use on low-voltage two-position control circuits.
- The unidirectional control circuits are made up of one holding and two starting circuits.
- Unidirectional motors will run in only one direction.
- Controllers used on unidirectional circuits include a single-pole double-throw (SPDT) switching action caused by either an open contact or a mercury switch.
- Because the unidirectional motor will run in only one direction, the linkage between it and the damper or valve must be arranged so its force is exerted in one direction during one half revolution and the other direction during the other half revolution.
- To achieve the two-position operation, the motor and its related circuit are designed so that once the motor is started it cannot stop until it has completed one half revolution.

# Service Call

A customer complains that his office building is not cooling properly. The unit is a hydronic heating and cooling system. The space temperature is above the thermostat setting. All the equipment is running. The water going to the coil is controlled by a unidirectional valve motor. The technician checks the position of the water valve and finds it half open to the coil and half open to the bypass. Because the thermostat is still demanding cooling, the motor and valve should be full open to the coil. The voltage to the motor is at 24 VAC. The tech checks the resistance of the motor winding and finds it to be infinite, and the motor winding is open. The tech replaces the motor and puts it back in operation. Then the valve moves to the full open to the coil position. More cool air is leaving the vents. The technician is satisfied that the system is repaired.

# Student Troubleshooting Problem

A customer complains that her office building gets too warm. The system is heated with hot water and cooled with chilled water. The technician checks the water valve and it is open to the coil and is operated by a unidirectional motor. The voltage to the motor terminals is at 24 VAC. The technician sets the thermostat below the space temperature, but the motor does not move; and when set above space temperature the motor does not move. A check of the motor windings resistance shows that one has continuity within the manufacturer's specifications and the other has infinite continuity. What could be the problem?

# Questions

1. How many starting circuits are used in unidirectional control circuits?
2. How do unidirectional motors operate?
3. When is the motor starting circuit established in a unidirectional control circuit?
4. In what direction does the unidirectional motor run?
5. Why does most of the current flow through the unidirectional motor holding circuit?
6. What is the purpose of the resistance heater in the unidirectional controller?
7. How can two unidirectional motors be successfully controlled with one controller?

# Unit 13: LINE-VOLTAGE TWO-POSITION CONTROL APPLICATIONS

## Introduction

The line-voltage two-position control circuits are quite similar to the unidirectional control circuit, except that the line-voltage two-position system is designed for line-voltage use. These control systems are useful for (1) industrial applications when line-voltage equipment is used; and (2) installations that require single-pole double-throw (SPDT) control of the line voltage.

The line-voltage two-position control circuit is not fail-safe. It is not recommended when a continued equipment operation would cause a hazardous situation to occur in a control power failure. The motors and any equipment under their control will remain in the same position as when the power failure occurred.

A basic line-voltage two-position control circuit can be formed by combining a line-voltage two-position controller and a line-voltage two-position motor. Any required limit controls can be added when and where needed.

**Line-voltage two-position equipment.**    The following describes the equipment used in line-voltage two-position control circuits.

*Controllers.*    The following controllers are designed for use on the line-voltage two-position control circuits.

1. Room thermostats
2. Insertion thermostats
3. Pressure controllers
4. Humidity controllers

*Motor Power Units.*    The motor units designed for use on line-voltage two-position control circuits are made up of a small line-voltage motor. This motor delivers its power through a gear train to a drive shaft that may be mechanically connected to either the dampers or the valves.

The result of this action is a two-position type control. Therefore, it is necessary that the movement of the crank arm be divided into two halves of each revolution, and between each half revolution the motor stops. The maintaining switch causes the motor to stop. This action is similar to that of the unidirectional control units.

**Line-voltage two-position control operation.**    The unidirectional and the line-voltage two-position control circuits have almost identical operating features. Therefore, a complete description of the operating features of the line-voltage two-position circuit will not be given at this time. Refer to Unit 12 on unidirection circuit operation.

The mechanical construction of the two types of equipment is, however, quite different because of the different types of service needed from each of them. Even with these exceptions, their basic designs are almost identical.

**Line-voltage two-position control combinations.**    The different combinations of line-voltage two-position controls are connected identically to the line-voltage unidirectional control combinations, except that line-voltage power rather than low-voltage power is used for the circuit. The controls may be the mercury bulb type or have snap-acting contacts to prevent contact arcing when switching line-voltage loads.

# Summary

- Line-voltage two-position control circuits are similar to the unidirectional control circuit, except that the line-voltage two-position system is designed for line-voltage use.
- A basic line-voltage two-position control circuit is not fail-safe and is not recommended where continued equipment operation would cause a hazardous situation if the control power should fail.
- A line-voltage two-position motor delivers its power through a gear train to a drive shaft that may be mechanically connected to either a damper or a valve.

- It is necessary that the crank arm be divided into two halves of each revolution. Between each half the motor stops turning. The maintaining switch causes the motor to stop.
- The unidirectional and the line-voltage two-position control circuits have almost identical operating characteristics.
- Mechanical construction of the unidirectional and the line-voltage two-position equipment is quite different because of the different types of service demanded from each.
- These controls must be the mercury bulb type or have snap-acting contacts to stop contact arcing.

# Service Call

A customer complains that her large building is not being cooled correctly. A check of the system reveals that it is a hydronic heating and cooling system. The dampers are actuated by a line-voltage two-position motor. There is no voltage at the load side of the relay terminals, but line voltage is found on the line side of the relay. The relay contacts are bad. The technician replaces the relay and puts the unit back in operation. The damper motor starts moving toward the full open to the cooling coil position. The tech sets the thermostat temperature lever above the space temperature. The dampers keep moving until one half rotation is made. Then the motor starts to make the other half revolution and the dampers start moving in the opposite direction. The technician then sets the thermostat temperature lever at the wanted temperature and is satisfied that the system is repaired.

# Student Troubleshooting Problem

A customer complains that the heating unit in his office building is not heating. A check of the system reveals that the system has face and bypass dampers controlled by a line-voltage two-position motor. The dampers are fully open to the cooling coil. Voltage at the line side of the relay is 240 VAC. The voltage at the load side of the relay is 240 VAC and the voltage at the motor terminals is 240 VAC and the water temperature is 145°F. The power is removed from the motor terminals and the motor moves to change the dampers to the heating position. What could be the problem?

# Questions

1. Are the line-voltage two-position control circuits completely safe?
2. What are the four types of controllers used in line-voltage two-position control circuits?
3. How are dampers and valves driven by a line-voltage two-position motor?
4. What is the crank arm action of a line-voltage two-position motor?
5. What causes the divided rotation of a line-voltage two-position motor?
6. What types of switches must be used with line-voltage two-position controls?

# Unit 14: PROPORTIONING (MODULATING) CONTROL APPLICATIONS

## Introduction

The proportioning control circuit is designed to provide modulating operation or proportioning control motion. This circuit is used on such controls as:

1. Motorized valves
2. Motorized dampers
3. Sequencing switching mechanisms

The proportioning control circuits are designed to place a controlled device at any point between the full open and the full closed position. In this way it controls how much medium is delivered to the space as needed by the controller.

Modulating type control circuits do not have the same limitations as the two-position or floating type control circuits. For example:

1. When the motor is energized in modulating type control circuits it will operate only long enough to move the final control element a distance proportional to the change in the controlled variable. A proportioning motor will move a definite amount for every change in the controller position.
2. A two-position power unit is limited in operation by the maintaining switch. When the motor is energized, it must advance to one of its extreme positions and stay there until the conditions at the controller have changed through its entire differential.
3. Once the floating control motor is energized, it is limited in operation by the time needed to have the change in its position detected at the controller location.

To form a complete proportioning control circuit, join a proportioning controller with a motor or a relay built to give the proportioning action. The required limit controls may be added where desired, as well as automatic temperature compensation.

**Proportioning control equipment.**    The following section describes the proportioning control circuit parts.

*Controllers.*    Following are the controllers used on proportioning control circuits.

1. Room thermostats
2. Insertion thermostats
3. Humidity controllers
4. Pressure controllers

The proportioning controllers are different from the other types. Proportioning type controllers use a variable potentiometer rather than a set of contacts or mercury bulb. It has a contact blade that moves across the potentiometer winding. This winding has exactly 135 $\Omega$ of resistance. The contact blade moves in response to temperature, pressure, or humidity changes in the controlled space.

*Motors.*    Following is a description of the proportioning motors, which are made up of these components:

1. Reversible capacitor motor unit
2. Balancing relay
3. Balancing potentiometer
4. Gear train

The motor unit is made up of a low-voltage capacitor type motor that turns the motor drive shaft using the gear train. Limit switches operated by the motor are used to limit motor rotation to 160 degrees.

The motor starts, stops, and is reversed by a single-pole double-throw (SPDT) set of contacts in the balancing relay. See Figure 14–1. The balancing relay uses two solenoid coils with a parallel axis. The legs of the U-shaped armature are placed into the center of the coils. The armature has a pivot at the center so that it can be tilted by a change of the magnetic field of the two coils.

An armature has a balancing potentiometer as a necessary part. This potentiometer is identical to the one in the proportional controller. It has a 135-$\Omega$ coil of wire and a movable

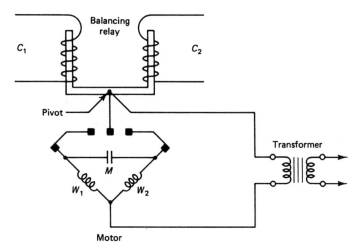

**Figure 14–1**
Balancing relay and a proportional, three-wire, low-voltage motor. (Courtesy of Honeywell Inc.)

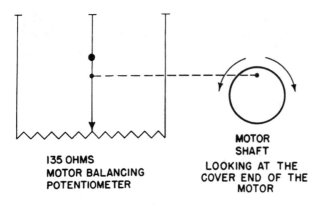

**135 OHMS**
**MOTOR BALANCING**
**POTENTIOMETER**

**MOTOR**
**SHAFT**
**LOOKING AT THE**
**COVER END OF THE**
**MOTOR**

**Figure 14–2**
Diagram of a proportional, three-wire,
low-voltage balancing potentiometer.
(Courtesy of Honeywell Inc.)

contact arm. The arm moves by the motor shaft so that it travels along the potentiometer coil. Contact occurs where it touches the coil. See Figure 14–2.

**Proportioning control operation.**    The following describes the operation of a proportioning control circuit.

*Balancing Relay Operation.*    The balancing relay is made of two coils equipped with a common armature mounted on a pivot. See Figure 14–1. The relay, as it is used in a proportioning control circuit, has electric current flowing through both coils. The amount of current flowing through each coil is regulated by the relative positions of the controller potentiometer and the motor balancing potentiometer. Therefore, when equal amounts of current are flowing through both balancing relay coils, the contact blade lies in the center of the space between the two stationary contacts. The motor is not turning in any direction. As the finger of the controller potentiometer moves in response to a change in the controlled variable, a greater amount of current flows through one of these coils than through the other, and the relay then becomes unbalanced. The relay armature is then moved so that the blade and movable contact touch one of the stationary contacts. The motor turns in the needed direction. See Figure 14–3.

Referring to Figure 14–1, if the relay coil $C_1$ receives more of the current than coil $C_2$ and has a stronger magnetic field, the contact blade will move to the left. The circuit between the transformer and the motor winding $W_1$ is then made. Electric current also flows through the capacitor M to the motor winding $W_2$. The motor will turn in the needed direction until the balancing relay is balanced (moves to the center position) and the circuit is broken.

If coil $C_2$ receives more of the current than coil $C_1$, the circuit is made to the motor winding $W_2$. The motor will turn in the opposite direction.

*Motor Balancing Potentiometer.*    The contact made by the balancing relay can be broken only if the current flowing through coil $C_1$ becomes equal to the current flowing through coil $C_2$. This is caused by the motor balancing potentiometer linked to the motor shaft. When the motor turns, the finger of the motor balancing relay is driven toward a position to equalize the resistances in both legs of the control circuit.

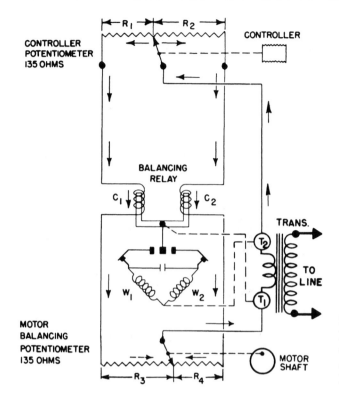

**Figure 14–3**
Proportional, three-wire, low-voltage control circuit in the balanced condition. (Courtesy of Honeywell Inc.)

Each position of the motor shaft has a corresponding position of the potentiometer contact finger for each degree of rotation through its complete 160 degrees of travel. The complete sequence of operation is shown in Figures 14–3, 14–4, and 14–5.

As shown in Figure 14–3, a balanced condition causes current to flow from the transformer, through the thermostat potentiometer finger, and down through both legs of the control circuit. In the positions shown, the thermostat potentiometer finger and the motor balancing potentiometer finger divide their respective coils so that the resistance on both sides of the circuit are equal. The coils $C_1$ and $C_2$ in the balancing relay have exactly the same amount of current flow. The armature of the balancing relay is balanced. The contact arm is floating between the two stationary contacts. No current is flowing to the motor and the motor is not running.

As shown in Figure 14–4, a condition can occur in which the temperature drops a small amount. Because of this drop, the thermostat potentiometer finger moves toward the right-hand end of the potentiometer coil. The amounts of resistance on both sides of the circuit are no longer equal. A larger part of the current is now flowing through the right-hand leg of the circuit. Coil $C_2$ makes contact with the side of the circuit that directs the current to the motor winding $W_2$. The motor then turns in the corresponding direction to move the motor balancing potentiometer finger to a new position.

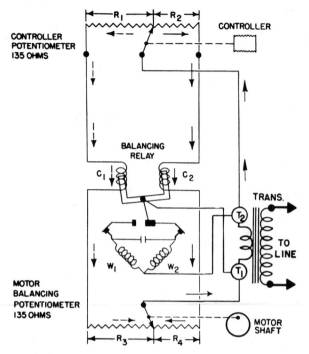

**Figure 14–4**
Proportional, three-wire, low-voltage control circuit on a drop in temperature. (Courtesy of Honeywell Inc.)

As shown in Figure 14–5, a new condition can cause the motor shaft to move the balancing relay potentiometer finger to a position that equalizes the current flow through the two legs of the circuit. In this position, the right-hand side of the thermostat potentiometer has a resistance equal to that of the left-hand side of the motor balancing potentiometer. Again, the current flowing through the two legs is equal, and the motor is not turning.

A careful analysis of these diagrams clearly shows that the motor turns until the finger of the motor balancing potentiometer reaches a position that corresponds to the finger position of the thermostat potentiometer.

**Proportioning control combinations.** Following is a description of the proportioning control combinations.

*Low-Limit Control.* The low-limit control is commonly used in proportioning control circuits used on central fan ventilating or air-conditioning systems in which the heating equipment is controlled from a space thermostat or return air duct controller. The space temperature can rise rapidly because of a rise in solar heat, a rise in occupancy, or any other type of condition that would suddenly result in a smaller heating load. When the room or return air controller is satisfied, it will close off the final control element.

Should the system be one that takes in some outside air, it is quite possible that air at too low a temperature could be blown into the space. When the supply air temperature drops below about 60°F, the space will probably feel drafty.

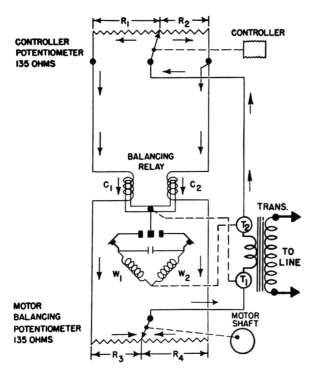

**Figure 14–5**
Proportional, three-wire, low-voltage
control circuit rebalanced in new
position. (Courtesy of Honeywell Inc.)

A low-limit control installed in the return air duct is often a wanted item in the control system. A controller regulates the valve feeding the coil to keep the discharge air temperature at a comfortable level. It will also limit the lower temperature of the discharge air.

Note that the low-limit temperature controller should be set at the lowest air temperature that can possibly be supplied to the space without drafts.

Usually, the low-limit controller has a potentiometer resistance of 135 Ω. This permits the controller to operate the final control element through 50 percent of its operating range.

Controllers equipped with double potentiometers can be used to give the final control 100 percent of its operating range. When double potentiometers are used they must be connected in series to give a resistance of 270 Ω. In most installations, however, the 50 percent range is more desirable; therefore, the following discussion will refer to the limit control equipped with the 135-Ω potentiometer. Figure 14–6 shows the external wiring connections for a proportional room thermostat, low-limit control, and motorized valve.

A typical proportional circuit with a low-limit control is shown in Figure 14–7. This diagram shows the low-limit control being satisfied, and the potentiometer finger on the W end of the potentiometer winding. This position shows that the temperature of the discharge air is above the setting of the low-limit control. When this happens, the controller operates the valve motor in the same way as in the proportioning circuit without the low-limit control described in the above heading "Proportioning Control Operation." As long as the low-limit controller remains satisfied, its potentiometer winding resistance will not

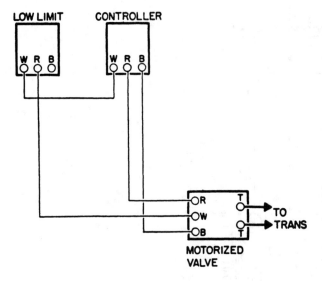

**Figure 14–6**
Proportional, three-wire, low-voltage valve, thermostat, and low-limit control diagram. (Courtesy of Honeywell Inc.)

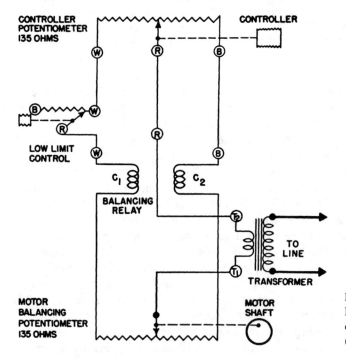

**Figure 14–7**
Proportional, three-wire, low-voltage circuit with low-limit control satisfied. (Courtesy of Honeywell Inc.)

affect the balancing relay. The relay coils are affected only by the controller potentiometer and the motor balancing potentiometer.

Figure 14–8 shows a typical proportioning circuit with a low-limit control when the room temperature is outside the modulating range of the room thermostat and the low-limit control controls the valve motor. In Figure 14–8 the finger of the room controller potentiometer has moved to the W end of the coil. The low-limit control remains satisfied. The motor is in the closed position of the motor balancing potentiometer.

If the system begins taking in outside air, the discharge air temperature may drop. As this temperature drops into the modulating range of the low-limit control and approaches its setting, the potentiometer finger will move away from the W end of the coil to a new position—see the dashed lines.

Assume that the finger has moved to a new position and that more of the resistance is placed into the left leg of the control circuit. The left and right legs of the circuit then become unbalanced. Most of the current is flowing through coil $C_2$ of the balancing relay. The armature of the balancing relay is moved so that it makes contact. The motor balancing potentiometer finger is moved toward the open end of the coil as the valve is opened. When

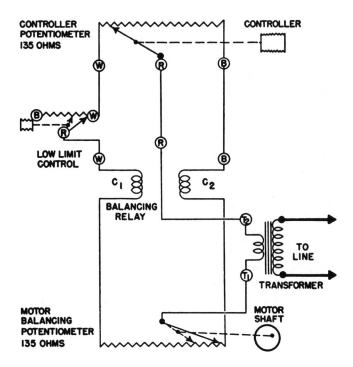

**Figure 14–8**
Proportional, three-wire, low-voltage circuit with low-limit controlling the valve or motor. (Courtesy of Honeywell Inc.)

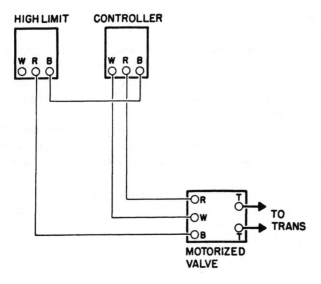

**Figure 14–9**
Wiring diagram for a proportional, three-wire, low-voltage circuit with a high-limit control. (Courtesy of Honeywell Inc.)

the potentiometer arm has moved far enough to cause the resistance of both sides of the circuit to be equal again, the circuit is broken by the balancing relay, and the motor stops.

*High-Limit Control.*    For best results, the high-limit control generally has a 135-$\Omega$ potentiometer. However, where 100 percent operating range is wanted, high-limit controls have dual potentiometers.

The external wiring diagram for a proportioning type high-limit controller and a motorized valve is shown in Figure 14–9. This type of circuit is used when a possibility exists that the space temperature will rise too high. The high-limit control will take control of the motorized valve when the discharge temperature enters its modulating range.

A complete proportional control circuit with a high-limit controller and a motorized valve is shown in Figure 14–10. This circuit is similar to the low-limit control circuit shown in Figure 14–8, except the high-limit control is connected into the B leg of the control circuit. The operation of the high-limit control is the same as that of the low-limit control.

When the temperature of the discharge air rises into the modulating range of the high-limit controller, the potentiometer finger starts moving toward the W side. This puts more resistance into the B leg of the circuit; therefore, coil $C_1$ becomes a stronger magnet than coil $C_2$. When contact is made by the relay armature, the motor turns toward the closed position until the motor balancing potentiometer finger reaches a final position so that the resistances of each circuit leg are equal again.

*Two-Position Limit Controls.*    In proportional control circuits using two-position limit controls when it is not necessary or not needed, the limit control operates as a proportioning control. This type limit control must always be of the snap-acting, single-pole double-throw (SPDT) type. Two-position limit controls must not be used in applications

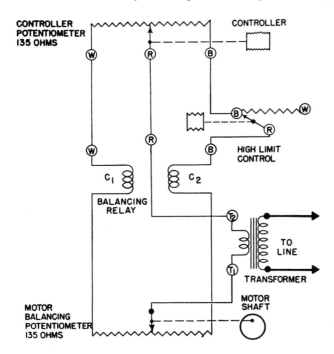

**Figure 14–10**
A complete proportional, three-wire, low-voltage circuit with a high-limit control. (Courtesy of Honeywell Inc.)

when the temperature of the medium in which their sensing elements are located affects the opening and closing of the valve or the damper it is controlling. As an example, suppose that a two-position high- or low-limit control is used in a proportioning control circuit to limit the temperature in a discharge air duct. Any time the air temperature enters the operating range of the limit control, the valve controlling the steam to the heating coil will cycle on and off continuously.

Two-position type limit controls are sometimes used in applications that require humidity high-limit control in a cooling application. This circuit uses a single-pole double-throw (SPDT) switching action. See Figure 14–11. When the humidity drops below the range of the humidity high-limit controller, the circuit is made from the thermostat to the motor. The cooling valve is controlled as usual. If the humidity rises into the range of the high-limit control, the limit control shorts terminals R to B at the motor, and the motor then opens the cooling valve to the full open position. When a power failure occurs, or when the fan shuts down, the cooling valve closes. Some type of reheat should be installed within a system of this type, because lowering the air temperature will result in a higher relative humidity condition. To successfully remove both the sensible and the latent heat, the air should be cooled below the dew point temperature and then be reheated to the wanted temperature.

When a reheat system is not a part of the installation, a second temperature controller should be used to act as a low-limit control. This is useful in high humidity climates.

The low-limit control should be wired into the circuit as shown in Figure 14–11, and it should be set a few degrees below the main temperature controller. Should the air

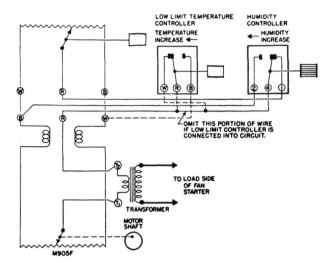

**Figure 14–11**
Proportional, three-wire, low-voltage
circuit with a two-position humidity
high limit. (Courtesy of Honeywell
Inc.)

temperature fall too low, the low-limit controller will take control away from the humidity
control and close the cooling valve.

*Manual and Automatic Switching.*    The use of manual switches and relays in proportioning control circuits is often wanted. Manual switches are shown in the diagrams
throughout the remainder of this unit. Relays that have the same switching action as the
manual switches are often used where automatic switching is wanted.

    Figure 14–12 illustrates a manual switch with a single-pole double-throw (SPDT)
switching action as used in a proportioning control circuit. When the switch is placed in the
automatic position, as shown in Figure 14–12, the R circuit to the motor is made. The motor turns in a normal manner and is controlled by the controller. When the switch is moved
to the closed position, the R wire from the controller is broken. The R to W wires are completed at the motor. The motor then moves to the closed position. This type system is often
used in forced air heating systems to manually shut off part of the controls when they are
not needed.

*Transfer of Motor from One Thermostat to Another.*    A double-pole double-throw
(DPDT) switch can be used to change the control of a single proportioning motor from one
thermostat to another. To prevent the two thermostats from affecting system operation, both
the W and B wires must be broken when the proportioning thermostat is taken out of control. The R wire does not have to be broken. See Figure 14–13.

*Reversing for Heating and Cooling Control.*    In applications when it is necessary
to use the same thermostat and the same final control element for both heating and cooling,
the wiring can be done as shown in Figure 14–14.  The double-pole double-throw (DPDT)
manual switch is used to reverse the wiring from the thermostat to the motor for cooling
control. This is done so the motor will operate in the opposite direction than when used in
the heating position. The thermostat and the motor are connected color to color for heating.

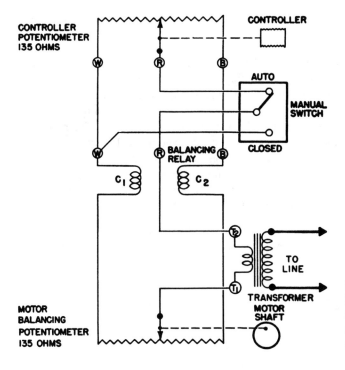

Figure 14–12
Proportional, two-wire, low-voltage circuit with single-pole double-throw manual switch for automatic or closed operation. (Courtesy of Honeywell Inc.)

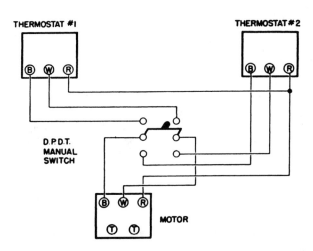

Figure 14–13
Diagram of a circuit for transferring control of a proportional, three-wire, low-voltage motor from one thermostat to another. (Courtesy of Honeywell Inc.)

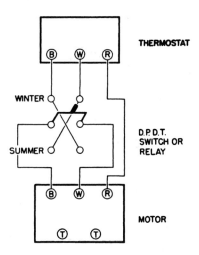

Figure 14–14
Wiring diagram arrangement used for reversing control so that the same proportional thermostat and final control element can be used for both heating and cooling. (Courtesy of Honeywell Inc.)

When the manual switch is placed in the cooling position, the connections are B to W for cooling.

*Transfer of Thermostat from One Motor to Another.*    In some installations, operating two proportioning motors—one at a time—from one controller is wanted. See Figure 14–15. A triple-pole double-throw (TPDT) manual switch is used to make the change. When the switch is placed in the cooling position, the control circuit is made from the thermostat to the cooling motor with the B to W and the W to B wires. The motor then operates normally as under the control of the thermostat. The manual switch, at the same time, is connected from R to W at the heating motor, causing it to close.

When the manual switch is placed in the heating position, the thermostat is connected color to color with the heating control motor that now operates normally inward. The cooling motor is closed when the circuit is wired this way.

*Sequence Control.*    A good method of controlling proportioning motors in sequence from one thermostat is shown in Figure 14–16. An auxiliary controller having two

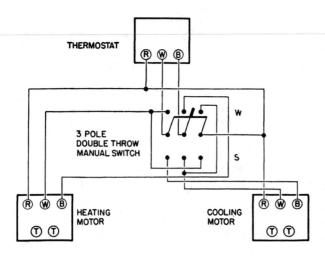

**Figure 14–15**
Diagram of arrangement for transferring a proportional thermostat to control one or the other of two motors. (Courtesy of Honeywell Inc.)

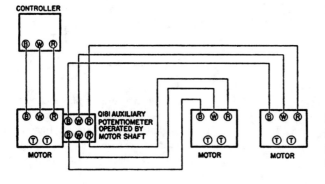

**Figure 14–16**
Diagram for controlling motors in sequence from a single controller. (Courtesy of Honeywell Inc.)

potentiometers is fastened to the shaft of the master control motor. Each potentiometer has control over one of the auxiliary motors. Therefore, the controller is in direct control of the master motor. The position of the motor shaft regulates the two potentiometer fingers, which in turn regulate the auxiliary motors.

The starting points and the proportional bands of the two auxiliary potentiometers can be adjusted to give the wanted sequence of operation.

***Unison Control.***    One method of controlling two motors with one controller at the same time is shown in  Figure 14–17. The master motor is a proportioning motor with a built-in dual potentiometer. The dual potentiometer operates the second motor. A third motor can also be added by installing more motors with dual potentiometers in the first and second places. Their motor can be controlled by the dual potentiometer in the second motor. Three motors are the maximum that should be used this way.

Another method that may be used for controlling several motors at the same time is shown in  Figure 14–18. Auxiliary potentiometers are available for controlling up to eight

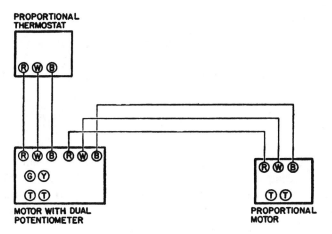

**Figure 14–17**
Diagram for unison control of two motors from one thermostat, using a master motor with built-in dual potentiometers. (Courtesy of Honeywell Inc.)

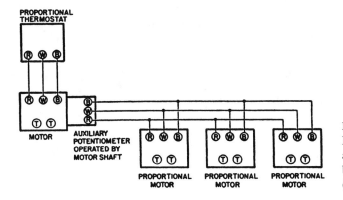

**Figure 14–18**
Diagram for unison control of three auxiliary motors and one master motor by means of an auxiliary potentiometer. (Courtesy of Honeywell Inc.)

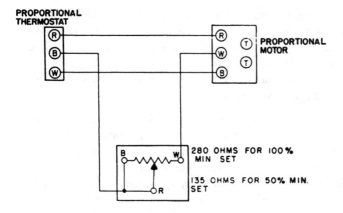

**Figure 14–19**
Diagram for manual minimum positioning of typical outdoor air control system. (Courtesy of Honeywell Inc.)

motors in parallel. Note that the correct potentiometer must be used to control the specific number of motors being used.

*Manual Minimum Positioning.*    A typical outdoor air control system using a manual potentiometer for minimum positioning is shown in  Figure 14–19. When the finger is turned so that it is at W, the potentiometer coil is shorted out of the circuit. The motor will operate normally, and the damper will go completely closed if the controller demands it. If the pointer is turned so that the finger is toward B, the resistance between B on the controller and W on the motor is increased. Thus, travel of the motor toward the closed position is limited. A manual potentiometer with 135 $\Omega$ of resistance will give up to 100 percent of the minimum position opening.

*Recycling Step Controllers.*    The wiring diagram of a proportioning step controller used to control five compressors in a large, central cooling system is shown in Figure 14–20. Step controllers are available with from five to ten auxiliary switches operated in sequence by a cam assembly turned by the control motor. The auxiliary switches are wired into the compressor start circuits. As the controller calls for more cooling, the proportioning motor rotates, turning on more compressors one at a time as needed. Thus, the correct amount of refrigeration for the load is always available.

When a power failure occurs, or even if the system is shut down, the control system must return to the starting point so the compressors will start one at a time rather than all at the same time, to prevent overloading the electrical system.

In the arrangement shown in Figure 14–20, the control motor remains in the position it occupied when the power went off. However, before any of the compressors can operate again, the control motor must travel back to the closed position. It then moves back to the position called for by the thermostat, starting the needed number of compressors one at a time as it turns.

To simplify the wiring schematic, the complete diagram is not shown in the figure. The power supply for the compressor coils is often taken from the fan circuit, thus making certain that the compressors will not run when the fan is shut down.

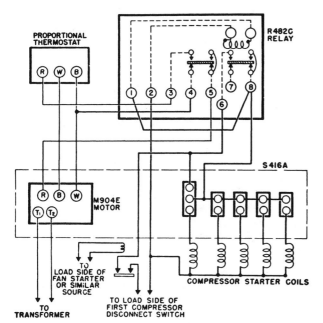

**Figure 14–20**
Wiring diagram of a Honeywell recycling step control system. (Courtesy of Honeywell Inc.)

This process is necessary in multiple-element control installations. It provides safe, economical compressor operation for all operating conditions with an added guarantee that the starting loads will never be great enough to overload the power line.

# Summary

- The proportioning control circuit is designed to provide modulating operation or proportioning control action.
- The proportioning control circuit is designed to operate and position the controlled device at any point between full open and full closed that will position the controlled device at any point between the full open and full closed positions. This will ensure control of the amount of medium delivered to the space as needed by the controlling device.
- Modulating type control circuits are not bothered by the same limitations as two-position or floating type control circuits.
- To form a complete proportioning control circuit, combine a proportioning controller with a motor power unit or relay built to give the proportioning action.
- The ohm rating of a potentiometer used in proportioning circuits is exactly 135 Ω.
- The motor is made up of a low-voltage, capacitor type motor that turns the motor drive shaft by using a gear train.
- Limit switches are operated by the motor to limit the rotation to 160 degrees.

- The motor is started, stopped, and reversed by a single-pole double-throw (SPDT) set of contacts in the balancing relay.
- The balancing relay is made of two coils using a common armature mounted on a pivot.
- The low-limit controller should be set at the lowest temperature that the air can possibly be supplied without causing drafts.
- Two-position controls used in proportioning control circuits must have snap-acting switches.

# Service Call

A customer complains that his air-conditioning system is getting too cold. A check of the system reveals that the system cools a large office building and the unit is equipped with three compressors controlled by a step controller. All three compressors are running even though the space temperature is below the thermostat setting. The proportioning motor has stopped in the full demand position and has not automatically adjusted for the space temperature. The technician checks the voltage at the motor terminals and finds none. A test of the voltage at the transformer secondary terminals shows none. Voltage testing at the transformer primary terminals shows correct voltage. The transformer is bad and is therefore replaced. The tech places the system back in operation. The proportioning motor turns to the full OFF position and restarts cycling the compressors on, one at a time, until the needed number to handle the building heat load are running. The technician is satisfied that the system is repaired.

# Student Troubleshooting Problem

A customer complains that her office building is too cold. A check of the system reveals that it is a hot water heating system with a modulating motor and a three-way valve. The valve is found open to the bypass. No hot water is flowing into the coil. The technician checks the voltage to the motor and finds it to be 19 VAC. What could be the problem?

# Questions

1. What does it take to make a complete proportioning control circuit?
2. In addition to room thermostats, what other type of temperature controller can be used in a proportioning control circuit?
3. What is different about a proportioning controller from other type controllers?
4. How much resistance does a potentiometer have?
5. Where are potentiometers used in a proportioning control circuit?
6. What is the limit of rotation of a proportioning motor?
7. What is the purpose of the balancing relay in a proportioning motor?
8. What is the lowest temperature recommended for a low-limit controller?

# 3 Electromechanical Controls

## Introduction

Electromechanical controls are those that use both electrical and mechanical components to do their job. They do not operate the same as electronic controls, although the industry is leaning toward just that. However, some knowledge about electromechanical controls is needed so the technician can service those remaining in the field and any that may be part of new installations.

# Unit 15: TRANSFORMERS
## Introduction

A *transformer* is an inductive device that transfers electrical energy from one circuit to another without the two being physically connected. When alternating current (ac) is applied to the primary coil, a current is induced into the secondary coil. A coupling transformer transfers energy at the same voltage; a step-down transformer transfers energy at a lower voltage. Step-up transformers transfer energy at a higher voltage.

**Purpose.**   The purpose of a transformer used in air-conditioning and refrigeration system control circuits is to reduce the line voltage to low voltage (24 volts). This is a step-down type transformer. It is made of two coils of insulated wire wound around an iron core. See Figure 15–1.

The coil connected to the line voltage is called the primary side. The coil connected to the control circuit (low voltage) is called the secondary side. These are step-down transformers that reduce the line voltage—either 120 VAC or 240 VAC—to 24 VAC.

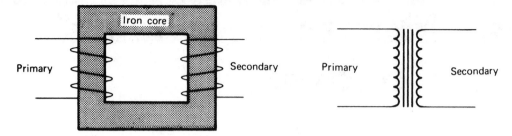

**Figure 15–1**
Symbolic sketch of a transformer and schematic symbol.

**Rating.**    Transformers are rated in volt-amps (VA) and are class 2 type. This VA rating, listed on the transformer in volt-amps, shows the amount of electrical power available at its secondary terminals. Power is found by multiplying the secondary volts by the amps drawn in the control circuit. For example, a circuit has 24 VAC and is drawing 0.9 amp. The power is 21.6 VA (watts). The transformer selected must be at least large enough to do the job. The rating of the transformer may be greater than needed, but not smaller, as it will only burn out in a short time. One that is too large will only cost more—it will not do a better job than one rated closer to the needed VA.

**Sizing.**    Because most control system parts are rated in amperes, knowing how to compare the VA rating with the ampere draw of the control system is necessary. Use the following formulas to make this comparison.

$$\text{Amps} = \text{VA output} \div \text{Secondary voltage}$$
$$\text{VA output} = \text{Secondary amps} \times \text{Secondary voltage}$$

Either of these formulas can be used when selecting a transformer.

The technician needs to check the ampere draw in the secondary circuit with all the equipment that will draw the greatest amount of current when operating. This is usually the cooling mode. On some special installations, however, the heaviest current draw will be in the heating mode. Be sure to check the mode having the most load. To check the amperage draw, start the unit and check the current draw in the power wire, which is usually the red wire, or the common wire to all loads connected to the secondary side of the transformer. For example, the amperage draw of the load on the secondary side of the transformer is 0.8 amp. The VA rating of the transformer would need to be at least

$$\text{VA output} = 0.8 \times 24$$
$$\text{VA} = 19.2$$

The situation requires a transformer with a minimum output rating of 19.2 volt-amps. Most equipment manufacturers equip their products with 40 VA transformers to handle the most common types of loads. Some heating-only units are equipped with 20 VA transformers because the heating load is usually smaller than the cooling load. These transformers must either be changed to one having a higher VA rating or to two transformers connected in parallel, or phased, to equal the needed power for the system.

Some transformers are equipped with fuses for short-circuit protection. Some fuses can be replaced but others cannot. When a nonreplaceable type is used, the transformer must be replaced when the circuit has been overloaded. Be sure to find and repair the problem in the circuit or the replacement transformer will burn out in a short time. When replacing a transformer be sure to turn off the power to the unit to prevent personal injury or damage to the unit, or both.

Several reasons exist for using transformers in control circuits. First, the low-voltage circuit is much safer than a line-voltage circuit, thus protecting the customer from electrical shock while protecting the equipment and the building from the greater fire potential of line voltage. Second, the controls used with low-voltage control circuits are made from much lighter material than the line-voltage type; therefore, they are much more sensitive to changes. They are more economical to buy than line-voltage controls. Third, the low-voltage wiring is greatly simplified when compared with line-voltage wiring.

**Phasing transformers.**    Equipment manufacturers sometimes phase transformers for various reasons, but generally phasing is done to increase the power for proper control circuit use. Transformer phasing guarantees that all the electrons are flowing in the same direction, and the process is sometimes referred to as *polarizing* the transformers.

To phase transformers, wire the primary sides in electrical parallel. See Figure 15–2. Connect one wire from each transformer secondary (B and C) together. See Figure 15–3. Then check the voltage across the other two leads (wires A and D) from the secondary sides. If the voltmeter shows 48 volts, the transformers are out of phase. The wires must be reversed. See Figure 15–4.

When the transformers are correctly phased, the voltmeter will show zero volts between the two disconnected wires on the secondary terminals of the transformer. At this point fasten the two wires on the secondary terminals of the transformers. Now fasten the two wires (A and C) together. The transformers are now in phase. The control circuit may now be connected to the two combinations of wires. See Figure 15–5.

This example illustrates the two wrong ways to connect the transformer secondaries. One way will result in burnout of either one or both transformers. The other way will result in too high a voltage. This high voltage will usually burn out the thermostat heat anticipator immediately and will cause other controls to burn out in a short time. Transformers supplied with the equipment by the manufacturer are usually properly phased with marked terminals. They are also supplied with wiring diagrams. Most transformers are supplied

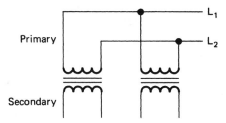

**Figure 15–2**
Transformers wired in parallel.

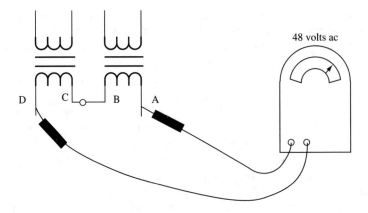

**Figure 15–3**
Checking voltage of improperly phased transformers.

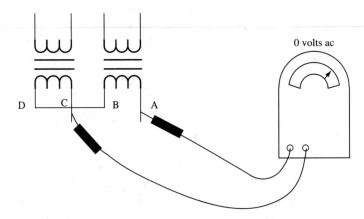

**Figure 15–4**
Checking voltage of properly phased transformers.

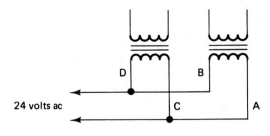

**Figure 15–5**
Control circuit connected to properly phased transformer.

with color-coded electrical leads to aid in proper phasing. Persons who install the transformer can be sure that proper phasing will follow by simply connecting like colors to like colors. Also, transformers equipped with screw terminals are color coded with one brass screw and the other is either silver or nickel in color.

***Transformer Phasing with Two Low-Voltage Control Components.***   Phasing is necessary when two low-voltage components having their own low-voltage power supply are to be used in the same control circuit. The components may be fan centers or similar devices. Generally, only one thermostat is used to control these devices. See Figure 15–6. In this example, the thermostat is supposed to be open. The + and − signs show instantaneous voltage polarity. The voltage measured across the secondary connections of either transformer is shown to be 24 volts. The voltage across the two wires from the relays is shown to be 48 volts.

There are two ways to properly phase transformers in this situation. The first method is as follows:

1. Wire the circuit, leaving one wire from the thermostat loose at the relay terminal.
2. Check the voltage between the unconnected wire and the relay terminal.
3. If the voltage is higher than 24 volts, reverse the two wires at the control relay.

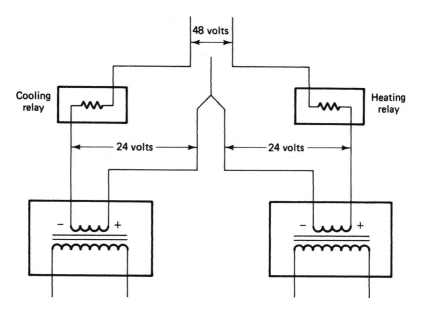

**Figure 15–6**
Phasing control components having separate power supplies.

The second method is as follows:

1. Wire the circuit, leaving the wires disconnected at the thermostat.
2. Check the voltage between one wire and the other two in turn.
3. If the voltage is greater than 24 volts, reverse the two thermostat wires at the control relay.

***Transformer Phasing in a Three-Phase Circuit.***   When phasing transformers in a three-phase circuit, the transformer primaries must be connected to the same two legs of the three-phase power supply. If not, exact phasing cannot be done. The secondaries are phased the same way as in the single-phase installations previously described.

# Summary

- A transformer is an inductive device that transfers electrical energy from one circuit to another without the two being physically connected.
- A step-down transformer transfers energy at a lower voltage than the supply voltage.
- The purpose of a transformer used in air-conditioning and refrigeration systems is to reduce the line voltage to low voltage (24 volts).
- The transformer coil connected to the line-voltage side is called the primary side and the coil connected to the control circuit is called the secondary side.
- Transformers are rated in volt-amps (VA) and are class 2 type.
- The VA rating shows the amount of electrical power that the transformer will give.
- To size a transformer, only one test needs to be made: check the amperage draw in the secondary side of the transformer with all the equipment operating or with all the equipment in the operating mode that will draw the greatest amount of current.
- Transformer phasing guarantees that all the electrons are flowing in the same direction. This is sometimes called polarizing transformers.

# Service Call

A customer complains that his heating unit is not operating. A check of the system reveals that it is completely off. The customer says that this has happened before. The technician checks the voltage at the transformer primary, which shows 120 VAC. The transformer secondary terminals show no voltage. The transformer is therefore bad. The tech replaces the transformer with a new one of the same size—a 40 VA transformer. A check of the current draw in the control circuit shows 1.75 amps. Calculation for the size of the needed transformer results in a minimum of 42 VA. The 40 VA transformer is too small; this is why the other transformers failed. At least a 45 VA transformer is needed; however, the available transformer nearest to this load is 50 VA. The tech therefore installs the 50 VA transformer and puts the unit back in operation. The technician is satisfied that the system is repaired.

# Student Troubleshooting Problem

A customer complains that her residential air-conditioning system is not cooling. The unit has an air-cooled condenser that operates on 240 VAC. The indoor unit is gas-fired with a matching evaporator. A transformer is located in both the indoor and outdoor units. The complete unit is off. The power supply terminals have 240 VAC, but no voltage is found at the contactor coil control circuit terminals. The technician checks the secondary side of the transformer in the outdoor unit and finds no voltage. The transformer is bad. There is no transformer on the service truck and it is Saturday afternoon so all the supply houses are closed. How can the unit be put in operation until the supply houses are open and the correct transformer can be bought and installed?

# Questions

1. What type transformers are used in air-conditioning and refrigeration systems?
2. What is the common secondary voltage of transformers used in air-conditioning and refrigeration systems?
3. To what transformer coil is the line voltage connected?
4. How are transformers sized?
5. Write the formulas for sizing transformers.
6. Describe how to check the amperage draw of the control circuit.
7. Regarding the control equipment, why are low-voltage control circuits preferred to line-voltage control circuits?
8. Why are transformers generally phased?
9. When phasing transformers, what are the possible voltages?
10. When phasing three-phase transformers, what must be done?

# Unit 16: CONTACTORS AND STARTERS

## Introduction

In the air-conditioning and refrigeration industry, the largest switching load for the control system is the compressor motor. The various fans, water pumps, and other needed machinery wired in electrical parallel to the compressor motor also add to the load requirements of the starter or contactor. Several methods of controlling the different loads are used, but this discussion will only involve the starter and contactor.

**Definitions.**    Following are definitions for both starters and contactors.

*Starter.* A starter is an electrical controller for accelerating a motor from rest to its normal speed.

*Contactor.* A contactor is a device for repeatedly closing and opening an electric circuit. It usually includes parts for equipment and circuit protection against overload.

Starters and contactors have many common features. For example, both have electromagnetically operated contacts that make or break an electric circuit on demand from a controller. Both may be used to make or break voltages that differ from the control circuit voltage. A single starter or contactor may be used to control more than one circuit. Thus, starters and contactors offer the following features for use in an electric circuit.

1. Complete isolation between the control circuit voltage and the controlled voltage
2. High power gain that can be used in the *controlled* circuit rather than in the *control* circuit
3. Ability to control most operations with one controller with a single control voltage; also, complete electrical isolation is possible between the controlled voltages

**Contactor.**    A contactor is used for switching heavy current, high voltage, or both. See Figure 16–1. Contactors generally do not have overload relays or any of the other features commonly found on starters.

**Starter.**    A motor starter may consist of a contactor used for switching the electric power to a motor; however, a starter usually has additional parts such as overload relays and holding contacts. These components may also include step resistors, disconnects, reactors, or other hardware needed for a more conclusive starter package for large motors. See Figure 16–2.

*Operation.*    Starters and contactors operate on the electromagnetic principle. When the coil is energized, the magnetic field that is generated pulls an armature into it. The armature carries movable contacts that either make or break a circuit.

Different manufacturers build starters and contactors with different types of armatures. Some connect the armatures to the relay frame with hinges or pivots. Others use slides to guide the armature into the magnetic field. Except for this difference their basic operation is the same; therefore, this discussion will include only the hinge type. See Figure 16–3.

The armature is the moving part of the starter or contactor. It is hinged on one end to the frame and has a movable contact assembly on the other end. A spring keeps the armature pulled out of the electromagnetic field when the coil is deenergized. The metallic armature is easily magnetized by the lines of flux from the electromagnetic coil when it is energized. The magnet is made of a coil wound around a laminated soft iron core. When energized, the coil becomes an electromagnet and pulls the armature toward it.

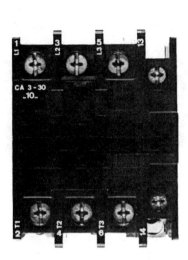

**Figure 16–1**
Three-pole contactor. (Courtesy
of Sprecher and Schuh.)

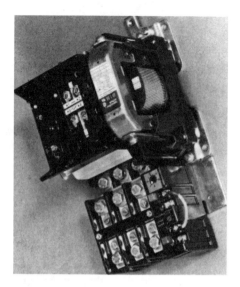

**Figure 16–2**
Three-pole starter. (Courtesy of Sprecher
and Schuh.)

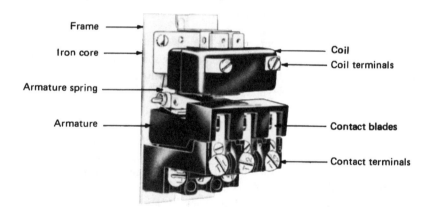

**Figure 16–3**
Contactor components. (Courtesy of Arrow-Hart, Inc.)

**Coil.**   Coil operation depends on the wire and the method used in winding it. Potential-wound type coils react to some value of voltage to pick up the relay armature. The current-wound type responds to some value of current when energized. Terminals connect the coil to the control voltage wiring through a controller such as a thermostat or pressure control. See Figure 16–4.

**Figure 16–4**
Contactor coil. (Courtesy of Sprecher and Schuh.)

The conditions that the coil would be used in determine the kind of insulation used. For example, those that will be exposed to high humidity conditions have water-resistant type insulation. Be sure to use the one recommended by the coil manufacturer.

The voltage rating of the coil must match the voltage of the control circuit. If the control circuit voltage is less than the coil rating, the starter or contactor will probably chatter and not operate properly, if at all. If the control circuit voltage is greater than the coil rating, the coil will burn out in a short time, requiring replacement.

**Contacts.**    Contacts are the current-carrying parts of the control circuit. The type of load to be switched is important when sizing starters and contactors. To match the starter or contactor correctly, the technician must consider the different characteristics of inductive (motors, for example) and noninductive (resistive) loads. An inductive load is a motor or some other device that draws a higher starting current than running current. A noninductive load is one in which the current draw never changes under normal operation, such as the heating elements in an electric heater, a relay coil, or a solenoid coil.

In an ac circuit with an inductive load, such as an electric motor, the initial inrush of current is always higher than the normal operating current. Before the motor reaches acceptable speed, the back EMF is quite low, allowing the maximum current to flow in the circuit. As the motor gains speed, the back EMF increases to oppose the high current flow, reducing it to the normal running current. Starter or contactor contacts, therefore, must not only withstand the normal running current but also the starting inrush current. A good table of contact ratings lists both full-load and locked-rotor current ratings. See Table 16–1.

Contacts are made of silver cadmium alloy. This material keeps contact-sticking to a minimum. The contacts are bonded to a strong backing member. They are cool operating even at loads of 25 percent above their ratings.

**Pole configuration.**    Starters and contactors may have from one to any number of contact poles. Generally the most popular sizes have from one to six poles with one, two, and three poles being the most popular. See Figure 16–5. The purpose of these

**TABLE 16–1**  COMPRESSOR MOTOR CURRENT RATINGS

| Single-Phase Motor | | Three-Phase Motor | |
| --- | --- | --- | --- |
| Compressor Size (Tons) | Motor Current (Amperes) | Compressor Size (Tons) | Motor Current (Amperes) |
| 2 | 18 | 3 | 18 |
| 3 | 25–30 | 4 | 25–30 |
| 4 | 30–40 | 5 | 30–40 |
| 5 | 40–50 | 7½ | 40–50 |

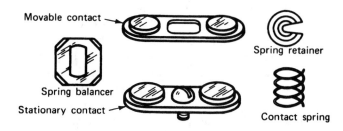

Movable contact

Spring retainer

Spring balancer

Stationary contact

Contact spring

**Figure 16–5**
Set of contacts consisting of one pole.

contacts is to connect the load to the power source. Starters and contactors may be bought with only the necessary number of poles or they may have extra sets, depending on the system needs. Any extra poles may be used as auxiliary contacts or left unused.

The contact size is determined by the current draw of the load they are expected to handle. Contactor or starter contacts may be larger than necessary, but they should never be smaller than the rated amperage of the load. When they are smaller, the contacts quickly pit and are ruined.

Auxiliary contacts are normally used to make an interlock circuit, such as another contactor connected to a water pump, fan motor, or similar part, or to make a circuit while the starter or contactor is in the deenergized position. These contacts are generally rated only for pilot duty (small current draw) and are not intended to withstand the heavy current used by the main load.

**Overload relays.**    Overload relays are primarily used to protect the equipment and circuit being controlled by the starter. Overload relays are usually mounted on the side of a contactor, thus completing the definition requirements of a starter. Overload relays may also be conveniently mounted elsewhere in the circuit, such as in the load side of the starter where the current draw can be sensed and the relays can be conveniently serviced. In case of an overload condition, the relays interrupt either the control circuit or one side of a single-phase power line. On three-phase circuits the control circuit is always broken.

***Operation.***     Overload relays operate on the principle that current produces heat. The main load is directed through a resistance wire in the relay. If the draw is greater than the rating of the resistance wire, additional heat is radiated either to a bimetal switch or to another type of material used to hold the overload contacts closed. When enough heat is caused, the overload contacts open the control circuit and deenergize the starter or contactor coil, thus breaking the electrical circuit to the load. The overload may be either manually reset or automatically reset before the overload will again complete the control circuit to the coil. See Figure 16–6.

### Two-speed compressor contactors.     Many manufacturers use two-speed compressors in their units, which requires a different type starting device. This is a combination contactor having two coils and nine or ten sets of contacts, depending on the equipment needs. See Figure 16–7.

Two-speed compressor contactors normally have two auxiliary switches in the contactors, one on the side of each contactor. These are normally closed (NC) contacts. Usually both contactors are mounted on one common base. They have both a mechanical and an electrical interlock to keep the compressor from operating in both speeds at the same time. The auxiliary contacts provide the electrical interlock. These interlocks must never be bypassed. To do so would result in a direct short circuit across the power line, causing damage to both the contactor and the circuit breaker. Also, any power surges created by this shorted condition could possibly damage the compressor motor windings. The tech should follow the wiring diagrams for each individual unit to

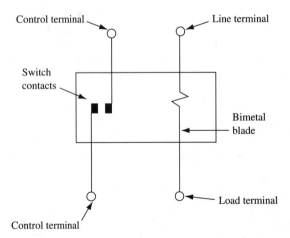

**Figure 16–6**
Representative overload relay.

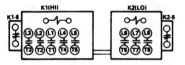

**Figure 16–7**
Two-speed contactor. (Courtesy of Lennox Industries, Inc.)

maintain correct operation of the equipment. Control relays (see Unit 17) are also sometimes used with two-speed contactors so the compressor can operate only in either high or low speed.

***Operation.***    The Lennox HP14 unit with a two-speed compressor will be used in describing the operation of the contactor—an important control. Figure 16–8 shows the compressor starting circuit, single phase, low speed. Each step number corresponds to the circled number in Figure 16–8.

1. The low-speed compressor contactor (K2) is energized by the timed OFF control circuit, thus closing contacts L6–T6, L7–T7, L8–T8, and L9–T9.
2. To energize the start windings, the start and run capacitors are wired in parallel to give maximum starting torque for the motor.

   L2 feeds power through the following:

   Low-speed capacitor (K2) contacts L9–T9

   Run capacitors (C1)

   Start capacitor (C3)

   Current-limiting device (RT-2)

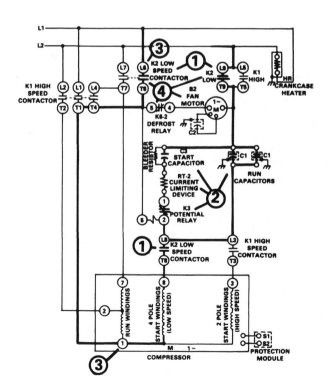

**Figure 16–8**
Compressor starting circuit, single phase, low speed. (Courtesy of Lennox Industries, Inc.)

Potential relay (NC) contacts (K3)

Low-speed contactor (K2) contacts L8–T8

The low-speed compressor start windings are energized at compressor terminal 8.

3. The start winding connects the common terminal 1 and completes the circuit to L1 through the low-speed contactor (K2) through contacts L6–T6.

4. The fan motor is also energized from L1 through contacts L6–T6 in K2 to L2.

Figure 16–9 shows the compressor-run circuit, single phase, low speed. Each step number corresponds to the circled number in Figure 16–9.

1. When the start windings are energized, the compressor-run windings that are in series are powered through the low-speed contactor (K2) contacts L6–T6 and L7–T7.

2. As the compressor comes up to speed, the potential relay coil (K3) is energized by voltage from the start-motor windings through contacts L8–T8 in (K2). This voltage is usually equal to or greater than the pickup voltage needed for the potential relay. The pickup voltage varies with each compressor model. Thus, a different potential relay is needed for each model.

The normally closed contacts in the starting relay K3 open, taking the start capacitor out of the circuit.

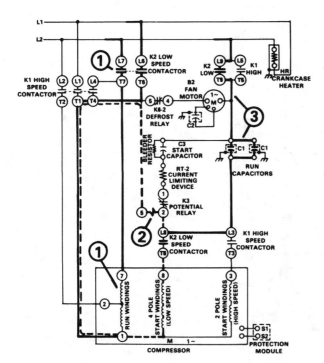

**Figure 16–9**
Compressor run circuit, single phase, low speed. (Courtesy of Lennox Industries, Inc.)

3. The run capacitor(s) remains connected to the motor start winding through contacts L9–T9 and L8–T8 in contactor K2. Run capacitors create the proper voltage phase shift to improve the power factor and increase the motor torque.

Note: If the run capacitor(s) fails, the compressor may not start. If it does start, it will run with very poor power factor causing high electric bills. The run capacitor(s) provides two functions:

1. Increasing the starting capacitance when connected in parallel with the start capacitor(s).
2. Improving the power factor and torque characteristics during the run mode.

Figure 16–10 shows the compressor starting circuit, single phase, high speed. Each step number corresponds to the circled number in Figure 16–10.

1. L2 feeds power through the following:

   High-speed contactor (K1) contacts L5–T5

   Run capacitor(s) C1

   Start capacitor(s) C3

   Current-limiting device RT-2

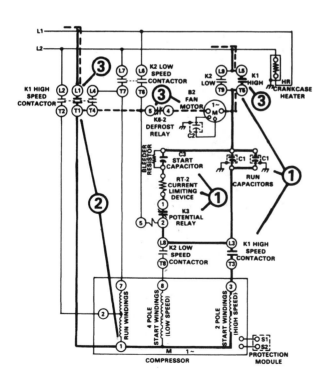

**Figure 16–10**
Compressor starting circuit, single phase, high speed. (Courtesy of Lennox Industries, Inc.)

Normally closed contacts in the potential relay K3

High-speed contactor (K1) contacts L3–T3.

High-speed start windings are energized at terminal 3

2. The start winding is connected to the common terminal 1 and completes the circuit through contacts L1–T1 in the high-speed contactor K1.

3. The fan motor is also energized from L1 through contacts L1–T1, and through contacts L5–T5 in K1.

Figure 16–11 shows the compressor-run circuit, single phase, high speed. When the start windings are energized, the compressor-run windings, wired in parallel, are powered through the high-speed contactor (K1), as follows:

1. L1 power is fed through contacts L1–T1 and L4–T4 in contactor K1 to the compressor terminals 1 and 7.

2. L2 power is fed through contacts L2–T2 in contactor K1 to compressor terminal 2.

3. As the compressor comes up to speed, the potential relay coil (K3) is energized by the voltage from the motor starting windings through contacts L3–T3 in contactor K1.

4. The run capacitor(s) remains connected to the start windings through contacts L5–T5 and L3–T3 in contactor K1.

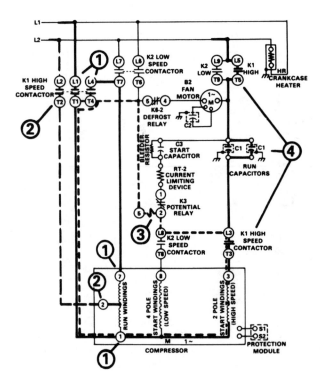

**Figure 16–11**
Compressor run circuit, single phase, high speed. (Courtesy of Lennox Industries, Inc.)

***Three-Phase Operation.***     The low-voltage circuits in three-phase units energize the compressor contactors K1 or K2 the same as they do in single-phase units. No starting components or run capacitors are needed or used on three-phase units.

Figure 16–12 shows the compressor circuit, three phase, low speed, series Y. Each step number corresponds to the circled number in Figure 16–12.

1. The K2–1 low-speed compressor contactor and the K3 fan relay are energized by the low-voltage circuit.

2. The fan motor B2 is energized from L1 through the normally open contacts in relay K3–2. The normally closed defrost relay contacts are in K6–2, and the normally open contacts are in K3–1 to L3. When the normally closed contacts in K3–1 and K3–2 open, they deenergize the crankcase heater.

3. The compressor terminals 1, 2, and 3 are energized through contacts L6–T6, L7–T7, and L8–T8 in contactor K2–1 to form a Y connection to the motor windings for low speed.

Figure 16–13 shows the compressor circuit, three phase, high speed, Parallel Y. Each step number corresponds to the circled number in Figure 16–13.

1. The K1–1 high-speed compressor contactor and the K3 fan relay are energized by the low-voltage circuit.

2. The fan motor B2 is energized from L1 through the normally open contacts in K3–2, the normally closed contacts in the defrost relay K6–2, and the normally open contacts in contactor K3–1 to L3. When the normally closed contacts in contactor K3–1 and K3–2 open, they deenergize the crankcase heater.

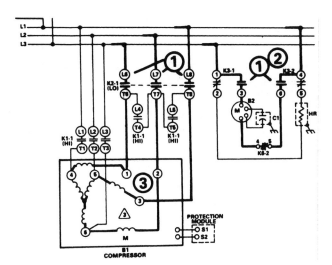

**Figure 16–12**
Compressor circuit, three-phase, low-speed series Y motor circuit. (Courtesy of Lennox Industries, Inc.)

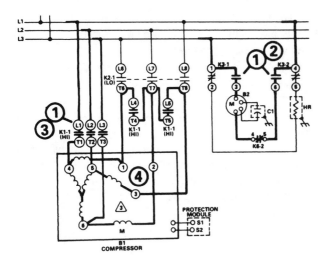

**Figure 16–13**
Compressor circuit, three-phase, high-speed parallel Y motor circuit.
(Courtesy of Lennox Industries, Inc.)

3. Compressor terminals 4, 5, and 6 are energized through contacts L1–T1, L2–T2, and L3–T3 in L1-1.

4. The compressor winding terminals 1, 2, and 3 are connected by contacts L4–T4 and L5–T5 in contactor K1 to complete the parallel Y connection to the motor high speed.

# Summary

- A starter is an electrical controller for accelerating a motor from rest to normal speed.
- A contactor is a device for repeatedly making and breaking an electric power circuit.
- Contactors are used for switching heavy current, high voltage, or both.
- Contactors generally do not have overload relays or any of the other features commonly found on starters.
- A starter usually has additional components such as overload relays, holding contacts, step resistors, disconnects, reactors, or other hardware needed for a more complete starter package.
- The armature is the moving part of a starter or contactor.
- The magnet consists of a coil wound around a laminated, soft iron core. This becomes an electromagnet and pulls the armature toward it when the coil is energized.
- The voltage rating of the coil must match the voltage applied to it.
- When sizing starters and contactors for a particular application, the type of load to be switched is quite important.
- Starter or contactor contacts must not only withstand the normal running current but also the starting inrush current.
- The purpose of contacts is to connect the load to the power source.

- Auxiliary contacts are normally used to complete an interlock circuit, such as another contactor connected to a water pump, fan motor, or similar component, or to complete a circuit while the starter or contactor is in the deenergized position.
- Overload relays are safety devices used to protect the main load from damage that could possibly be caused by a too high current draw.
- Overload relays operate on the principle that current produces heat.

# Service Call

A customer complains that the outside unit keeps tripping the circuit breaker. A check of the system reveals that the three-ton residential system will sometimes trip the circuit breaker when it tries to start. The voltage to the line-side terminals of the contactor is 240 volts when the unit is trying to start. The amperage is shown as normal starting amperage when the unit is trying to start. Voltage at the compressor terminals was 210 volts when starting. The voltage at the load side of the contactor is also 210. The technician checks the contactor contacts and finds them to be pitted and burned too bad for use, so he replaces the contactor. The resistor on the capacitor is open, so the tech also replaces the resistor. The tech starts the unit, and the starting and running amperage and voltage are now normal. The breaker does not trip. The technician is satisfied that the system is repaired.

# Student Troubleshooting Problem

A customer complains that a commercial refrigeration unit is not cooling. A check of the system reveals that it has an air-cooled condenser. Nothing in the condensing unit is running. The unit uses 240 VAC current. The voltage to the unit is at 240 VAC. The thermostat and pressure switches are demanding cooling. The technician pulls in the double-pole contactor to close the contacts. The line side of the contactor terminals has 240 VAC. The load-side terminals have no voltage. The contactor contacts are bad. The technician has only a three-pole on the service truck. How can the technician wire the three-pole contactor so that the system will operate?

# Questions

1. What is the largest switching load in air-conditioning and refrigeration systems?
2. How many circuits can a starter or contactor control?
3. What devices do starters have that contactors do not?
4. What is the moving component in a starter or contactor?
5. What causes the armature to change positions in a contactor or starter?
6. What are the two types of coils used in starters and contactors?
7. Is it safe to use a 24-volt coil on a 240-volt control system?
8. Does the type of load affect the type of contacts used?
9. When does the most current flow in a motor circuit?

10. How many poles are used in a starter for a three-phase single-speed motor?
11. How are auxiliary contacts rated?
12. For what are the overload relays used on starters?
13. What line is interrupted by an overload in a three-phase system?
14. How many auxiliary switches are generally used in two-speed contactors?
15. What is the best method to use when wiring a two-speed compressor?

# Unit 17: ELECTROMAGNETIC RELAYS
# Introduction

The electromagnetic relay is used for many purposes in the air-conditioning and refrigeration industry. Each year new and imaginative needs are specified by equipment manufacturers.

**Definition.**    A *relay* is a switching device that operates from an electrical input to the coil to operate the armature that opens and closes the contacts used to control the same or other circuits.

Relays are switching controls designed for the automatic control of single- or multi-speed motors in heating, refrigeration, and air-conditioning. They are also used for controlling heating and cooling system operation. The use of relays is limited only by the imagination. Covering all the uses for relays at one time would be almost impossible; therefore, discussion will involve only a small representation for the theory of operation.

*Operation.*    The operation of electromagnetic relays is almost the same as that of starters and contactors. If necessary, review the operation of those devices at this time. The main differences are the temperature of the relay coil, especially when the relay is enclosed; and the current-carrying capacity of the contacts. A coil is limited by how much heat it can radiate in a given amount of time to the ambient air or to the core of the magnet. Increasing the operating voltage causes an increase in temperature.

The outer windings of a coil operate cooler than the inner windings because they are closer to the surrounding air and give up more heat. Temperature differences cause hot areas inside the coil. These hot areas cause a breakdown of the coil insulation, resulting in a shorted winding.

**Pole configuration.**    Relays are available in almost any pole arrangement imaginable. The main arrangements are normally open (NO); normally closed (NC); single-pole double-throw (SPDT); double-pole single-throw (DPST); double-pole double-throw (DPDT); single-pole single-throw (SPST); and variations involving these arrangements. For symbols of these arrangements, see Table 17–1.

**Contacts.**    Most relay contacts consist of silver cadmium oxide mounted on beryllium copper blades for longer life and low resistance. When contacts become pitted and dirty, replacing the entire relay is usually best. If it is necessary to file them make

**TABLE 17–1**  RELAY SYMBOLS FOR
ELECTRICAL DIAGRAMS

| Pole Form | Symbol |
| --- | --- |
| SPST, NO | |
| SPST, NC | |
| SPDT | |
| DPST, NO | |
| DPST, NC | |
| DPST, NO, and NC | |

certain that the matching surfaces are square to each other, otherwise the contacts will pit and become useless in a short time. When contacts are filed, the thin layer of silver is removed from the contact. Contacts will not usually last long after filing. Filing should be done only as a temporary repair. The relay should be replaced as soon as possible. Be sure to turn off all electricity to the relay before attempting to work on the contacts, connections, or terminals.

**Electromagnetic relays.**   When the control circuit is made and the relay coil is energized, the armature is pulled into the electromagnetic field. As the NO contacts are closed (made), the electrical circuit to the other devices is made. When the control circuit to the relay is broken, the electromagnetic field collapses. This allows the armature spring to pull the armature away from the coil, moving the contacts, and either makes or breaks the circuit to the controlled parts.

**Fan relays and fan centers.**   *Fan relays and fan centers* are controls designed for the automatic control of one- or two-speed fan motors used in heating, refrigeration, and air-conditioning systems. Their primary purpose is to cause the fan or blower motor to operate at the correct speed at the correct time. Some original equipment manufacturers make their own fan centers using a relay and transformer.

*Fan Relay.*   The electromagnetic fan relay is primarily used to bypass the heating fan control to energize the fan motor for air-conditioning or ventilation. This relay is available alone or with other parts inside the housing for protection and safety, and is usually referred to as a fan center. The contacts of this relay should be heavy enough to handle the fan motor current. This may be either an SPST or an SPDT relay. The SPST relay has a set of normally open (NO) contacts; the SPDT relay has a set of normally open (NO) and a set of normally closed (NC) contacts. They are wired into the circuit according to their use. See Figure 17–1.

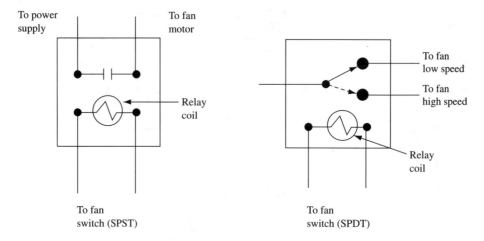

**Figure 17–1**
Typical fan relay schematics.

*Operation.*    When the thermostat demands cooling or when the fan switch on the thermostat is turned to the ON position, the relay coil is energized, thus causing the armature to be pulled into the magnetic field of the coil. The armature causes the NO contacts to close and the NC contacts to open, directing electric current to the blower motor. When the SPDT relay is used, the opening of the NC contacts keeps the fan motor from being energized in both the high and low speed at the same time when two-speed motors are used. SPDT relays are used in heating and cooling applications. When the NC contacts are opened and the NO contacts are closed, electricity can flow through the fan control as needed by the temperature inside the heat exchanger and make the fan motor run.

*Fan Center.*    Fan centers are used for the same purposes as fan relays, but they include a low-voltage transformer and a terminal board for wiring convenience. They also include an electromagnetic relay. Transformers are used to provide low-voltage power for the control system and its parts. The transformer is sized with enough capacity to operate the fan relay, the compressor contactor, and any auxiliary equipment controls as long as their combined load does not exceed the transformer VA rating.

*Operation.*    Fan centers operate the same as the fan relay. The only difference is that the transformer and terminal board are used in fan centers.

**Lockout relay.**    The lockout relay is a compressor safety device, but not a compressor overload. The relay coil voltage is the same as the contactor coil voltage. Lockout relay contacts are usually for pilot duty only, as they have only one set of NC contacts. It is wired into the control circuit with the NC contacts in electrical series with the starter or contactor coil. The coil is wired in parallel with the control circuit. See Figure 17–2.

*Operation.*    If any of the other safety controls such as pressure switches or compressor overloads stop the compressor motor, the lockout relay coil is energized and opens the NC contacts that keep the unit from restarting until the lockout relay has been reset. When the circuit through the NC contacts is broken by a safety device in the control circuit, the voltage is directed through the relay holding coil. This pulls in the armature, opening the NC contacts in the control circuit. The system is now locked out and will not operate until it is reset. The control voltage is not strong enough to cause both the lockout relay and the contactor coil to pull in their armatures simultaneously when they are wired in series. The lockout relay coil needs less voltage than the contactor coil; therefore, its magnetic field will pull in the armature and open the NC contacts. To reset the lockout relay, turn the electric power off to the control circuit at the main disconnect, the thermostat selector switch, or any other place that will break the power to the control circuit voltage.

**Heating relay.**    Heating relays are used in the heating units to energize the heat strips or the gas valve as the thermostat demands. The coil is energized by the 24 VAC control circuit. The NO contacts may energize more than one gas valve or step relays for electric heating elements. See Figure 17–3.

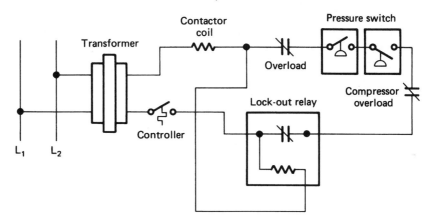

**Figure 17–2**
Typical lockout relay diagram.

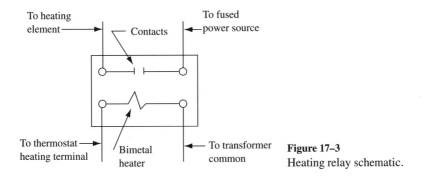

**Figure 17–3**
Heating relay schematic.

*Operation.*    When the thermostat demands heat, the control circuit energizes the heating relay coil. The coil then becomes an electromagnet and pulls in the relay armature. At this time the NO contacts are closed, completing the circuit to the gas valve, electronic ignition system, or the heat strips. The heating relay remains energized until the thermostat is satisfied. When the control circuit is broken, the coil will lose its magnetic field. The relay contacts will return to their NO position, stopping the flow of power to the gas valve, ignition circuit, or the heat strips. The contact rating of these relays must be heavy enough to handle the current and voltage of the heating equipment.

**Electric heat relay.**    These relays are used to control conventional ON-OFF electric heating elements and to control fan operation in electric furnaces. They have two sets of NO contacts that close when the coil is energized. Electric heat relays may be used on furnaces that use either line-voltage or pilot-duty limit controls. The contact rating must be high enough to switch the load. Electric heat relays may be wired to control up to four heating elements. See Figure 17–4.

*Operation.*    The electric heat relay coils are energized when the thermostat demands heat. When the thermostat makes the control circuit to the heating relay coil, the NO contacts close. One set of the contacts makes the circuit to the fan motor. The other set makes the circuit to the heating element or elements. When used this way, the fan stays energized as long as the thermostat is demanding heat. All the elements are energized together.

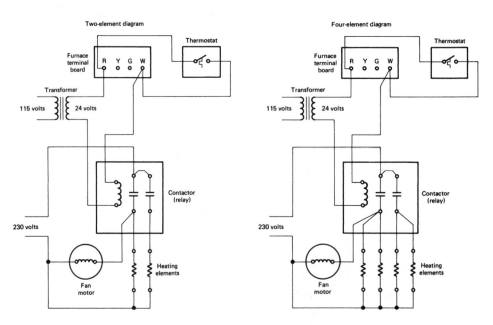

**Figure 17–4**
Wiring diagrams for electric heat relays.

**Figure 17–5**
Blower control. (Courtesy of Lennox
Industries, Inc.)

**Blower control.**    A blower control is used to control operation of the blower motor
on electric heating units. This control will start the blower motor any time the first heating element in the unit is energized. These relays have normally open (NO) contacts. When they
close, the blower motor is energized, moving air into the conditioned space. These line-voltage
contacts are opened and closed by the low-voltage control circuit. See Figure 17–5.

# Summary

- A relay is a switching device that operates from an electrical input to cause the
  operation of the controls in the same or other circuits.
- The operation of electromechanical relays is almost the same as that of starters and
  contactors.
- The coil is limited by how much heat it can radiate in a given amount of time to the
  ambient air or to the core of the magnet.
- Relays can be purchased in almost any type of pole arrangement imaginable.
- Most relay contacts are silver cadmium oxide mounted on beryllium copper blades
  for longer life and low resistance.
- When the control circuit is completed and the relay coil is energized, the electromagnet pulls the armature toward it. As the NO contacts are closed (made), the
  electrical circuit to the other devices is completed.
- Fan relays and fan centers are controls used for the automatic control of one- or
  two-speed fan motors used in heating, refrigeration, or air-conditioning systems.
- The electromagnetic relay is primarily used to bypass the heating fan control to
  operate the fan motor that provides air-conditioning or ventilation.
- Heating relays are used in the heating unit to energize the heat strips, electronic
  ignition module, or the gas valve in response to demands from the thermostat.

# Service Call

A customer complains that his outdoor unit is operating but no cold air is coming from the vents. A check of the system reveals that the fan motor is not running but the condensing unit is. The tech checks the fan relay and finds the contacts open to the fan motor high speed. The coil is then energized and a magnetic field is shown by the attraction of a screwdriver placed close to the coil core. Also, there is an audible click inside the relay when the control circuit power is turned off and on. The fan relay is bad. The technician replaces the fan relay, deices the evaporator coil, and puts the system back in operation. The temperature of the air leaving the grills inside the building is about 20°F cooler than the return air to the evaporator coil. The technician is satisfied that the system is repaired.

# Student Troubleshooting Problem

A customer complains that her small commercial building is not as warm as wanted. The system is an electric heating unit. It operates with 240 VAC, single-phase power. The thermostat is found to be demanding heat. The voltage to the unit is 240 VAC, single phase. The voltage from the transformer secondary is 24 VAC and the voltage to the inlet terminals of the heating relay coil is 24 VAC. There are 120 VAC to one side of the heating elements to ground and 120 VAC from the other side of the heating elements to ground. What could be the problem?

# Questions

1. To what device do relays operate similarly?
2. Why is the inside of a relay coil warmer than the outside?
3. Draw the electrical symbol for an SPDT relay.
4. How often should relay contacts be filed?
5. Draw the electrical diagram for a NO DPST relay.
6. For what are fan centers used?
7. What device is used to bypass the heating fan control on furnaces?
8. Draw the electrical symbol for a DPDT fan relay.
9. When replacing a fan center, what two things must be considered?
10. What relay is a compressor safety device, but not an overload?
11. Where in the control circuit are the contacts of a lockout relay located?
12. When a system is stopped by the lockout relay, how can it be restarted?
13. When is a heating relay coil energized?
14. In what position are the contacts of an electric heat relay when the coil is deenergized?
15. How are electric heat relays sized?

# Unit 18: THERMAL RELAYS
# Introduction

The thermal relay contacts and terminals are the same as the electromagnetic relay. The method of pulling the movable contact against the stationary contact is the major difference between them. The thermal relay uses a bimetal blade with a heater coil wound around it. The bimetal heater leads are connected to the control circuit power supply through the controller. When the controller demands operation, the bimetal heater warms the bimetallic element, causing it to bend toward the stationary contact to make the circuit. When the controller opens the control circuit, the heating element cools and moves the movable contact away from the stationary contact, breaking the circuit. These contacts open and close with a snap action to prevent excessive arcing. The heater element voltage rating must be the same as the control circuit voltage or the coil will either not operate or will burn out. See Figure 18–1.

**Fan relay.** The main purpose of the thermal type fan relay is to bypass the winter fan control and energize the fan motor for air-conditioning and ventilation. The contacts of this relay should be heavy enough to handle the fan motor current. Fan relay contacts may be either SPST or DPDT. The SPST relay has a set of NO contacts. The SPDT relay has a set of NO and a set of NC contacts. The SPDT model is necessary when a two-speed fan motor is used. They are wired according to their use. See Figure 18–2.

*Operation.* When the thermostat demands cooling or when the fan switch on the thermostat is turned to the ON position, the relay bimetal heating element is energized. After a given amount of time, the bimetal element causes the NO contacts to close and the NC contacts to open, making the circuit to the blower motor high speed. When the SPDT relay is used, the opening of the NC contacts keeps the fan motor from being energized in both high and low speed at the same time. The NC contacts make the circuit to the heating fan control.

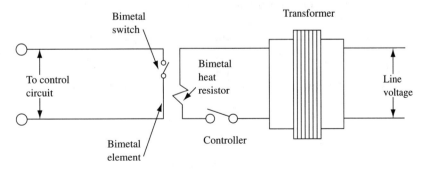

**Figure 18–1**
Thermal relay circuit.

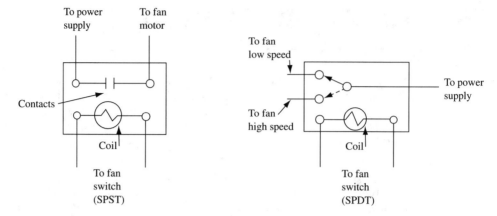

**Figure 18–2**
Typical fan relay schematics.

**Heating relays.**    Heating relays are used to make the circuit to the heat strips, or the gas valve, when the thermostat demands heat. The heating coil is energized by the 24-volt control circuit. The NO contacts may energize more than one gas valve or step relay for electric heating elements. They are operated by the heat which is generated by an electric coil wrapped around a bimetal strip that bends to either make or break the contacts. See Figure 18–3.

*Operation.*    When the thermostat demands heat, the heater relay coil is energized. After the bimetal has heated sufficiently, its NO contacts close and then energize the heating elements or gas valve, whichever is used. The heating relay stays in this position until the thermostat is satisfied. When the control circuit is broken, the relay contacts will open after the bimetal blade has cooled, breaking the circuit to the heating equipment. The contact rating of these relays must be heavy enough to handle the current and voltage of the heating elements or gas valve.

**Time-delay sequencing relays.**    Time-delay sequencing relays are bimetal type relays that use a heating element wrapped around a bimetal strip. These relays are normally used with electric heating units. They are popular in furnaces equipped with several electric heat strips for heating the building. Their purpose is to reduce the input of current to the unit during start-up by staging, or sequencing, the elements to come on at different times. This method may be done by using 24-volt heaters in the relays or a combination of 24 volts and line voltage to the relay heating elements, depending on the manufacturer's design and needs. Sequencing relays that have different time-delay operations and have the wanted time delay between each element on demand from the thermostat must be selected. A selection may be made so that the first one energized has a low-voltage heating element and the remaining relays have line-voltage heating elements, which are energized from a second set of NO contacts in the relay. Depending on the use, they may be either single relays or stacked relays. See Figure 18–4.

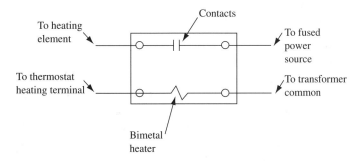

**Figure 18–3**
Thermal type heating relay.

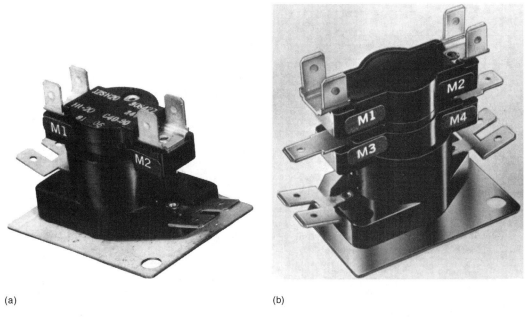

(a)                                          (b)

**Figure 18–4**
(a) Single and (b) stacked time-delay relays. (Courtesy of White-Rodgers Division,
Emerson Electric Co.)

*Operation.*    When the first relay has a low-voltage heater and the other relays have
line-voltage heaters, the operation is as follows: When the thermostat demands heat, the
first relay is energized. After the specified amount of time has passed, the relay closes its
contacts, energizing (1) the first heating element, (2) the blower motor that usually has its
own relay, and (3) the line-voltage to the bimetal heater of the second relay. After the needed
amount of time has passed, the second relay closes its contacts, energizing the second

heating element. When it closes its NO contacts, line voltage is directed to the third relay bimetal heater, which it energizes. This sequence continues until all heating elements are energized. See Figure 18–5.

When the thermostat is satisfied, the low voltage to the first relay is opened. This allows the bimetal heater to cool and open the contacts to break the circuit to the first heating element and to the bimetal heater of the second relay. After enough time has passed for the second relay heater to cool, the contacts open, breaking the circuit to the third relay, and so on until all the relays have been deenergized. Remember that the second, third, and following relays must have a very short deenergizing time, and also that the fan relay may have a longer deenergizing time.

When all the relays have low-voltage bimetal heaters, the operating sequence may be as just described or as follows: When the thermostat demands heat, all the bimetal heaters in all the relays are energized. However, because each has a different time before the contacts close, the heating elements will be sequenced automatically. See Figure 18–6.

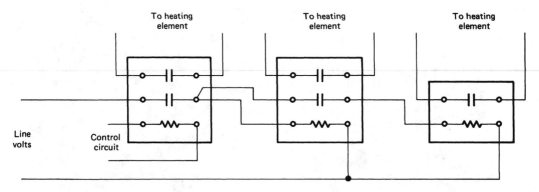

**Figure 18–5**
Series-wired time-delay sequencing relay wiring diagram.

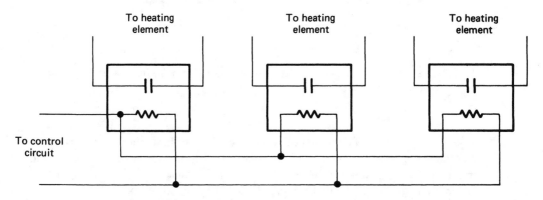

**Figure 18–6**
Parallel-wired time-delay sequencing relay wiring diagram.

# Summary

- The thermal relay uses a bimetal blade with a heater coil wound around it. The heater leads are connected to the control circuit power supply through the control circuit controller.
- Relay contacts are snap acting to stop excessive arcing of the contacts.
- The main purpose of the thermal type fan relay is to bypass the winter fan control and operate the fan motor for air-conditioning and ventilation.
- The contacts of this relay should be heavy enough to handle the current used by the fan motor.
- The SPDT type relay is needed when a two-speed fan motor is used.
- These relays are used in the heating unit to energize the heat strips, electronic ignition, or the gas valve in response to the temperature demands of the thermostat.
- Time-delay sequencing relays are bimetal type relays that use a heating element wrapped around the bimetal blade and are normally used in electric heating units.
- The purpose of time delay is to reduce the input current to the unit during start-up by staging, or sequencing, the elements to come on at different times.
- When bimetal type relays are used in a sequence, all those following the first must have very short off-time intervals.

# Service Call

A customer complains that the fan is off but the heating unit closet is very hot. The unit is an electric heating unit. The fan motor is off and the heat strips are cycling on the over temperature overload. After checking the unit, the technician learns that the fan relay is not operating. The contacts are found open and will not close after the specified time has passed. A check of the continuity of the heater coil shows it has resistance, and the contacts in the fan relay have infinite resistance. The fan relay is bad. The tech replaces the relay and puts the unit back in operation. The fan relay and all the electric strip relays operate correctly. After checking the voltage and amperage to all the heating elements, the technician is satisfied that the system is repaired.

# Student Troubleshooting Problem

A customer complains that her residential unit is not heating. The system is a gas-fired horizontal heating unit. Only the fan is running. The voltage to the unit is 120 VAC. The thermostat is demanding heat. There are 24 VAC at the transformer secondary terminals. The unit has a thermal fan relay. There is no voltage at the gas valve terminals. The unit is warmer than usual. What could be the problem?

# Questions

1. What is used to operate the contacts on thermal relays?
2. What is the purpose of the thermal fan relay?

3. In what position are the contacts of a NO relay when the coil has been energized for the given period of time?

4. In the thermal type fan relay, when is the bimetal heating element energized?

5. What causes time-delay heating relays to operate?

6. In what application are time-delay sequencing relays popular?

7. When using time-delay relays for heating element sequencing, how should they be selected?

# Unit 19: MOTOR STARTING DEVICES

## Introduction

Motor starting devices are operated either by current or voltage (potential). Current relays can be operated with either an electromagnetic coil or a thermal heating element. Voltage relays are operated only with voltage-sensitive electromagnetic coils.

**Voltage (potential) relay.** The voltage relay may be recognized by its high resistance coil wound with small gauge wire. The coil of the voltage (potential) relay is connected in electrical parallel with the starting winding of the compressor motor.

Potential relays are generally used on high-starting torque motors, but they may also be used on low-starting torque motors to help in starting the motor. They have a single set of contacts that are normally closed.

Potential relays have three terminals used to wire the relay into the circuit. The relay may sometimes have as many as six terminals, but the extras are only used as binder or auxiliary terminals, and have nothing to do with normal relay operation. Terminals 1, 2, and 5 are the operating terminals and terminals 3, 4, and 6 are the binding terminals—if they are on the relay housing. See Figure 19–1.

When wiring the relay into the starting system, terminal 5 and the run winding terminal of the compressor motor are connected to the same electric supply line. See Figure 19–2. Terminal 2 on the relay and the compressor motor start terminal are connected. Terminal 1 on the relay is connected to one of the motor starting capacitor terminals.

Potential relays are easily sized. If the correct size is not known, start the compressor motor manually and check the voltage between the start and common terminals of the compressor motor while the motor is running at full speed.

To start the motor manually, use a test cord that can be connected to the line voltage and to the common and run compressor motor terminals. Connect a wire to a spring-loaded switch from the run terminal to the start terminal. Place a start capacitor in this line. Turn on the electricity and press the spring-loaded switch to energize the capacitor circuit. When the compressor motor has started, release the spring-loaded switch. The compressor motor should keep running. Now the running voltage of the compressor motor can be read with an ammeter. See Figure 19–3.

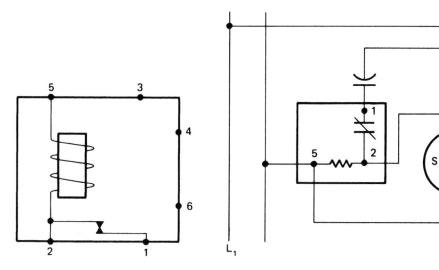

**Figure 19–1**
Symbolic voltage relay.

**Figure 19–2**
Typical voltage relay wiring diagram.

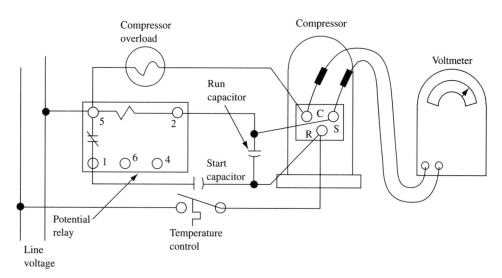

**Figure 19–3**
Checking voltage between start and run terminals.

Then multiply this voltage reading by 0.75 to find the pickup voltage of the relay. The pickup voltage is the voltage that will cause the magnetic field of the relay coil to be strong enough to pull the armature into its magnetic field. For example, a compressor motor is running at full speed and the voltage between the start and common terminals of the compressor motor is 300 VAC.

To find the pickup voltage of the needed potential relay:

$$E = 300 \times 0.75$$
$$E = 225 \text{ volts}$$

Thus, a potential relay with a pickup voltage of 225 volts is needed. It is best to check with the compressor motor manufacturer to find out exactly which relay should be used.

Table 19–1 shows the specifications of different potential relays. The table shows that the Mars 66 potential relay has a wide range of applications. It can operate with a continuous voltage of 432 volts. The minimum voltage pickup is 215 volts, with a maximum pickup voltage of 225 volts. The maximum dropout voltage is 120 volts. This would be a good replacement for our compressor motor.

*Operation.*    When electric power is first applied to the compressor motor, the contacts of the potential starting relay are closed. In the closed position an electrical connection is made between the start and run motor windings. The difference in the size of the wiring used to make the main and auxiliary (starting) motor windings causes an out-of-phase condition in the motor.

During the initial start-up, no back EMF is present and the maximum current is flowing through the motor windings. As the motor gains speed, a back EMF is generated in the auxiliary (start) winding. As the speed increases, the back EMF also increases. When the back EMF reaches the pickup voltage of the relay, the armature will pull in, opening the NC set of contacts. At this time the out-of-phase condition is removed by the starting relay, allowing the motor to run normally.

**Amperage (current) relay.**    The coil of the amperage relay is connected electrically in series with the running winding of the compressor motor. The amperage relay can be recognized by the low-resistance, heavy wire with which the coil is wound. This

**TABLE 19–1**  CALIBRATION SPECIFICATIONS: MARS—
GENERAL ELECTRIC—POTENTIAL RELAYS

| Mars Relay # | | Continuous Volt | | Pickup Min. | Pickup Max. | Drop Out Max. |
|---|---|---|---|---|---|---|
| Mars 63 | ¼-⅓-½-¾ | 200 | 115v | 139 | 153 | 55 |
| Mars 64 | 1-1½-1¾ | 432 | 230v | 260 | 275 | 120 |
| Mars 65 | ½-¾-1-1½ | 332 | 115v | 168 | 182 | 90 |
| Mars 66 | 3-4-5 | 432 | 230v | 215 | 225 | 120 |
| Mars 67 | 1¾-2-3-4-5 | 457 | 230v | 295 | 315 | 125 |
| Mars 68 | 2-3-4-5 | 502 | 230v | 325 | 345 | 135 |
| Mars 69 | ¾-1-1½ | 378 | 115v | 180 | 195 | 105 |
| Mars 70 | ¾-1 | 253 | 230v | 285 | 305 | 177 |

(Reprinted by permission of Prentice-Hall, Englewood Cliffs, N.J.)

is an electromagnetic relay. Amperage relays are normally used on 0.5-horsepower and smaller units with low-starting torque motors. The contacts of this relay are normally open. Amperage relays are positional and must be installed so that the weighted armature carrying the movable contacts will open the starting circuit when the current draw drops enough to lower the magnetism of the coil. Gravity will then open the contacts.

Both three- and four-terminal relays are available. Some are equipped with binding terminals. Three-terminal current relays have switch connections from L to S and the coil is between terminals L and M. See Figure 19–4. Four-terminal relays have switch connections from 3 to S and the coil is between terminals 2 and 4 and M. See Figure 19–5. Four-terminal relays are often used with three-terminal overloads and are a convenient method of connecting starting capacitors into the circuit.

An overload must be used with this type relay for motor protection in case of high current draw.

*Operation.*    When power to the motor is first turned on, the contacts of the coil type current relay are open. The high inrush of current through the relay coil causes the contacts to close and thus energizes the start winding, and therefore causes the electric phase shift in the start winding needed to start the compressor motor. As the motor gains speed, the current draw drops because of the back EMF, thus lowering the electromagnetic field in the relay coil. This allows gravity to pull the armature out of the relay coil and open the starting contacts. The relay takes the starting winding out of the circuit when the motor reaches approximately 75 percent of its full running speed. The motor then runs normally.

Amperage relays must be sized for each motor horsepower and amperage rating. When a relay is rated too large, the current draw may not be enough to close the relay contacts, leaving out the much needed starting circuit. The motor probably will not start under these conditions. A relay rated too small for the motor may keep the contacts closed at

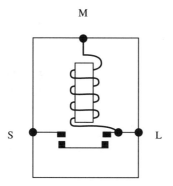

**Figure 19–4**
Symbolic three-terminal current relay.

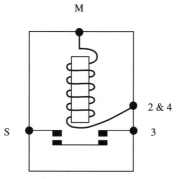

**Figure 19–5**
Symbolic four-terminal current relay.

all times when electric power is applied, leaving the starting circuit engaged continuously. The starting winding will probably be damaged. A motor protector must be used with current starting relays.

**Hard-start kit.**    The purpose of the hard-start kit is to help the compressor motor start under hard starting conditions. Such conditions include when the electrical power to the unit is below normal, when it fluctuates too low to give the necessary power for starting the compressor motor, or when a PSC motor is used on a system that uses rapid cycle operation. Hard-start kits are designed to convert PSC motors to capacitor-start capacitor-run (CSCR) motors. These kits consist of the correct starting relay and starting capacitor, and the necessary wiring for installation. The individual parts may also be bought to make up a hard-start kit. To install a hard-start kit, turn off the electrical power and complete the electrical connections as shown in Figure 19–6. Usually both a run and a start capacitor are used in these systems to help in starting and to increase unit efficiency. The two capacitors are connected in electrical parallel with each other. The run capacitor is connected between the start and run terminals and the run capacitor is connected between the line and the run terminal of the compressor motor. See Figure 19–7. When a potential starting relay is used, both it and the capacitors should be sized according to the motor manufacturer's recommendations. Some hard-start packages available are designed to be used on almost any size unit.

*Operation.*    When the compressor motor is first energized, the starting relay contacts are closed. The motor receives power through both the start and run capacitors. Capacitors produce the out of phase in the motor starting (auxiliary) winding to help it start

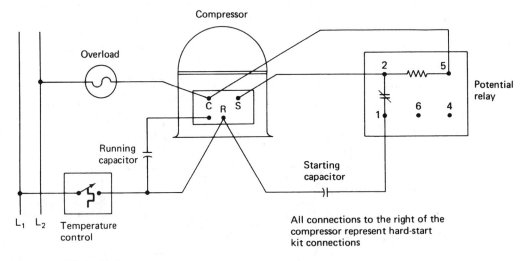

**Figure 19–6**
Diagram showing hard-start kit connections. (Prentice-Hall, Englewood Cliffs, N.J.)

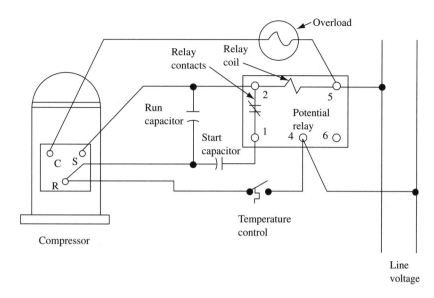

**Figure 19–7**
Compressor wiring diagram using run and start capacitors.

turning. As the motor reaches about 75 percent of its full rpm, the back EMF from the motor increases to the point that the relay coil will pull in the armature and open the contacts to the start capacitor. The run capacitor remains in the circuit to cause some phase shift in the start winding and to allow the motor to run more efficiently than when one is not used. The correct run capacitor will reduce the running amperage of the motor; however, if the wrong one is used, the amperage will increase. This increase in amperage will occur whether the run capacitor is too small or too large.

The system will operate as long as the motor is running. When the motor stops, the relay contacts close and the system is ready for the next start.

*Kickstart TO-5®.*    As with other hard-start devices, the Kickstart TO-5® also connects the start capacitor in parallel with the run capacitor through the normally closed contacts of the potential relay. See Figure 19–8. This hard-start kit differs from other hard-start controls because the back EMF is read across both the start and run windings of the motor. For every compressor motor, the back EMF across both the start and run windings is nearly identical. See Figure 19–9. Therefore, this type of hard-start kit uses only one potential starting relay that can be used on all PSC and CSCR motors. A larger than normal starting capacitor is used to make the Kickstart TO-5® a maximum torque hard-start unit. It is a two-wire device that uses a potential relay and a start capacitor all in one enclosure.

For wiring diagrams of different type installations see Figures 19–10 through 19–13 on pages 119–121.

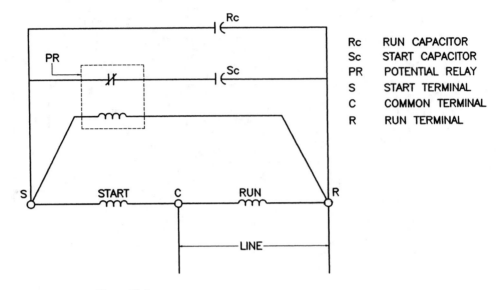

| | |
|---|---|
| Rc | RUN CAPACITOR |
| Sc | START CAPACITOR |
| PR | POTENTIAL RELAY |
| S | START TERMINAL |
| C | COMMON TERMINAL |
| R | RUN TERMINAL |

**Figure 19–8**
Wiring connections for the Kickstart TO-5®. (Courtesy of Kickstart®.)

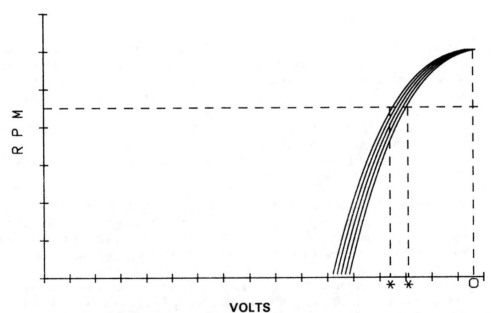

**VOLTS**
**BACK E.M.F. ACROSS START AND RUN WINDINGS FOR DIFFERENT COMPRESSORS**
**✳ POTENTIAL RELAY PICK-UP VOLTAGE RANGE AT DESIGNATED RPM FOR DIFFERENT COMPRESSORS**
**O POTENTIAL RELAY CONTINUOUS VOLTAGE**

**Figure 19–9**
Graph of back EMF across both start and run windings. (Courtesy of Kickstart®.)

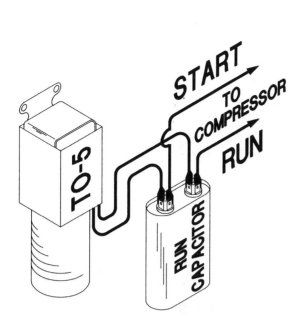

## STANDARD CONNECTION

**Figure 19–10**
Standard connection. (Courtesy
of Kickstart®.)

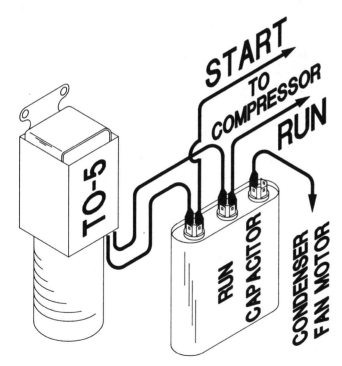

## DUAL CAPACITOR

**Figure 19–11**
Dual capacitor. (Courtesy of Kickstart®.)

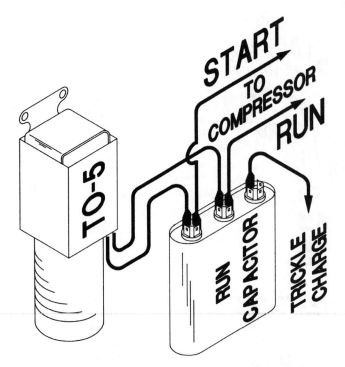

# SPLIT CAPACITOR

**Figure 19–12**
Split capacitor. (Courtesy of Kickstart®.)

## Summary

- The voltage (potential) relay may be recognized by its high-resistance coil wound with small gauge wire.
- The coil of the voltage (potential) relay is connected in electrical parallel with the start and run windings of the compressor motor.
- Contacts of a potential relay are NC.
- Potential relay terminals 1, 2, and 5 are the relay operating terminals.
- Relay terminal 5 and the run winding of the compressor motor are connected to the same electrical power line. A start capacitor may be installed in this line.
- Relay terminal 2 and the compressor motor start terminal are connected to the same electrical power line.
- Terminal 1 is connected to one of the motor start capacitor terminals.
- The pickup voltage is the voltage that will cause the relay coil electromagnetism to be strong enough to pull the armature into its magnetic field.

# CSIR COMPRESSOR

**Figure 19–13**
CSIR compressor. (Courtesy of
Kickstart®.)

- The coil of the amperage relay is connected electrically in series with the running winding of the motor.
- The amperage relay can be recognized by the low-resistance, heavy wire with which the coil is wound.
- The amperage relay is an electromagnetic relay.
- Amperage relays are normally used on 0.5 horsepower units and smaller with low-starting torque motors.
- The contacts of an amperage relay are NO.
- Amperage relays are positional and must be installed so that the weighted armature carrying the movable contacts will open the starting circuit when the current drops low enough to reduce the magnetism of the coil and allow gravity to open the contacts.
- Amperage relays need the use of a motor overload to give motor protection in case of high current draw.
- Amperage relays must be sized for each motor horsepower and amperage rating.
- The purpose of the hard-start kit is to help the compressor motor start under hard starting conditions.
- Hard-start kits are used to convert PSC motors to capacitor-start capacitor-run (CSCR) motors.

# Service Calls

*Service Call 1.*    A customer complains that the compressor will hum but will not usually start. The system is using a potential start relay. The technician checks the compressor starting amperage and finds it near the locked rotor amperage rating of the motor. The overload breaks the electric circuit to the compressor motor after a few seconds of trying to start. The electric power to the unit is turned off. The coil of the potential relay has continuity, and the contacts are in the closed position. The electric power is turned back on. When the compressor tries to start, the technician taps the relay with a screwdriver handle. The compressor motor starts and runs with the correct amperage. The starting relay is bad, so it is replaced with one of the correct size. The tech starts the system and finds the voltage and amperage to the compressor motor within the rating of the motor. The technician is satisfied that the system is repaired.

*Service Call 2.*    A customer complains that a domestic refrigerator sometimes hums before the compressor will start. The refrigerator's compressor uses an amperage starting relay. The refrigerator is turned on and the compressor starts humming but does not start. The technician taps the relay with a screwdriver handle and the unit starts running. The amperage is within the rating of the manufacturer. The tech removes the electric power cord from the wall plug and the relay from the compressor. A check of the contacts shows that they sometimes stick in the open position. The technician replaces the relay with the exact replacement and plugs the electrical cord back into the wall outlet. The compressor starts immediately. The voltage and amperage of the compressor motor are now within the rating of the compressor motor. The technician is satisfied that the system is repaired.

# Student Troubleshooting Problem

A customer complains that an air-conditioning unit will not always cool the building. A check of the system reveals that the compressor is warm and will not start at this time. The unit is located in a rural area. The unit is turned off. The voltage is 218 VAC at the line side of the compressor contactor terminals. The compressor is a PSC type. When the unit is turned on, the compressor tries to start. The voltage at the contactor contacts drops to 195. The voltage is too low for satisfactory operation. The compressor will not start. What could be done to help the compressor start every time?

# Questions

1. Are potential relays positional?
2. In what applications are potential relays mostly used?
3. How many terminals may be on a potential relay?

4. On a potential relay, to what does terminal 2 connect?

5. To what are potential relay terminal 5 and the common motor terminal connected?

6. On a potential relay, where is the start capacitor placed?

7. What value of voltage is used when sizing a potential starting relay?

8. If a compressor motor is producing 370 volts between the start and common terminals, what would be the pickup voltage of a replacement potential starting relay?

9. Can a potential relay with a pickup voltage of 195 volts be used on a 240-volt system?

10. What are two outstanding characteristics of an amperage relay?

11. Must an overload be used with an amperage relay?

12. In an amperage relay, what causes the contacts to close?

13. How are amperage relays sized?

14. On what type motors are hard-start kits used?

15. How are start and run capacitors connected into the motor circuits?

16. Does the start relay remove both the start and run capacitors from the circuit?

17. Will a larger than necessary run capacitor reduce the amperage draw?

# Unit 20: COMPRESSOR MOTOR OVERLOADS

## Introduction

Compressor motor overloads are used to protect the compressor motor during over temperature or over current conditions or both. Motor overloads may be mounted internal or external of the motor housing, depending on the compressor design. In most cases they are mounted near or in the hottest part of the motor winding. When replacing any type of compressor motor overload, be sure to use an exact replacement to ensure the correct protection. Overloads not properly sized can cause many problems. If sized for a high current draw, the compressor motor will not have the needed protection. If sized for too small a current draw, the compressor will shut down unnecessarily.

**Externally mounted overloads.**    The two types of external compressor motor overloads are the bimetal operated type and the hydraulic operated type. See Figure 20–1. The bimetal types react in response to both the temperature of the windings and the current draw of the motor. The hydraulic type reacts to the temperature of a hydraulic fluid being heated by the current flowing through the overload. These overloads are manufactured with two, three, or four electrical terminals.

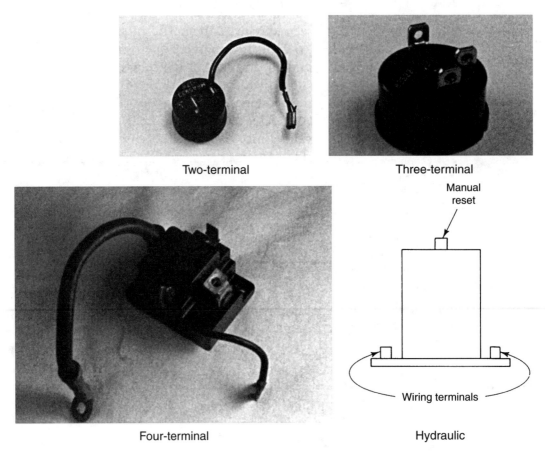

Two-terminal                        Three-terminal

Four-terminal                          Hydraulic

**Figure 20–1**
External compressor overloads.

*Bimetal Type Overload.*    The bimetal type overloads are designed with the following two methods of stopping the compressor motor in case of an overload: (1) the bimetal breaks the line voltage to the motor when an overload condition is sensed, and (2) the bimetal breaks the control voltage when an overload is sensed. This deenergizes the contactor or starter holding coil. Its contacts will then open to stop the line voltage to the compressor motor.

Some overloads are automatically reset but some must be manually reset before the motor will run again. The type used depends on the wants of the equipment manufacturer. An exact replacement must be used to maintain the correct protection.

*Operation.*    When the compressor motor is first energized, a slight time delay occurs before the overload will open. This delay is due to the cool bimetal. The bimetal takes a certain amount of time before it has warmed to the contact opening temperature. By the time this temperature is reached, the compressor motor amperage will have, under normal operating conditions, dropped to within its normal operating range. The overload contacts will remain closed until the compressor is cycled off normally. If the overload senses enough heat rise caused by the over current or an over temperature condition, its contacts will automatically open, breaking the electrical circuit to the compressor motor. The compressor will then stop. It will stay off until the overload has cooled enough to automatically reset or until it is manually reset. This cycling will continue until the overload condition has been corrected or until either a fuse is blown or a circuit breaker is tripped. The cause of the compressor motor overload must be found and corrected or the condition will only worsen.

**Hydraulic type overload.**    Hydraulic type overloads are generally mounted away from the compressor in a control panel. They may be either automatically or manually reset, depending on the equipment design. This type senses the current draw to the compressor motor only. The motor temperature does not affect these overloads, as they are generally used only on larger compressor motors. It uses the temperature of the hydraulic fluid in the overload to open and close the circuit.

*Operation.*    When the compressor motor starts, because the hydraulic fluid is cool, it takes a few seconds before it will sense an overload. Therefore, a slight time delay occurs to prevent nuisance shutdowns. All the current to the motor is directed through a heater that warms the hydraulic fluid. The warming of the fluid causes a rise in the fluid pressure that in turn causes a set of contacts to open and break the electrical circuit to the compressor motor. After the hydraulic fluid has cooled, the overload may be manually reset or it may reset automatically, depending on which type is used, before the compressor motor will restart. This cycling will continue until the cause of the compressor motor overload has been corrected.

**Internal overloads.**    Internal overloads may be of the line-break or the thermostatic type. Both types are exactly in the heat sink (the hottest place) of the motor windings to keep the motor windings from getting too hot and to sense high current draw and high winding temperature. See Figure 20–2.

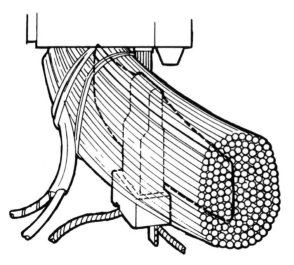

**Figure 20–2**
Internal overload location. (Courtesy of Lennox Industries, Inc.)

***Operation.***    The line-break type breaks the electric power to the common terminal of the motor winding when an overload occurs. See Figure 20–3. The thermostatic overload protector is also mounted in the heat sink of the motor windings. However, the switch breaks the control circuit to the starter or contactor holding coil. Notice that two additional terminals are on the compressor shell when this type is used. The electrical connections are shown in Figure 20–4. If either of these overloads goes bad, the compressor must be replaced. Therefore, before condemning the compressor motor be sure to let it cool enough so the overload will reset. Sometimes a slight tap on the compressor housing will jar the contacts closed. When the thermostatic type has opened, the two overload terminals located on the outside of the compressor housing can be jumped to complete the control circuit. The compressor motor should run and the suction gas will cool the compressor. Then the overload must be placed back in the control circuit and the cause of the overload corrected, otherwise the overload will stop the compressor again. If the overload contacts will not reclose after the compressor has cooled, the compressor must be replaced. Do not leave this overload jumpered because more damage will occur and thus make the repair more costly.

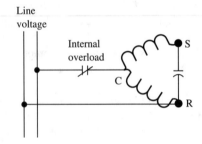

**Figure 20–3**
Line-break internal overload connection.

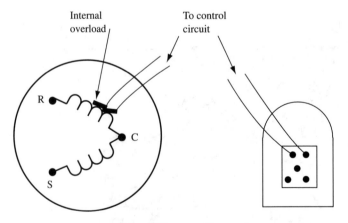

**Figure 20–4**
Internal overload connections.

# Summary

- Compressor overloads give protection for the compressor motor during over temperature or over current conditions or both.
- When replacing any type of compressor motor overload be sure to use an exact replacement to avoid improper motor protection.
- Two types of external compressor motor overloads are the bimetal operated type and the hydraulic operated type.
- Hydraulic overloads are generally mounted away from the compressor in a control panel.
- Internal overloads may be of the line-break or the thermostatic type.
- Both types are located exactly in the heat sink of the winding to protect the motor from high temperatures and high current draw.

# Service Calls

*Service Call 1.*    A customer complains that the compressor will start and then stop. A check of the system reveals that the compressor is cycling on and off. The unit has an external bimetal overload. The compressor housing feels quite warm to the touch. The technician removes the overload from the compressor housing and allows it to cool. After the overload has cooled enough for the contacts to close, the technician attempts to start the compressor motor but it only trips the overload. He jumps the overload and the compressor starts and continues to run. The voltage and amperage are within the normal operating range. The tech replaces the overload with an exact replacement and starts the compressor. It continues to run. Both the amperage and voltage stay within the normal operating range. The technician is satisfied that the system is repaired.

*Service Call 2.*    A customer complains that a refrigeration system is not cooling the product. The system's compressor is not running. The unit is equipped with an internal line-break motor overload. The compressor body is cool to the touch. There is voltage at the contactor load terminals. The technician checks the voltage at the compressor motor terminals and it is correct. The internal overload is open but will not close. The compressor is cool enough for it to reset. The tech jolts the compressor very hard, but the overload contacts do not close. The tech replaces the compressor using the correct procedures and turns the unit on. It starts and continues to run. The voltage and amperage are now within the normal operating range for the unit. The technician is satisfied that the system is repaired.

# Student Troubleshooting Problem

A customer complains that her air-conditioning system is not cooling. A check of the unit reveals that the compressor is not running and the contactor is pulled in. There is voltage at the compressor terminals. The compressor is hot to the touch. A PSC compressor motor with an internal thermostatic overload is used. The power to the unit is turned off. The technician removes all wiring from the overload terminals and checks the continuity of the overload. The overload is found open. The overload terminals are jumpered and the power to the unit is turned back on. The compressor starts running. The voltage and current are within the normal operating range. After a few minutes of operation the compressor has cooled down. The technician turns off the unit again and places the overload back in the control circuit. With the electricity on, the compressor starts running. A check of the voltage and amperage show that the voltage is in the normal operating range, but the amperage is just at the upper limits of the normal operating range. These readings show that the overload is probably good, but something has caused the compressor to get too hot. The suction pressure is at 70 psig. The return air is 80°F. What could cause this problem?

# Questions

1. Where are motor overloads usually placed?
2. What must be done when replacing a compressor motor overload?
3. What two things operate a bimetal type overload?
4. What causes the hydraulic type overload contacts to open?
5. What are the two types of bimetal overloads?
6. Why is a compressor motor usually allowed time to start before the overload trips?
7. Do hydraulic overloads sense motor temperature?
8. How does the fluid in a hydraulic overload open its contacts?
9. Where are internal overloads located?

# Unit 21: SOLENOID VALVES
# Introduction

Solenoid valves are used in refrigeration systems to control the flow of the refrigerant. A solenoid valve is a refrigerant control made by putting two separate parts into one component. A solenoid is a coil of wire, which, when energized by an electric current, becomes an electromagnet. See Figure 21–1. When this coil which is fastened to the stem of a valve is energized, a magnetic field pulls the plunger into the coil. The plunger either opens or closes the valve, depending on the needs of the system.

**Solenoid valves.**   Solenoid valves have many different uses in the refrigeration and air-conditioning industry. Because they are electrically operated and perform similarly to a manually operated valve, they are useful for automatic control of the refrigeration equipment.

Solenoid valves can be made to do many things merely by turning on or off an electrical circuit, which is usually done by a thermostat or another controller. When the electrical circuit to a solenoid coil is made, a magnetic field is built up around it and a plunger is pulled into the center of the coil. See Figure 21–2. By connecting this plunger to a valve stem, the valve can be made to open or close by making or breaking the electrical circuit to the coil.

*Operation.*    In the direct-acting type, when the circuit is made to coil A, the energized coil pulls plunger B upward, lifting valve disc C from the valve seat D. See Figure 21–3. This allows the fluid to pass through the valve until the circuit is broken to coil A. The plunger B then drops, and valve seat C returns to seat D, stopping the fluid flow. The valve spring E makes sure that the valve closes completely.

Valves equipped with manual operation have a knob on the bottom of the valve body. In case of power failure the valve may be opened by sliding the instruction sleeve off the knob and pushing the knob upward with the fingers. Then the knob is turned one-quarter turn in either direction to hold valve seat C in the open position.

To close the valve manually, rotate the knob one-quarter turn and the manual operator will return the valve to its normally closed position. The force of the valve spring E causes the valve to go fully closed.

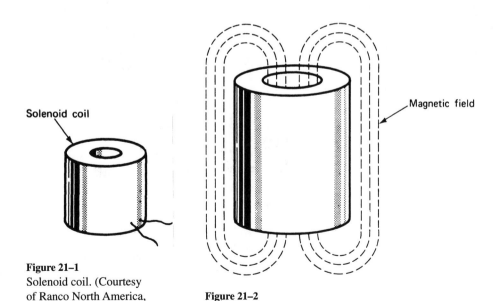

Solenoid coil

Magnetic field

**Figure 21–1**
Solenoid coil. (Courtesy of Ranco North America, Division of Ranco, Inc.)

**Figure 21–2**
Magnetic field.

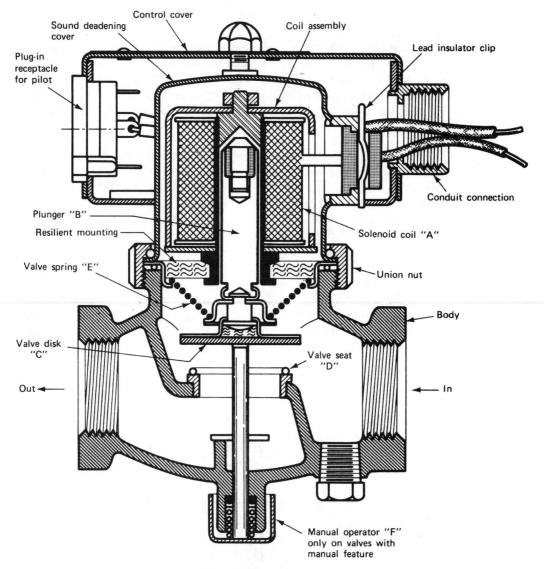

**Figure 21–3**
Direct-acting solenoid valve.

*Note:* On manual operation, the control does not return to automatic operation when the power is turned back on. It must be manually returned to the closed position before automatic operation will continue.

Pilot-operated solenoid valves are usually made in the larger sizes. See Figure 21–4. The plunger does not directly open the main valve seat in this type valve. When coil A is energized, plunger B, which has two seats, moves from seat C to seat D. Pilot port E is

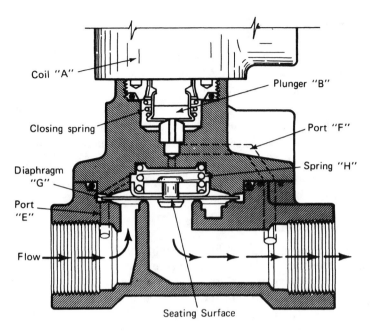

Coil "A"

Closing spring

Diaphragm
"G"

Port
"E"

Flow

Plunger "B"

Port "F"

Spring "H"

Seating Surface

**Figure 21–4**
Pilot-operated solenoid valve.

closed and bleed port F is opened, allowing the fluid to escape to the outlet side of the valve. The pressure is now the same on both sides of diaphragm G, and spring H can open the main seat and the fluid will flow straight through the valve.

When the electric circuit to the solenoid coil is broken, the pilot valve closes and allows pressure to build up above the diaphragm. The pressure, the closing spring, and the weight of the diaphragm close the main seat to stop the flow through the main valve.

All solenoid valves are made on these two basic principles. However, a few exceptions apply in the mechanical construction of these valves. Some examples are the lever valve, the general purpose two- or three-way valve, and the four-way valve. A correct valve should be chosen for the application.

For a solenoid valve to operate properly, the following general rules should be kept in mind when installing the valve.

1. Observe the arrows on the valve body—they show the direction of flow through the valve.

2. Use new pipe, properly chamfered and reamed, when making connections. Be careful when using pipe dope—keep it from getting inside the pipe or valve.

3. Ensure that the ambient air temperature around the solenoid valve does not exceed 125°F.

# Summary

- Solenoid valves are used in refrigeration systems to control the flow of refrigerant.
- A solenoid is a coil of wire, which, when energized by an electric current, will operate like an electromagnet.
- These valves can be made to do many things by turning on or off an electric circuit, which is usually done by a thermostat or another controller.
- In pilot-operated solenoid valves, the plunger does not open the main valve seat.
- When installing solenoid valves, make certain that the arrow on the valve body is pointing in the direction of fluid flow.

# Service Call

A customer complains that his air-conditioning system does not cool the building. A check of the system reveals that the compressor is off and that the system is operating with a pump-down cycle. The voltage to the solenoid coil is at 240 VAC. The technician removes one wire from the coil but hears no click. It is believed that the solenoid valve is stuck in the closed position. The tech reconnects the wire to the coil and taps the valve lightly, but it does not open. The solenoid valve is stuck closed. The tech removes all the refrigerant from that part of the system, replaces the solenoid valve, and puts the system back in operation. All refrigerant pressures are in the normal operating range. The tech again removes a wire from the solenoid coil and now hears a click. When the wire is back on the coil, the technician can again hear a click. The valve also clicks when the controller is turned on and off. The technician is satisfied that the system is repaired.

# Student Troubleshooting Problem

A customer complains that his refrigeration system is freezing everything in the case. The system is air-cooled and uses a pump-down cycle. The evaporator coil is iced over. All the equipment is running. A check of the thermostat shows that its contacts are open. There is no voltage to the pump-down solenoid. What could be the problem?

# Questions

1. How is a solenoid valve made?
2. How are solenoid valves controlled?
3. What causes the plunger to be pulled into the solenoid coil?
4. What component makes certain that a solenoid valve closes?
5. In what type solenoid does the plunger not directly open the valve seat?
6. When installing a solenoid valve, in what direction should the arrow point?

# Unit 22: REVERSING (FOUR-WAY) VALVES

## Introduction

Reversing valves are designed for various nominal tonnage capacities and for the automatic operation of heat pump and air-conditioning systems. They are also used on commercial refrigeration systems that use hot gas for defrosting the evaporator coils. Reversing valves should be sized according to the equipment manufacturer's specifications. These valves are hermetically built and pressure-differential operated. See Figure 22–1.

*Operation.* Reversing valve operation is controlled by energizing and deenergizing the solenoid coil. The coil is fastened on top of a three-way pilot valve with a locknut. The pilot valve is part of the main valve. When used on a heat pump system, the coil may be energized either during the cooling or heating cycle. One way is no more correct than the other—the choice depends on the needs of the equipment manufacturer. When used for hot gas defrost on refrigeration systems, the coil is deenergized during the cooling cycle. In the following discussion assume that the coil is energized during the heating cycle.

Reversing valves instantly reverse running system refrigerant pressures and operate completely on the pressure differential between the high and low sides of the system, within the valves listed capacities.

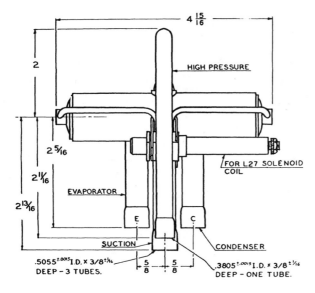

**Figure 22–1**

Reversing (four-way) valve. (Courtesy of Ranco North America, Division of Ranco, Inc.)

The refrigerant path is shown schematically through the main valve. This diagram shows the sliding port at a position over two tube openings as it transfers both refrigerant coils between the operating phases of cooling, deicing, and heating.

The solenoid coil is not energized in the normal cooling cycle, which allows the refrigerant to flow through its normal cycle. See Figure 22–2.

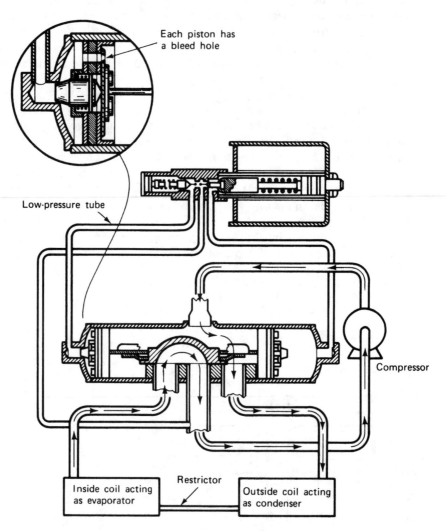

**Figure 22–2**
Deenergized solenoid coil. (Courtesy of Ranco North America, Division of Ranco, Inc.)

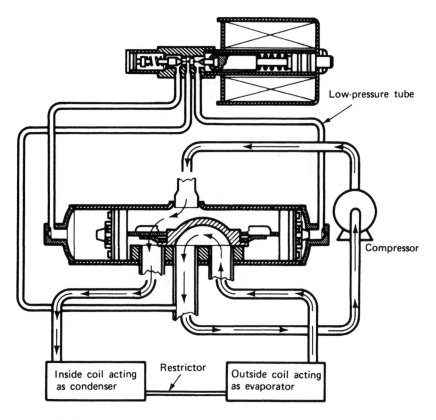

**Figure 22–3**
Energized solenoid coil. (Courtesy of Ranco North America, Division of Ranco, Inc.)

In the heating cycle, the solenoid coil is first energized to operate the pilot-valve plunger. See Figure 22–3. This causes the left port to open with one needle valve and keeps the right port open with the other. The pressure differential, created in the two main valves and the chambers by action of the pilot valve, instantly causes the two pistons to move the sliding port. Both end chambers soon have equal pressures. However, this condition is instantly changed by the pilot valve on demand from the temperature control.

When replacing a bad reversing valve, be sure to protect the valve from being overheated by the welding torch. This is best done with some type of heat shield or wet rags wrapped around the valve body. Be sure to make the tube connections exactly as they were or the valve will not operate correctly.

For touch testing steps for reversing valves see Table 22–1.

**TABLE 22—1**  TOUCH TEST CHART

| Valve Operating Condition | DISCHARGE TUBE from Compressor | SUCTION TUBE to Compressor | Tube to INSIDE COIL | Tube to OUTSIDE COIL | LEFT Pilot Back Capillary Tube | RIGHT Pilot Front Capillary Tube | Possible Causes | Corrections |
|---|---|---|---|---|---|---|---|---|
| | 1 | 2 | 3 | 4 | 5 | 6 | Possible Causes | Corrections |
| Normal Operation of Valve | | | | | | | | |
| Normal Cooling | Hot | Cool | Cool, as (2) | Hot, as (1) | *TVB | *TVB | | |
| Normal Heating | Hot | Cool | Hot, as (1) | Cool, as (2) | *TVB | *TVB | | |
| Malfunction of Valve | | | | | | | | |
| Valve will not shift from cool to heat | Check electrical circuit and coil. | | | | | | No voltage to coil. Defective coil. | Repair electrical circuit. Replace coil. |
| | Check refrigeration charge. | | | | | | Low charge. Pressure differential too high. | Repair leak, recharge system. Recheck system. |
| | Hot | Cool | Cool, as (2) | Hot, as (1) | *TVB | Hot | Pilot valve okay. Dirt in one bleeder hole. | Deenergize solenoid, raise head pressure, reenergize solenoid to break dirt loose. If unsuccessful, remove valve, wash out. Check on air before installing. If no movement, replace valve, add strainer to discharge tube, mount valve horizontally. |
| | | | | | | | Piston cup leak. | Stop unit. After pressures equalize, restart with solenoid energized. If valve shifts, reattempt with compressor running. If still no shift, replace valve. |

(Courtesy of Ranco North America, Division of Ranco, Inc.)

**TABLE 22–1** (Continued)

| Valve Operating Condition | DISCHARGE TUBE from Compressor 1 | SUCTION TUBE to Compressor 2 | Tube to INSIDE COIL 3 | Tube to OUTSIDE COIL 4 | LEFT Pilot Back Capillary Tube 5 | RIGHT Pilot Front Capillary Tube 6 | Possible Causes | Corrections |
|---|---|---|---|---|---|---|---|---|
| | Hot | Cool | Cool, as (2) | Hot, as (1) | *TVB | *TVB | Clogged pilot tubes. | Raise head pressure, operate solenoid to free. If still no shift, replace valve. |
| | Hot | Cool | Cool, as (2) | Hot, as (1) | Hot | Hot | Both ports of pilot open. (Back seat port did not close.) | Raise head pressure, operate solenoid to free partially clogged port. If still no shift, replace valve. |
| | Warm | Cool | Cool, as (2) | Warm, as (1) | *TVB | Warm | Defective compressor. | |
| Start to shift but does not complete reversal | Hot | Warm | Warm | Hot | *TVB | Hot | Not enough pressure differential at start of stroke or not enough flow to maintain pressure differential. | Check unit for correct operating pressures and charge. Raise head pressure. If no shift, use valve with smaller parts. |
| | | | | | | | Body damage. | Replace valve. |
| | Hot | Warm | Warm | Hot | Hot | Hot | Both ports of pilot open. | Raise head pressure, operate solenoid. If no shift, replace valve. |
| Start to shift but does not complete reversal | | | | | | | Body damage. | Replace valve. |
| | Hot | Hot | Hot | Hot | *TVB | Hot | Valve hung up at mid-stroke. Pumping volume of compressor not sufficient to maintain reversal. | Raise head pressure, operate solenoid. If no shift, use valve with smaller ports. |

*(continued)*

Notes:

*Temperature of Valve Body.

**Warmer than Valve Body.

**TABLE 22–1** TOUCH TEST CHART

| Valve Operating Condition | DISCHARGE TUBE from Compressor | SUCTION TUBE to Compressor | Tube to INSIDE COIL | Tube to OUTSIDE COIL | LEFT Pilot Back Capillary Tube | RIGHT Pilot Front Capillary Tube | Possible Causes | Corrections |
|---|---|---|---|---|---|---|---|---|
| | 1 | 2 | 3 | 4 | 5 | 6 | Possible Causes | Corrections |
| Start to shift but does not complete reversal | Hot | Hot | Hot | Hot | Hot | Hot | Both ports of pilot open. | Raise head pressure, operate solenoid. If no shift, replace valve. |
| Apparent leak in heating | Hot | Cool | Hot, as (1) | Cool, as (2) | *TVB | **WVB | Piston needle on end of slide leaking. | Operate valve several times then recheck. If excessive leak, replace valve. |
| | Hot | Cool | Hot, as (1) | Cool, as (2) | **WVB | **WVB | Pilot needle and piston needle leaking. | Operate valve several times then recheck. If excessive leak, replace valve. |
| Will not shift from heat to cool | Hot | Cool | Hot, as (1) | Cool, as (2) | *TVB | *TVB | Pressure differential too high. | Stop unit. Will reverse during equalization period. Recheck system. |
| | | | | | | | Clogged pilot tube. | Raise head pressure, operate solenoid to free dirt. If still no shift, replace valve. |
| | Hot | Cool | Hot, as (1) | Cool, as (2) | Hot | *TVB | Dirt in bleeder hole. | Raise head pressure, operate solenoid. Remove valve and wash out. Check on air before reinstalling. If no movement, replace valve. Add strainer to discharge tube. Mount valve horizontally. |

(Courtesy of Ranco North America, Division of Ranco, Inc.)

138

**TABLE 22—1** (Continued)

| Valve Operating Condition | DISCHARGE TUBE from Compressor | SUCTION TUBE to Compressor | Tube to INSIDE COIL | Tube to OUTSIDE COIL | LEFT Pilot Back Capillary Tube | RIGHT Pilot Front Capillary Tube | Possible Causes | Corrections |
|---|---|---|---|---|---|---|---|---|
| | 1 | 2 | 3 | 4 | 5 | 6 | | |
| Will not shift from heat to cool | Hot | Cool | Hot, as (1) | Cool, as (2) | Hot | *TVB | Piston cup leak. | Stop unit, after pressures equalize, restart with solenoid deenergized. If valve shifts, reattempt with compressor running. If it still will not reverse while running, replace valve. |
| | Hot | Cool | Hot, as (1) | Cool, as (2) | Hot | Hot | Defective pilot. | Replace valve. |
| | Warm | Cool | Warm, as (1) | Cool, as (2) | Warm | *TVB | Defector compressor. | |

Notes:

*Temperature of Valve Body.

**Warmer than Valve Body.

VALVE OPERATED SATISFACTORILY PRIOR TO COMPRESSOR MOTOR BURN OUT—caused by dirt and small greasy particles inside the valve. To CORRECT: Remove valve, thoroughly wash it out. Check on air before reinstalling, or replace valve. Add strainer and filter-dryer to discharge tube between valve and compressor.

# Summary

- Reversing valves should be sized according to the equipment manufacturer's specifications.
- Reversing valves are hermetically made and are pressure-differential operated.
- These valves are controlled by an energized or deenergized solenoid coil fastened over a three-way pilot valve with a locknut. The pilot valve is part of the main valve.

# Service Call

A customer complains that his heat pump system is not heating his building. A check of the system reveals that all the equipment is running, but it is not heating as it should. The reversing valve is energized during the heating season. A check of the temperature of the different tubes on the reversing valve shows the compressor discharge tube to be warm to the touch. The compressor suction tube is cool. The tube to the inside coil is only as warm as the compressor discharge tube. The tube to the outside coil is cool. The left pilot back capillary tube is the temperature of the valve body.

From this information the tech learns that the reversing valve is not changing the direction of all the refrigerant, and recovers the refrigerant from the system. Taking the proper precautions to keep from ruining the new valve, the technician replaces the reversing valve. Evacuation of the system is complete, the system is recharged with refrigerant and put back in operation. The temperature of the air leaving the registers inside the building is 110°F. The temperature of the reversing valve tubes is correct. The technician is satisfied that the system is repaired.

# Student Troubleshooting Problem

A customer complains that a heat pump is not heating. The system is an air-cooled unit and everything is running. The indoor temperature is below the thermostat setting. The air coming from the indoor vents is cool. The reversing valve is energized during the cooling season. There is no voltage across the defrost termination switch contacts. What could cause this problem?

# Questions

1. In what applications are reversing valves used?
2. For what are reversing valves designed?
3. How are reversing valves operated?
4. When installing a reversing valve, what must be done to protect it?

# Unit 23: PRESSURE CONTROLS
## Introduction

Any time a motor stalls or is overloaded, the motor draws from more than its rated amperage to about six times its full-load amperage rating. If the overload continues, the motor windings get overheated. The least that will happen is that the insulation on the motor windings will be ruined, shorting out the motor.

A compressor motor can be protected from over current damage or possible damage to the bearings from the lack of oil by using pressure controls. These controls are used to sense high or low refrigerant pressures.

**Definition.**    *Pressure controls* are switching controls used to stop the compressor motor when the refrigerant pressures reach a predetermined value.

*Operation.*    A low-pressure control is usually connected into the suction (low) side of the refrigeration system and is set to stop the compressor motor if the low-side pressure drops to a certain value depending on the type of refrigerant used. See Figure 23–1. Suction-pressure controls, or low-pressure controls, are those whose contacts open on a fall and close on a rise in refrigerant pressure. They have NO contacts.

High-pressure controls are connected to the high (discharge) side of the system and are set to stop the compressor motor if the high-side refrigerant pressure rises to a certain value. See Figure 23–2. High-pressure controls, or discharge-pressure controls, are those whose contacts open on a rise and close on a fall in refrigerant pressure. They have NC contacts.

Both the high- and low-pressure controls are available in either automatic or manual reset models. The manual reset models are used to keep the equipment from running until it is inspected. The manual reset type will not allow the unit to run until the problem is repaired to avoid damage to the compressor or motor. They are available in either adjustable or nonadjustable modules. See Figure 23–3.

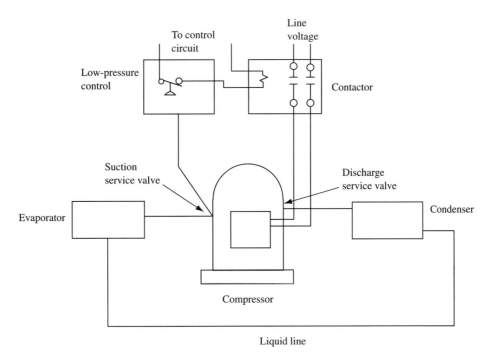

**Figure 23–1**
Low-pressure control connections into a refrigeration and electric system.

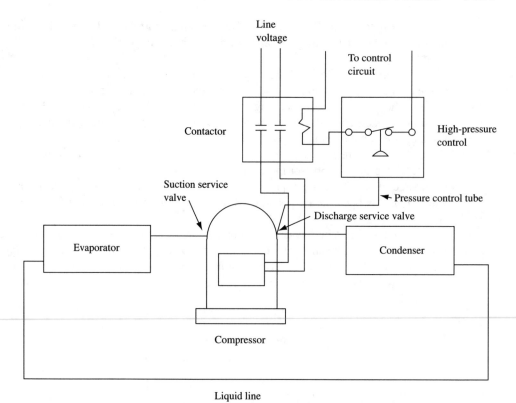

**Figure 23–2**
Refrigerant and electrical connections for a high-pressure control.

Pressure controls are available in either single-function or dual-function models. The type used depends on the equipment design, needs, and cost considerations. See Figure 23–4.

Pressure-control contact ratings have a range of about 24 full-load amperes and 102 locked-rotor amperes on 120 VAC or 240 VAC. Following are most of the uses for pressure controls.

Suction-pressure sensing for temperature control
Suction-pressure sensing for pump-down control
Suction-pressure sensing for capacity control
Suction-pressure sensing for low-pressure limit control
Suction-pressure sensing for alarm control
High-pressure sensing for high-pressure control
High-pressure sensing for condenser fan control

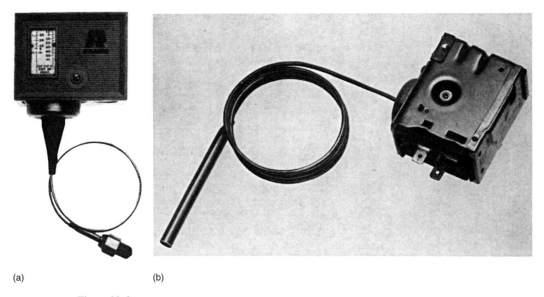

(a)               (b)

**Figure 23–3**
(a) Adjustable and (b) nonadjustable pressure controls. (Courtesy of Ranco North America, Division of Ranco, Inc.)

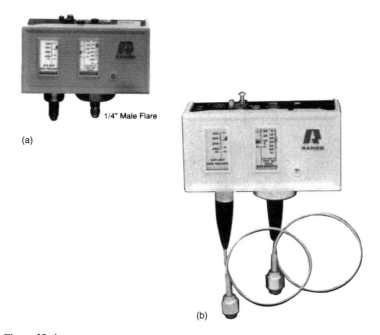

**Figure 23–4**
(a) Single-function and (b) dual-function pressure controls. (Courtesy of Ranco North America, Division of Ranco, Inc.)

Suction-pressure sensing control contacts open on a fall and close on a rise in pressure. They are normally used on thermostatic expansion valve systems and control the temperature by sensing the suction pressure and the resulting temperature of the evaporator. See Figure 23–5.

The low-pressure control is adjusted to start (cut in) the compressor motor when the suction pressure rises to a chosen value, which is determined by the required temperature of the stored product. It will then stop (cut out) the compressor motor when the suction pressure drops to a chosen value. See Table 23–1.

**TABLE 23–1** APPROXIMATE PRESSURE CONTROL SETTINGS

| Vacuum: Italic Figures | | | Gage Pressure: Light Figures | | | |
|---|---|---|---|---|---|---|
| | | | Refrigerant | | | |
| | 12 | | 22 | | 502 | |
| Application | Out | In | Out | In | Out | In |
| Ice cube maker, dry type coil | 4 | 17 | 16 | 37 | 22 | 45 |
| Sweet water bath, soda fountain | 21 | 29 | 43 | 56 | 52 | 66 |
| Beer, water, milk cooler, wet type | 19 | 29 | 40 | 56 | 48 | 66 |
| Ice cream, hardening rooms | 2 | 15 | 13 | 34 | 18 | 41 |
| Eutectic plates, ice cream truck | 1 | 4 | 11 | 16 | 16 | 22 |
| Walk-in, defrost cycle | 14 | 34 | 32 | 64 | 40 | 75 |
| Vegetable display, defrost cycle | 13 | 35 | 30 | 66 | 38 | 77 |
| Vegetable display case, open type | 16 | 42 | 35 | 77 | 44 | 89 |
| Beverage cooler, blower, dry type | 15 | 34 | 34 | 64 | 42 | 75 |
| Retail florist, blower coil | 28 | 42 | 55 | 77 | 65 | 89 |
| Meat display case, defrost cycle | 17 | 35 | 37 | 66 | 45 | 77 |
| Meat display case, open type | 11 | 27 | 27 | 53 | 35 | 63 |
| Dairy case, open type | 10 | 35 | 26 | 66 | 33 | 77 |
| Frozen food, open type | −7 | 5 | 4 | 17 | 8 | 24 |
| Frozen food, open type, thermostat | 2°F | 10°F | — | — | — | — |
| Frozen food, closed type | 1 | 8 | 11 | 22 | 16 | 29 |

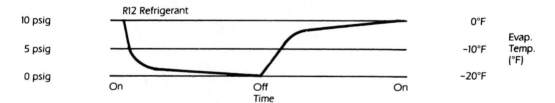

**Figure 23–5**
Suction-pressure sensing for temperature cycle chart. (Courtesy of Ranco North America, Division of Ranco, Inc.)

As discussed, the distance between the cut-in and cut-out pressure is known as the differential. When the control is adjusted for a narrow differential, the temperature changes will also be narrow. However, a narrow differential may cause short cycling (rapid on and off) of the compressor motor. A wide differential allows the compressor to operate with longer ON-OFF cycles, causing wide temperature differences inside the refrigerated area.

**Suction-pressure sensing for pumpdown control.**    Suction-pressure sensing controls open on a fall and close on a rise in pressure. This type control is generally used on systems that have the evaporator some distance from the condensing unit and on systems that the condensing unit is outside the building.

The advantage of using a pump-down control is to remove the refrigerant from the low side of the system during the OFF cycle, which prevents liquid slugging of the compressor and a possible loss of lubricating oil on the next system start-up. In these type systems, a temperature control energizes and deenergizes the liquid-line solenoid valve in response to the temperature in the controlled space, thus causing the system to either stop or start because of either a rise or fall in suction pressure at the pressure control. See Figure 23–6.

When the space requires cooling, the temperature control contacts close to energize the liquid-line solenoid coil. The coil then opens the valve and allows refrigerant to flow into the low side of the system. As a result, the low-side pressure rises to the cut-in setting of the low-pressure control. The electric circuit is then made to the compressor motor causing it to start. Small compressor motors may be started directly with line current through the pressure-control contacts. Larger compressor motors must be started with a contactor or starter in which the holding coil is energized through the pressure-control contacts.

When the space has cooled down and the temperature control is satisfied, the solenoid valve coil is deenergized, allowing the valve to close and stop the flow of refrigerant. The compressor continues to run until the refrigerant has been pumped from the low side and the pressure falls to the cut-out setting of the low-pressure control. The electrical circuit to the compressor motor is broken and it stops.

Pressure settings used on controls for pump-down control are quite important. The cut-in setting measures how high the refrigerant pressure in the low side of the system must go before the contacts close to start the compressor motor. Therefore, the cut-in setting must be carefully chosen. Equipment manufacturers generally make recommendations for this setting.

The cut-in setting for condensing units placed outside is determined by selecting either the coldest unit operating temperature or the coldest anticipated outdoor ambient temperature, whichever is colder. The equipment manufacturer generally makes recommendations for this setting also.

**Suction-pressure sensing for capacity control.**    Suction-pressure sensing controls are designed to close on a rise and open on a fall in pressure. They are used to energize and deenergize electric solenoid-controlled, compressor capacity controls. As the load decreases (the suction pressure falls), the unloader solenoid is deenergized, unloading one or more compressor cylinders and causing a reduced compressor capacity. Generally one suction-pressure sensing control is used for each compressor unloader. See Figure 23–7.

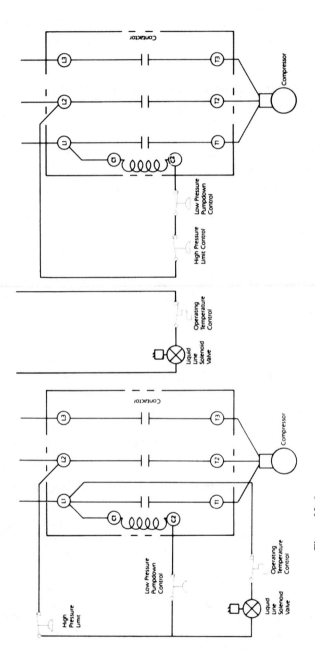

**Figure 23-6**

Suction-pressure sensing for pump-down control connections. (Courtesy of Ranco North America, Division of Ranco, Inc.)

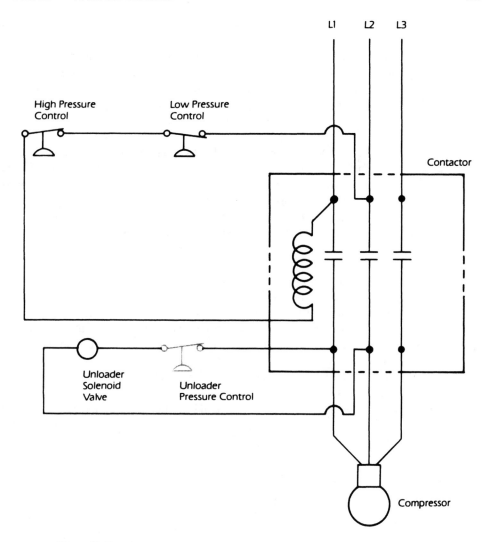

**Figure 23–7**
Suction-pressure sensing for capacity control connections. (Courtesy of Ranco
North America, Division of Ranco, Inc.)

The setting for pressure controls used as unloader controls is higher than low-pressure control settings for other uses. They are set to keep the evaporator temperature recommended by the equipment manufacturer.

**Suction-pressure sensing for alarm control.**    This type of suction-pressure control closes its contacts on a rise and opens them on a fall in pressure. These are popular on systems with remote evaporators. A high evaporator temperature is indicated by a high suction pressure, which causes the contacts in the pressure control to close and energizes the alarm. See Figure 23–8.

Generally, the low-pressure control is used with an automatic reset time delay. The low-pressure control energizes the time-delay relay. If the suction pressure stays too high for a given time, the relay contacts close to energize the alarm, alerting the personnel that the temperature is too high. The purpose of the time delay is to prevent nuisance alarm signals. When a defrost cycle is used, the time delay may be from 45 to 60 minutes.

### Suction-pressure sensing for low-pressure limit control.

This suction-pressure sensing control opens its contacts on a fall and closes them on a rise in pressure. Systems not equipped with a pump-down control use this control to give compressor protection when low suction pressure exists, showing a possible loss of refrigerant charge or an extremely low evaporator temperature. Sometimes these controls are used on air-conditioning systems to keep from frosting the evaporator and to keep from freezing the evaporator tubes in a water-chiller system. When the suction pressure falls to the set point, these controls stop the compressor to keep from damaging the evaporator tubes. See Figure 23–9.

Usually the manual reset type control is used for low-pressure control on these systems. However, when perishable products are being stored, an automatic reset type may be used to keep product loss to a minimum when the problem exists for only a short time.

Another method using this type control is to use a manual reset, time-delay relay with an automatic reset, low-pressure control with contacts that open on a rise and close on a fall in pressure. This system almost eliminates a system shutdown due to short-term problems. They also give the equipment the same protection as that from a manual reset control.

In this type system, when the control contacts close, the time-delay device is energized. If the suction pressure does not rise to the cut-out setting of the control, the time delay

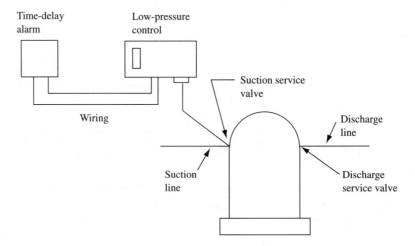

**Figure 23–8**
Connections used for low-pressure sensing alarm control. (Courtesy of Ranco North America, Division of Ranco, Inc.)

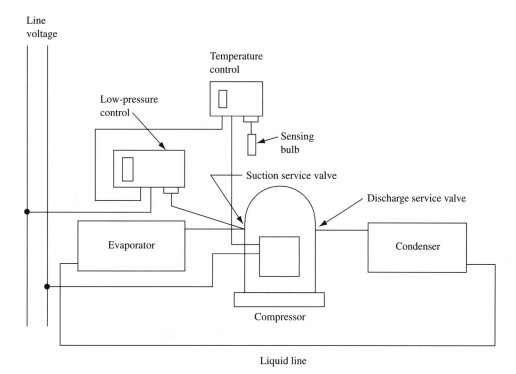

**Figure 23–9**
Refrigerant and electrical connections for low-pressure control.

will shut the system down. The time-delay relay must be reset before the system can run again. Typically the time delay is two to five minutes. See Figure 23–10.

**High-pressure sensing for high-limit control.**   High-pressure sensing pressure controls have contacts that open on a rise and close on a fall in pressure. High-pressure controls are used to sense the discharge pressure and to stop the compressor motor when a high discharge pressure is sensed. See Figure 23–11.

The pressure-control settings vary with the different type refrigerants and different system designs. The equipment manufacturers generally recommend the settings for their equipment. Most manufacturers prefer that manual reset, high-pressure controls be used to give better protection for their equipment.

In some installations an SPDT type switch is used. The NO contacts are used to energize an alarm circuit, showing that the pressure-control contacts have opened and need resetting. The NC contacts control compressor motor operation.

An open high-pressure control contact suggests a problem with the equipment that must be corrected, or damage to the compressor and motor. For this reason the manual reset control is chosen over the automatic reset type.

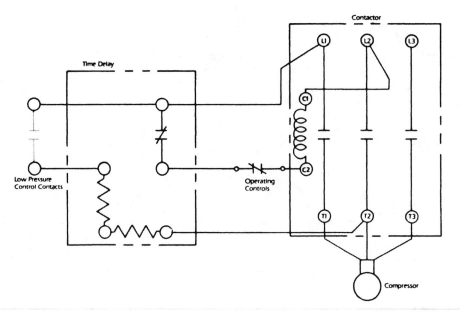

**Figure 23–10**
Suction-pressure limit with time delay. (Courtesy of Ranco North America, Division of Ranco, Inc.)

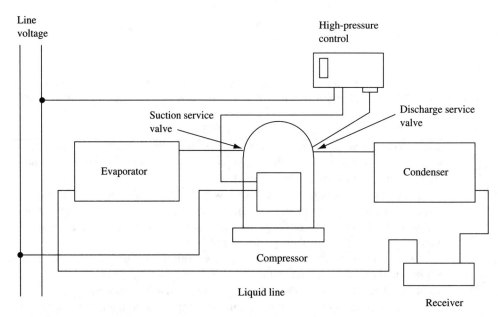

**Figure 23–11**
Refrigerant and electrical connections for high-pressure sensing.

**High-pressure sensing for condenser fan control.**    High-pressure sensing for condenser fan controls have contacts that close on a rise and open on a fall in pressure. They are used on air-cooled systems where the discharge pressure needs to be kept at a certain value in low ambient temperatures. The higher pressure is needed to ensure that the right amount of refrigerant is fed into the evaporator.

In this type of application the control contacts close when the discharge pressure drops to a chosen value to start the condenser fan motor. They open when the discharge pressure rises to a chosen value to stop the condenser fan motor. This keeps the discharge pressure at a near normal operating value. See Figure 23–12.

These controls should be adjusted correctly for proper operation. Too wide a differential will allow the fan motor to stay off too long, causing a wide differential in the discharge pressure and incorrect feeding of the refrigerant flow control device. A differential that is too narrow will cycle the fan too rapidly, causing a shortened fan motor life. The equipment manufacturer will generally make recommendations for these control settings.

When more than one fan is to be cycled, a separate control exists for each fan motor. Usually multiple controls are set with approximately 10 pounds per square inch gauge (psig) difference between them. For example, an air-conditioning system using refrigerant HCFC-22 has four condenser fans. Three of them are to be cycled to control the head pressure. The first fan to cycle off should have the pressure control set to open at 250 psig. The second control should be set to open at 240 psig, and the third control should be set to open at 230 psig. The last fan to turn off will be the first to come back on. This difference allows for fan staging and better control of the discharge pressure.

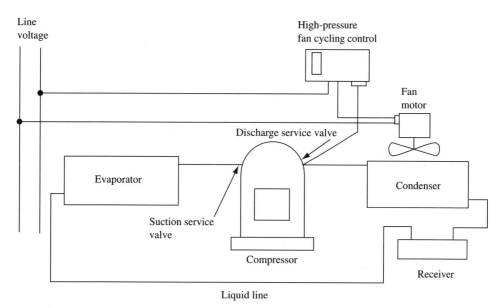

**Figure 23–12**
Fan controlled by a high-pressure control.

**Proportional pressure controls.**   Proportional pressure controls are used to control actuators, which control the dampers, valves, and other similar devices. Some models have a throttling adjustment so that on-the-job selection of the correct throttling range for stable system control is possible. These controls may be used with others to control the flow of water through a water-cooled condenser, to position face and bypass dampers on multizone systems, and to control the position of outdoor air dampers for energy conservation. They should be used with the manufacturer's recommended accessory equipment for correct system control.

*Operation.*   These controls operate the same as the modulating motors to be described in Unit 38.

# Summary

- A compressor motor can be protected from over current damage or possible damage to the bearings from the loss of oil by using pressure controls that will sense high or low refrigerant pressures.
- Pressure controls are switching devices used to stop the compressor motor when the refrigerant pressures reach a chosen value.
- Usually, the low-pressure control is connected into the suction (low) side of the refrigerant system and is set to stop the compressor motor if the low-side pressure drops to a chosen value.
- Low-pressure controls are considered those that open on a fall and close on a rise in refrigerant pressure. They have NO contacts.
- High-pressure controls are generally connected into the high (discharge) side of the refrigeration system and are set to stop the compressor motor if the high-side refrigerant pressure rises to a chosen value.
- High-pressure controls are considered those that open their contacts on a rise and close on a fall in refrigerant pressure. They have NC contacts.
- Pressure controls are available in either single-function or dual-function models.
- The distance between the cut-in and cut-out pressure is known as the differential.
- The pressure control settings vary with the different types of refrigerants and different system needs.
- Some high-pressure controls have SPDT switching. When a high pressure is sensed, the NC contacts open to stop the compressor motor. The NO contacts close to energize an alarm circuit.
- Proportional pressure controls are used to control actuators for controlling dampers, valves, and other similar devices.
- Proportioning pressure controls may also be used along with other accessories to control the flow of water through a water-cooled condenser, to position face and bypass dampers on multizone systems, and to control the position of outdoor dampers for energy conservation.

# Service Call

A customer complains that a refrigeration unit is going on and off and not cooling the cabinet. A check of the system reveals that the compressor will run for a short time and then stop. It is short cycling. A check of the voltage and amperage shows that the amperage draw is below the rating of the compressor and the voltage is 240 VAC. The refrigerant pressures show that both the discharge and the suction pressures are lower than normal. The tech decides that the system has a refrigerant leak. The system undergoes a thorough leak test, and the leak is found on the flare nut at the inlet of the TXV. The technician tightens the flare nut and performs another leak test. The tech repairs the leak and puts the system back in operation. The tech then adds the necessary refrigerant to bring the system up to normal charge. Both the discharge and suction pressures are now normal and the system is cooling the box. The amperage draw is normal. The technician, now satisfied that the system is repaired, records the exact amount of refrigerant used.

# Student Troubleshooting Problem

A customer complains that an air-conditioning system is not cooling the building properly. A check of the system shows that the system is running but not cooling as it should. The building is almost completely glass on the south side. The outdoor ambient temperature is 95°F. The unit uses HCFC-22 refrigerant. A check of the refrigerant pressures show that the suction pressure is 55 psig and the discharge pressure is 170 psig. The unit has pressure control to stop one of the condenser fans when the discharge pressure drops to a given value. What could be the problem?

# Questions

1. What are pressure controls?
2. What is the purpose of a low-pressure control?
3. In what position are the low-pressure control contacts normally?
4. In what position are the high-pressure control contacts normally?
5. What is the purpose of a manual reset on a pressure control?
6. What temperature does the low pressure show?
7. What is pressure-control differential?
8. What type of compressor operation is possible with a wide pressure-control differential?
9. What are the advantages of a pump-down cycle?
10. What controls the solenoid valve on a pump-down system?
11. On a system using a pump-down cycle, why does the compressor run after the temperature has been reached?
12. What is determined by selecting either the coldest unit operating temperature or the coldest anticipated outdoor ambient temperature, whichever is colder?

13. What is the purpose of using SPDT contacts in a pressure control?

14. What type of pressure control is used to cycle the fans on an air-cooled condensing unit during low ambient temperatures?

# Unit 24: OIL-FAILURE CONTROLS
# Introduction

Oil-failure controls are designed to protect compressors from bearing damage in case the oil pressure drops below that recommended by the compressor manufacturer or does not develop on compressor start-up.

**Oil-failure controls.**    The oil-failure control is a differential pressure control used to monitor the effective oil pressure and stop the compressor motor in case of oil failure. See Figure 24–1.

These controls have two opposing pressure-sensing elements that sense the *net* oil pressure. Net oil pressure is the difference between the crankcase pressure and the oil pump outlet pressure. For example, the suction pressure is 70 psig on HCFC-22 refrigerant. The oil pump outlet pressure is 110 psig. The net oil pressure is:

$$\text{Net oil pressure} = \text{Oil pump outlet pressure} - \text{Suction pressure}$$
$$= 110 - 70$$
$$= 40 \text{ psig}$$

The compressor crankcase is usually operating at a pressure other than atmospheric; therefore, the net pressure must be measured to ensure proper lubrication. These pressure-sensing elements operate a set of NC contacts that open on a fall in pressure difference.

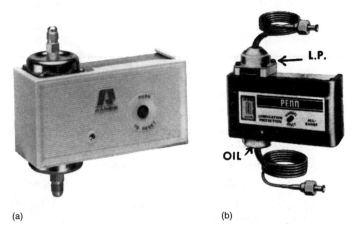

(a)                                              (b)

**Figure 24–1**
Oil-failure control. (Courtesy of Control Products Division, Johnson Controls, Inc.)

A built-in time-delay switch allows for oil-pressure pickup on starting to avoid nuisance shutdowns on short-term oil-pressure drops during the running cycle. See Figure 24–2.

***Operation.***    In operation the total oil pressure is the combination of the crankcase pressure and the pressure generated by the oil pump. The net oil pressure available to circulate the oil is the difference between the total oil pressure and the refrigerant pressure in the crankcase.

$$\text{Total oil pressure} - \text{Refrigerant pressure} = \text{Net oil pressure}$$

This control measures that difference in pressure, or net oil pressure.

When the compressor starts, a time-delay switch is energized. If the net oil pressure does not rise to the control cut-in setting within the required time limit, the time-delay switch trips to stop the compressor motor. If the net oil pressure rises to the cut-in setting within the required time after the compressor starts, the time-delay switch is automatically deenergized and the compressor continues its normal operation. If the net oil pressure falls below the control cut-out setting during the running cycle, the time-delay switch is energized, and unless the net oil pressure rises to the cut-in point within the time delay period, the compressor stops.

The time-delay switch is a trip-free thermal control. It is compensated to reduce the effect of ambient temperatures from 32°F to 150°F. The timing is also affected by voltage changes. The technician should follow the equipment manufacturers' recommendations in the installation and adjustment of these controls. The oil-pressure line is connected to the pressure connection labeled OIL. The crankcase line to the pressure connection is labeled LOW. Wire as suggested for correct equipment protection. See Figure 24–3. Oil-failure controls are electrically rated for pilot duty only.

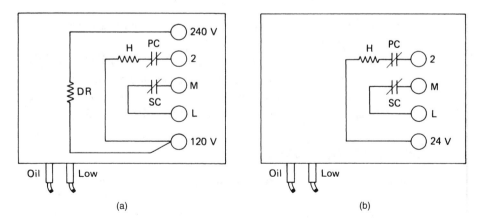

**Figure 24–2**
Internal diagram for Penn line-voltage and low-voltage controls. (Courtesy of Control Products Division, Johnson Controls, Inc.)

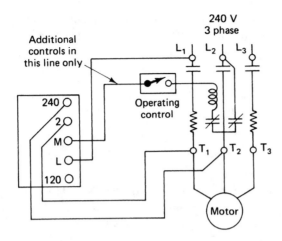

**Figure 24–3**
Penn control wiring diagram used on 240-volt system. (Courtesy of Control Products Division, Johnson Controls, Inc.)

# Summary

- Oil-failure controls are used to protect the compressor from bearing damage in case the oil pressure drops below that recommended by the compressor manufacturer or does not develop on compressor start-up.
- Oil-failure controls are switching controls that give protection against major breakdowns on pressure-lubricated refrigeration compressors by guarding against low lubricating oil pressure.
- The oil-failure control is a differential pressure control used to monitor the effective oil pressure and stop the compressor motor in case of oil failure.
- These controls have two opposing pressure-sensing elements that sense the net oil pressure.
- Net oil pressure is the difference between the crankcase pressure and the oil pump outlet pressure.
- Total oil pressure minus refrigerant pressure equals net oil pressure.
- The oil-pressure line is connected to the pressure connection labeled OIL and the crankcase line to the pressure connection is labeled LOW.
- Oil-failure controls are electrically rated for pilot duty only.

# Service Call

A customer complains that a large air-conditioning system is not cooling. A checking of the system reveals that the compressor is not running. The unit has an oil-failure control. The reset button on the control is out and needs to be reset. The tech installs a set of refrigeration gauges on the compressor with one line connected to the suction valve and the other line connected to the oil pump outlet. The tech resets the control and the compressor starts running. The total oil pressure is now 100 psig. The suction pressure drops to 80 psig. The net oil pressure is 20 psig. The tech checks the control and finds that the control differential is set at 35 psig. The heater remains energized and the compressor is again stopped by the oil-failure control. The tech adjusts the oil-failure control for a differential of 20 psig, and puts the unit back in operation. The compressor continues to run and cool the building. The technician is satisfied that the system is repaired.

# Student Troubleshooting Problem

A customer complains that a large air-conditioning system is not cooling. A check of the system reveals that the system has an oil-failure control, a high-pressure control, and a low-pressure control. Everything is running except the compressor. The technician checks the three-phase voltage, which is 240 VAC. The starter is not pulled in. There is no control voltage at the starter coil terminals. The thermostat is demanding cooling. The compressor oil level is at the bottom of the oil sight glass. What could be causing the problem?

# Questions

1. Are the contacts in an oil-pressure control NO or NC?
2. Why are oil-failure controls used?
3. What type of control is the oil-failure control?
4. What allows the compressor to start and run until the lubricating oil pressure rises to the correct pressure?
5. Define *total oil pressure.*
6. With a total oil pressure of 60 psig and a suction pressure of 40 psig, what is the net oil pressure?
7. If a compressor is running and the oil pressure drops for a short time, will the compressor shut down on the oil-failure control?

# Unit 25: TEMPERATURE CONTROLS
# Introduction

Any building, regardless of location, age, or architecture, can now be heated and cooled with year-round comfort and safety with a well-designed, correctly installed, and accurately controlled air-conditioning system. Keeping perishable foods at the necessary temperature would also be a problem if automatic temperature controls had not been designed and manufactured.

Temperature controls are temperature sensing devices used to keep a wanted temperature by switching on or off the necessary equipment in response to space temperature demands.

Generally, temperature controls are divided into two different types: refrigeration temperature control and air-conditioning temperature control. Air-conditioning includes both cooling and heating.

# Refrigeration Temperature Controls

Refrigeration temperature controls are used to start the equipment when the space needs cooling. The process is the same whether it is a commercial refrigeration case or a domestic refrigerator.

In a commercial refrigeration case, the temperature control is used to energize a motor contactor, or starter holding coil, to either start or stop the compressor motor as the temperature control demands. See Figure 25–1.

Refrigeration temperature controls are made with the temperature-sensing vapor or liquid-filled sealed power elements. See Figure 25–2.

The vapor-pressure element is based on the principle that the boiling temperature of a liquid depends on the vapor pressure at the liquid surface. By partially filling the bulb with a liquid and connecting the vapor space to a pressure-sensitive element, a closed system is formed in which the vapor pressure will depend on the temperature of the bulb. In this type of control the pressure-sensitive mechanism is usually a bellows. See Figure 25–3.

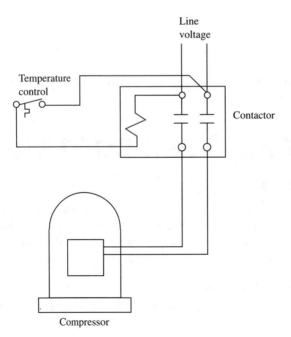

**Figure 25–1**
Wiring diagram for refrigeration temperature control.

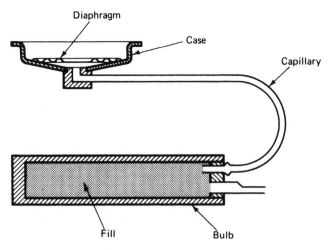

Diaphragm

Case

Capillary

Fill

Bulb

**Figure 25–2**
Liquid-filled remote bulb controller.

**Figure 25–3**
Refrigeration thermostat. (Courtesy of Control Products Division, Johnson Controls, Inc.)

**Methods of refrigeration temperature control.**   The most popular methods of refrigeration temperature control are (1) space and return-air sensing, (2) evaporator temperature sensing, (3) product-temperature sensing, and (4) ice thickness sensing.

*Space and Return-Air Sensing for Temperature Control.*   This method is used on most walk-in refrigeration cases. The contacts in this control close on a rise and open on a fall in temperature. See Figure 25–3. Thermostats that are used as refrigeration controls have a second set of NO contacts that close when the space temperature has risen a few degrees above the normal thermostat cut-in setting. These NO contacts, when closed, energize an alarm circuit to notify the user that the space temperature is too high and product loss is possible.

The major advantage of this method of control is the close control of the space temperature. Temperature controls that sense the space temperature by use of air coils are popular on systems such as walk-in coolers and freezers. See Figure 25–4.

Refrigeration thermostats should be installed where they will not be easily damaged. The stored product should not prevent free air circulation over the sensing element. The sensing element should not be affected by the opening of doors or airflow directly from the evaporator coil. The sensing bulb or capillary tube must be placed so that it will sense the return air to the evaporating coil. The thermostat body may be installed either in the conditioned space or outside the case. When used in display cases the thermostat body is generally mounted on a wall close to the case. See Figure 25–5.

*Note:* Do not connect the sensing bulb to the evaporating coil or any other surface that may affect its sensing of the return air temperature.

***Operation.*** As the temperature inside the conditioned space rises to the thermostat cut-in setting, its contacts close, making the electrical circuit to the compressor motor contactor or starter holding coil. The contactor or starter holding coil is energized to close its contacts. The compressor starts, and the space is cooled to the thermostat cut-out setting. The thermostat contacts open and deenergize the compressor motor contactor or starter holding coil. When the coil is deenergized, the contactor or starter contacts open and the compressor stops. The unit remains stopped until the temperature has risen enough to close the thermostat contacts again.

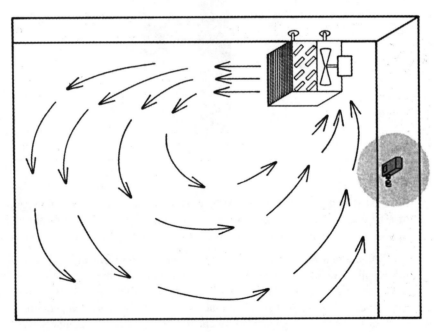

**Figure 25–4**
Location of refrigeration thermostat for a walk-in cooler, (Courtesy of Ranco North America, Division of Ranco, Inc.)

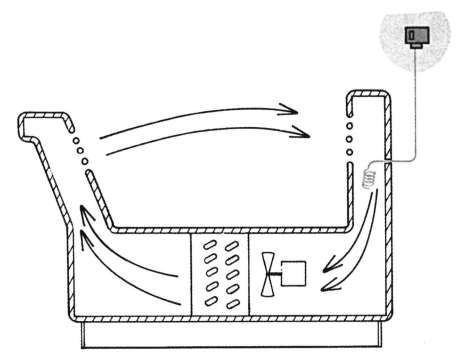

**Figure 25–5**
Thermostat location on a display case. (Courtesy of Ranco North America,
Division of Ranco, Inc.)

If an alarm signal is included, it is not used until the temperature rises a chosen number of degrees above the thermostat cut-in setting. Then the alarm contacts close and either
start a bell ringing or turn on a light to alert the user that equipment is in need of service.
For example, the temperature of a refrigerated cabinet has risen to 3°F above the thermostat cut-in setting. The alarm contacts close and energize the alarm system to alert the user
that the temperature is rising too high for the stored product. The user, or maintenance technician, will check the cabinet and either make the needed repairs or call a service company
to make them.

***Product-Temperature Sensing for Temperature Control.***    Product-temperature
sensing is a popular control method used on liquid chillers and soft ice cream, slush ice, and
yogurt machines. The contacts in these type controls close on a rise and open on a fall in
temperature. Close temperature control is done by directly sensing the product temperature.
See Figure 25–6. Product-sensing thermostats usually have a narrow differential to give the
needed close product-temperature control.

When sensing the temperature of a liquid, such as wine or another liquid such as in a
bulk milk tank, the control element is usually inserted into a well surrounded by the refrigerated produce. The well is securely fastened to the tank and is insulated from the outside
air temperature. See Figure 25–7.

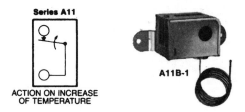

**Figure 25–6**
Thermostat used for product-temperature sensing. (Courtesy of Control Products
Division, Johnson Controls, Inc.)

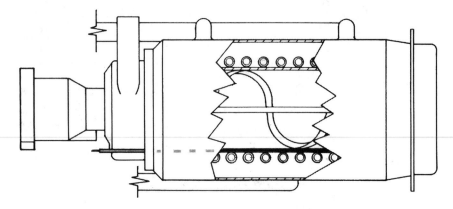

**Figure 25–7**
Thermostat installation in a soft ice cream, slush drink, or yogurt machine.
(Courtesy of Ranco North America, Division of Ranco, Inc.)

In a soft ice cream, slush drink, or yogurt machine, the sensing element may be placed
in a well, if provided, or it may be wrapped around and fastened to the freezing drum. See
Figure 25–7.

*Operation.*    In operation, if the product temperature rises to the cut-in setting, the
electrical circuit is made to the compressor motor starter or contactor holding coil. The contactor or starter contacts are made. The compressor motor is started and the product cooled
down to the thermostat cut-out temperature. At this time, the thermostat contacts open,
breaking the electrical circuit to the compressor motor contactor or starter holding coil, and
the compressor motor is stopped. As the product temperature rises to the thermostat cut-in
temperature, the cycle is repeated.

*Evaporator Temperature Sensing for Temperature Control.*    Evaporator temperature sensing to control temperature is popular on medium- and low-temperature
self-contained refrigeration systems using a capillary tube as the flow control device.
This method of control also ensures that the evaporator is completely defrosted during

the OFF cycle. The evaporator coil will not reach the thermostat cut-in setting until all the frost is melted. In this type thermostat the contacts close on a rise and open on a fall in temperature. Temperature of the evaporator can be sensed with either a capillary tube or a capillary tube and bulb. Sensing elements may be installed in many ways to sense the evaporator temperature—on a return bend of the coil, on a plate secured to several return bends, wrapped around some of the evaporator tubes, or inserted into a well used for this purpose. If it is placed into a well, a sealant must be used to prevent moisture from entering the well and causing corrosion. At least 6 inches of the capillary tube must be in direct contact with the surface sensed to maintain proper control. See Figure 25–8.

Thermostats used for this method of control have a wide differential. When the correct differential is chosen and used, the compressor will not short cycle.

*Operation.*    When the surface temperature of the evaporator rises to the cut-in setting, the thermostat contacts will close, thus making the electrical circuit to the compressor motor contactor or starter holding coil, and start the compressor motor. The evaporator surface temperature falls until the cut-out setting of the thermostat is reached. Then the contacts open, deenergizing the compressor contactor or starter holding coil. The contactor or starter contacts open and the compressor stops. The evaporator surface temperature begins to rise to the thermostat cut-in setting and the cycle is started again.

*Ice Thickness Sensing for Temperature Control.*    Ice thickness sensing for temperature control is popular on ice bank jobs. The contacts in these controls close on a rise and open on a fall in temperature of the water in the tank. Ice bank controls are designed to control the refrigeration compressor so that the wanted volume of ice is built up on the evaporator coil. See Figure 25–9. Water is used to fill the bulb of this control. The sensing bulb is mounted close to the evaporator coil so that it can sense the buildup of ice. See Figure 25–10.

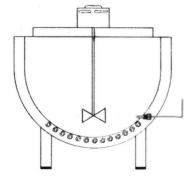

**Figure 25–8**
Thermostat installation in a bulk milk tank. (Courtesy of Ranco North America, Division of Ranco, Inc.)

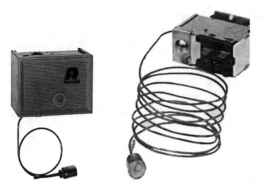

**Figure 25–9**
Thermostat used on ice bank systems. (Courtesy of Ranco North America, Division of Ranco, Inc.)

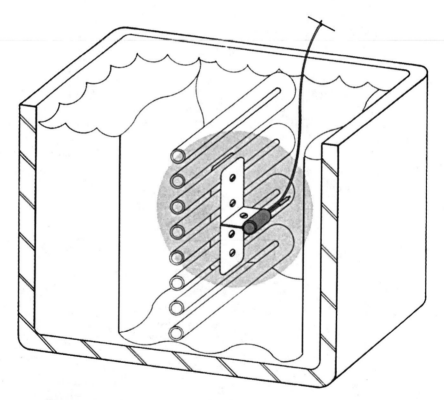

**Figure 25–10**
Thermostat installation in an ice bank. (Courtesy of Ranco North America, Division of Ranco, Inc.)

*Operation.*    When the temperature of the water in the ice storage tank warms to the thermostat cut-in setting, the electrical circuit to the compressor motor contactor or starter holding coil is made. The compressor motor contactor or starter contacts are closed and the motor is started.

When the water temperature has fallen to the thermostat cut-out setting, the water in the thermostat bulb has also frozen and expands. This expansion pressure is transferred to the control through the liquid-filled capillary to open the thermostat contacts and deenergize the contactor or starter holding coil circuit. The contactor or starter contacts open, stopping the compressor motor.

# Air-Conditioning (Room) Thermostats

The most familiar temperature control is the room thermostat, which is used to sense the room temperature and signal the air-conditioning or heating equipment either to operate or to stop in response to the room temperature.

In its simplest form, a room thermostat is a control that responds to changes in air temperature and causes a set of electrical contacts to open or close. This is the basic function of a room thermostat. Many variations are available which are designed to do a variety of operations in heating, cooling, or heating and cooling models. Some are designed for multiple stages for both heating and cooling operation.

One early type of heating system capable of some degree of automatic control was the hand-fired coal furnace. Thermostatic control of this system was made possible with an SPDT thermostat and damper motor. This was a long way from the completely automatic control for residential and commercial building heating and cooling systems in use today.

**Types of room thermostats.**    The three types of electric thermostats used in heating and cooling systems today are the bimetal type, the bellows type, and the proportioning (modulating) type. The bimetal type is by far the most popular.

*Bimetal Thermostat.*    The bimetal thermostat gets its name because it uses a bimetal to open and close a set of contacts on a rise or fall in the room air temperature. See Figure 25–11. This set of contacts may be the open type or they may be enclosed in a mercury tube.

Bimetals are made of two pieces of metal, which, at a given temperature, are the same length. If the temperature of these two pieces of metal is raised, one gets longer than the other because they are different types of metal, having different rates of expansion. These two metals are welded together so they become one solid piece, but each piece keeps its individual characteristics. See Figure 25–12.

When heat is applied, one piece expands at a faster rate than the other. For one piece to become longer than the other, it must bend the entire bimetal into an arc. See Figure 25–13.

By anchoring one end of the bimetal to something solid, the free end will move up or down with a change in temperature. By attaching contacts to the free end and placing a stationary contact nearby, different switching actions result with changes in temperature. See Figure 25–14.

The first bimetal thermostat was not satisfactory due to unstable contact action. Because of the small changes in room air temperature, the bimetal could not develop enough contact pressure to make a good electrical connection. With the development of the permanent magnet, convenience of a control system using the best features of modern control circuits was possible. See Figure 25–15.

***Snap-Action Versus Mercury Switch.***    Room thermostats are available with either snap-action or mercury switches. Snap-action switches are made with a fixed contact securely

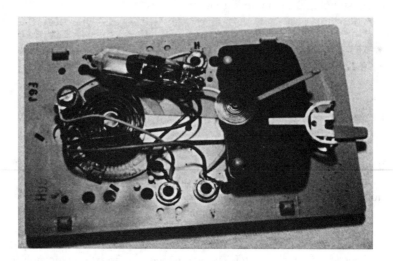

**Figure 25–11**
Bimetal thermostat with cover removed.

Dissimilar metals

**Figure 25–12**
Bimetal.

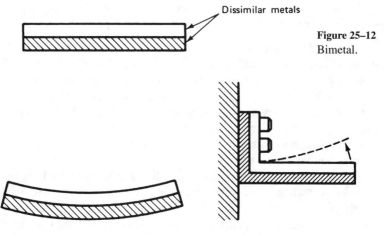

**Figure 25–13**
Bimetal in heated position.

**Figure 25–14**
Anchored bimetal.

attached to the base of the thermostat. This contact is mounted inside a round, permanent magnet, which produces a magnetic field in the contact. The movable contact is attached to the bimetal and when the temperature falls (on heating models) it moves slowly toward the fixed contact. See Figure 25–16.

As the movable contact enters the magnetic field around the fixed contact, it pulls the movable contact against the fixed contact with a positive snap. Because the movable contact has a floating action, it closes with a clean snap, that is, without contact bounce. This floating action also eliminates any tendency for the contacts to "walk" while opening. Either the walking action or a lack of the positive snap will cause arcing between the contacts, which in time will burn and pit them, reducing the continuity between them. This will eventually ruin the contacts and the thermostat must be replaced.

As the bimetal gets warmer (on heating models), it wants to pull the movable contact away from the fixed contact. Nevertheless, because the movable contact is in the magnetic

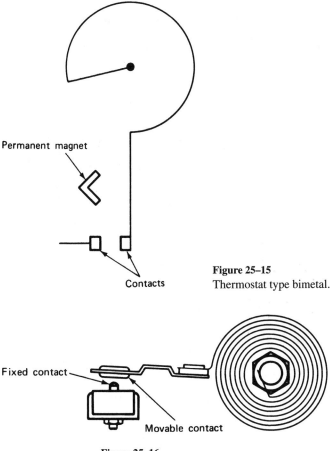

Permanent magnet

Contacts

**Figure 25–15**
Thermostat type bimetal.

Fixed contact

Movable contact

**Figure 25–16**
Typical thermostat bimetal.

field around the fixed contact, the bimetal does not, at this time, have enough force to pull away from the magnetic field. As the bimetal continues to warm, it soon develops a force stronger than the magnetic field. At this time the movable contact moves away from the stationary contact with a snap. In the SPDT switch, another fixed contact is used. This contact is placed so that when one contact is broken, another is made. See Figure 25–17.

All snap-action thermostats have dust covers to prevent dirt and other contaminants from getting on the contacts. If it is necessary to clean these contacts, never use a file or sandpaper. A clean business card or smooth cardboard should be placed between the contacts. With gentle pressure on the movable contact, pull the card back and forth to clean the dirt or film from the contacts. Filing the contacts will ruin them and the thermostat must be replaced.

Mercury switches do the same switching action as snap-acting switches. However, the switching action is done by a puddle of mercury sealed inside a glass tube that moves between two or three fixed probes. Two probes are used on SPST switches. See Figure 25–18. The SPDT models have three probes. See Figure 25–19.

Mercury switches are attached to the thermostat bimetal and do the wanted operation.

**Thermostat anticipators (resistors).**   The two types of thermostat anticipators (resistors) are heating anticipators and cooling anticipators. Anticipators cause artifi-

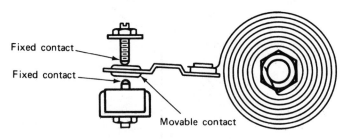

**Figure 25–17**
Single-pole double-throw thermostat.

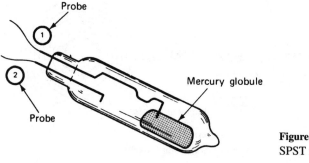

**Figure 25–18**
SPST mercury switch.

cial heat inside the thermostat to help in temperature control. Without them, the space temperature swings would be quite wide and would not be considered comfortable. The following is a description of the anticipators and how they affect the overall system operation. Anticipators are resistance heaters that produce heat when current passes through them.

***Heating Anticipator (Resistor).***    The heating anticipator is placed in the thermostat circuit in electrical series with the control circuit, which causes all the current flowing in the control circuit to pass through the anticipator. Heat is produced by the anticipator in an amount equal to the current flowing through it. This heat is released inside the thermostat and causes it to cycle more often than it would without anticipation. See Figure 25–20.

Heat anticipators are of two different types—fixed and adjustable. A fixed heat anticipator can be either wire wound or a carbon resistor type. Earlier models used the wire-wound fixed anticipators. Fixed anticipators were dipped in an insulating material and color-coded to show the primary control (main gas valve) current they were to be used with. When an anticipator was the wrong size it had to be replaced.

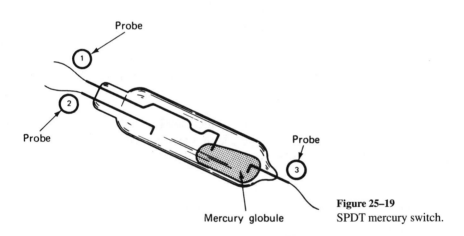

**Figure 25–19**
SPDT mercury switch.

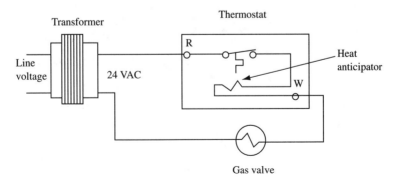

**Figure 25–20**
Location of heating anticipator.

Present-day thermostats with fixed anticipators use tubular resistors, color coded to show the current draw of the primary control. These are nonremovable (riveted in place) or removable (attached with a screw). Fixed anticipators must match the current draw of the control circuit. Various ranges are available. The most versatile heat anticipator (resistor) is the adjustable type. See Figure 25–21.

*Setting an Adjustable Heat Anticipator.*    The primary purpose of an adjustable thermostat anticipator is to provide a single thermostat to match almost any type of current draw that might be experienced in the field. Before installing the thermostat, check the amperage draw of the temperature control circuit to the gas valve or relay. Make sure that the voltage of the control circuit and the thermostat are the same. Otherwise the thermostat could be damaged or the system does not operate correctly. After making sure that the voltages are the same and the amperage draw has been learned, set the heat anticipator.

The current draw of the circuit can be learned by using a current multiplier with the ammeter. Current multipliers can be bought at the supply house or they can be made. To make one, simply wrap ten turns of the control wire around the tong of the ammeter. Read the current and divide by ten to find the current draw.

For example, if the amperage draw has been measured with a multiplier and found to be 4.5 amps, divide 4.5 by 10 to get 0.45 amp. Set the adjustable anticipator to 0.45. See Figure 25–21. By matching the adjustable anticipator to the current draw of the control circuit, this guarantees the best heat anticipation for the system and temperature for the space.

**Cooling anticipators.**    The cooling anticipator is a 0.25-watt carbon type resistor, which is nonadjustable. This is generally known as OFF-cycle anticipation. To understand better the operation of this type of anticipation, look first at how it is used in a typical

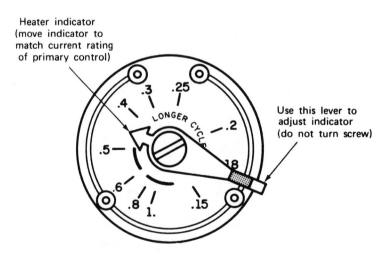

**Figure 25–21**
Adjustable heat anticipator.

cooling system. See Figure 25–22. The cooling anticipator is wired in electrical parallel with the thermostat contacts and in series with the starter or contactor holding coil. When the thermostat contacts close, a low resistance path for the current in the thermostat circuit pulls in the cooling relay or contactor or starter. When the thermostat is satisfied, its contacts open. Now there is a higher resistance path from the transformer, through the cooling anticipator and the winding of the relay coil, and back to the transformer. Because of the high resistance in the cooling anticipator, the voltage drops to a value that will not close the contactor or starter contacts.

This is just the opposite of heat anticipation—in a cooling thermostat the contacts close and a low resistance path for the current in the thermostat circuit pulls in the compressor contactor or starter holding coil. A higher resistance path from the transformer is now through the cooling anticipator and the winding of the starter or contactor holding coil, and back to the transformer. Because of the high resistance in the cooling anticipator, the voltage drops to a value that the starter or contactor will not pull in.

The current flowing through the cooling anticipator during the compressor OFF cycle heats the bimetal and causes it to be warmer than the surrounding room air, which is also rising (the system is off). The false heat actually causes the thermostat contacts to close before the room temperature reaches the cut-in point. This way the cooling system is started sooner than without cooling anticipation, reducing system lag to a minimum and giving a narrow space temperature differential. Cooling anticipators are used on all low-voltage cooling thermostats.

**Forced warm-air system—nonanticipated thermostat.**    If the temperature selector is set on 75°F and the furnace has been off for some time, the temperature

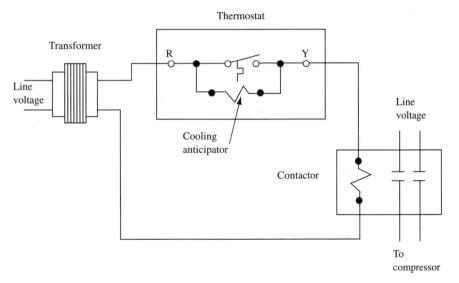

**Figure 25–22**
Off-cycle anticipation schematic diagram.

in the room drops slowly. The bimetal element follows the air temperature change and closes at 75°F. This causes the heating system to start. At this moment no warm air is being blown into the space because the heating system must warm up to the fan control ON setting.

While the heating system is warming, the space air temperature continues to fall slowly. Depending on the type of heating system, the space air temperature falls to about 74.5°F or below before the blower comes on and the warm air is sensed by the thermostat bimetal. This difference in temperature between the point at which the thermostat contacts close and the point the air temperature at the thermostat begins to rise is known as system lag. The amount of system lag in degrees Fahrenheit depends on the thermostat location and type, the size of the furnace, and the design of the air distribution system.

With the furnace on and the blower running, the space air temperature continues to rise. If the thermostat has a mechanical differential of 2°F, the electrical contacts open at 77°F to close the gas valve or deenergize the heating relay on the furnace. However, the furnace is still hot and the blower continues to run and deliver warm air to the space until the furnace temperature drops to the fan control OFF setting. The additional heat delivered to the space after the thermostat contacts have opened is called overshoot. Overshoot can carry the space air temperature to 77.5°F or higher.

**Forced warm-air system—anticipated thermostat.** To reduce the wide differential resulting from a nonanticipated thermostat, simply add a small amount of heat to the bimetal element so that it gets warmer than the surrounding air. Place a resistor in the thermostat close to the bimetal element. This resistor is in electrical series with the contacts. See Figure 25–20. When the thermostat contacts close, the current flowing through the temperature control circuit must also flow through the resistor. The current flowing through the resistor causes it to heat up, which in turn heats the bimetal element. Thus the temperature that the thermostat contacts should open is "anticipated" to give a narrow space temperature differential.

Because the furnace has been on for a shorter time, the heat left in the heat exchanger will still be blown into the space, but the thermostat contacts opened at a lower temperature than the surrounding space temperature. This results in less overshoot in the space temperature. The bimetal element cools faster because it is no longer being heated by the anticipator. The contacts close and the cycle begins again.

System lag and overshoot have not been completely eliminated, but by using anticipation these factors are reduced to a small amount. Through proper system design and thermostat location, keeping a space temperature differential no greater than 0.5°F by using heat anticipation is not unusual.

**Bellows-operated thermostat.** The bellows-operated thermostat has the same function as the bimetal thermostat. The biggest difference between the two is that a bellows filled with some type of fluid that expands when heated and contracts when cooled is used instead of a bimetal element. Bellows-operated thermostats do the same things as bimetal thermostats.

*Operation.*    When the space temperature begins rising, the fluid temperature inside the bellows also begins rising. As its temperature rises, the bellows begins to expand in direct relation to the temperature and builds a pressure inside. When the thermostat cut-in setting is reached, the pressure inside the bellows causes the contacts to close and start the system. As the space temperature drops, the temperature of the fluid and the pressure inside the bellows also drop. When the cut-off temperature is reached, the thermostat contacts open the control circuit and the system stops.

### Staging thermostats.
Because of the increased demand for greater comfort and efficiency in indoor heating and cooling systems, staging thermostats have become more popular. See Figure 25–23.

The staging thermostat is designed to be used on systems that have more than one-stage heating and cooling or any combination of heating and cooling stages.

In a typical system with two-stage heating and two-stage cooling, when the thermostat is in the heating position the heating system operates at reduced Btu input capacity during mild weather, or when the heating demand is low. As the weather gets colder, or the heating demand rises, this reduced capacity is not enough to keep the wanted comfort level. The thermostat automatically brings on the additional capacity of the heating system.

In the cooling position the situation would be similar, except that we would be bringing on more levels of cooling rather than heating. These thermostats are also available with an automatic changeover. In the automatic changeover position it is only necessary to set

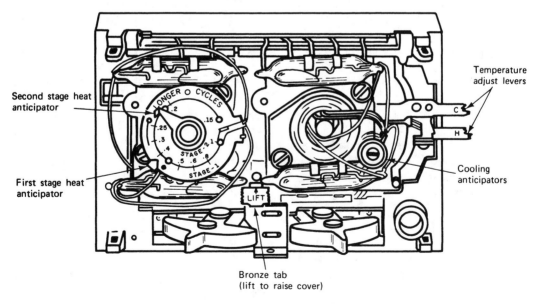

**Figure 25–23**
Multistage thermostat.

the temperature to the wanted temperatures. The thermostat will automatically switch from heating to cooling and back to heating, based on the temperature settings on the thermostat and the space needs. Two separate temperature settings are used on this type thermostat— C is for cooling and H is for heating. The cooling cannot be set at a temperature below what the heating lever is set. Nor can the heating be set at a temperature higher than what the cooling lever is set.

*Circuit Wiring Diagram.*    The circuit wiring diagram with staging thermostats varies with each installation because of the number of possible stages that can be used. Here, study of thermostat operation will involve the two-stage heating–two-stage cooling wiring diagram. See Figure 25–24. This diagram illustrates that the thermostat system switch is in the heat position and the fan switch in the auto position. From one side of the transformer, the circuit goes to thermostat terminal RC, through the jumper wire to RH, through the internal wiring to the bar contact on heat, and to the stage 1 and stage 2 heat anticipators. Also note the circuit to terminal B, which will keep another circuit energized continuously. When there is a call for heat, the stage 1 heating switch closes, providing a circuit to $W_1$, then to the first stage of the heating system, and back to the other side of the transformer. If the temperature continues to drop, the thermostat stage 2 heating switch closes, making a circuit through $W_2$ to the second stage of the heating system and back to the other side of the transformer.

When the selector switch is in the cool position, the heat bar breaks contact and the cool contacts are made. This makes a circuit from RC to the cool contact, through the internal jumper, to the auto contact, and to stage 1 and stage 2 cooling contacts. When stage 1 contacts close, a circuit through $Y_1$, stage 1 cooling, and back to the other side of the transformer is made. If this is not enough cooling, stage 2 closes and the circuit is made through $Y_2$, stage 2 cooling, and back to the other side of the transformer. When the temperature inside the space drops to the chosen degree, stage 2 will open and lower the cooling capacity. If the temperature continues to drop, stage 1 will also open and stop the cooling completely. This is just the reverse of the steps taken during a rise in space temperature. When the thermostat again demands cooling, the unit starts to give the needed cooling.

**Fan switches.**    The fan switch on a thermostat has two positions—one for automatic operation and one for continuous operation. When the fan switch is in the auto position, the fan runs only on demand by that part of the system in use at the time (heating or cooling). If the switch is placed in the ON position, the fan runs all the time regardless of system demand. The fan will also run in this position when the system switch (heat or cool) is moved to the OFF position.

**Modulating thermostats.**    Modulating control systems are built around the Wheatstone bridge principle. Both the controller and the controlled part, usually a modulating motor operating a damper or water valve, use this principle. Both the thermostat (the controller) and the motors have potentiometers in them.

The operation of this type of circuit is as follows. See Figure 25–25. If the temperature of the controller rises, the following occurs:

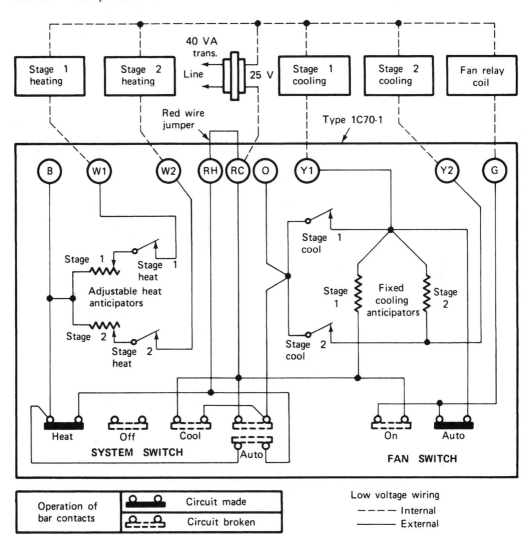

**Figure 25–24**
Multistage system wiring diagram.

1. The wiper blade on the controller potentiometer moves toward the W terminal lowering the resistance between terminals W and R at the controller.
2. Current flow from the transformer through the W terminal of the controller is increased, and this increased current in the corresponding relay coil pulls the DPST relay switch to contact 1.
3. Current from the transformer then flows through the relay switch 1 to the motor.

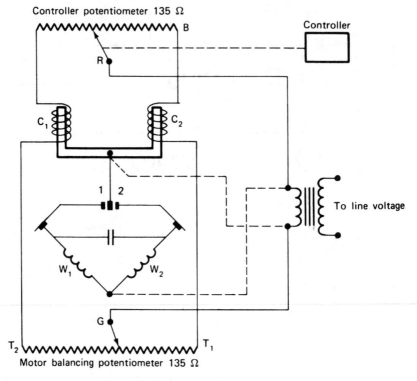

**Figure 25–25**
Motor balancing potentiometer (135 ohms).

4. The motor runs counterclockwise to reposition the controlled part. As the motor moves, the wiper blade on the motor potentiometer moves toward the G terminal.
5. When the wiper blade on the motor potentiometer reaches a point where the resistance between T and G on the motor potentiometer equals the resistance between R and W on the controller potentiometer, the current to the relay coil is equalized.
6. The relay contacts break, stopping the motor. The circuit is again in balance.

On a drop in temperature, the current flows to the other side of the relay and the motor runs in the opposite direction. The controller potentiometer wiper blade is moved through a series of levers and springs by a gas-filled bellows.

**Thermostat location.**    As mentioned earlier, the thermostat location is important for successful operation of the total system. For your guidance, the following are suggestions for correct thermostat location.

1. Always locate the thermostat on an inside wall. If it is on an outside wall, space over-heating during cold weather is likely because the thermostat will always feel cold.

2. Avoid the false sources of heat such as lamps, television sets, warm air ducts, or hot water pipes in the wall. Also avoid places where heat-producing appliances such as ranges, ovens, or dryers are on the other side of the wall. Locations near windows may allow direct sun to reach the thermostat.

3. Avoid sources of vibration such as sliding doors and room doors. Always locate the thermostat at least 4 feet from such sources of vibration and near a wall support if possible.

4. Mount the thermostat level.

5. Mount the thermostat about 5 feet from the floor.

**Thermostat voltage.**    Thermostats are available for all common voltages and must be used with the stated voltage. If used with another voltage, they will either be damaged or will not operate correctly. A thermostat will be damaged if it is used on a voltage that is greater than its rated voltage. If the thermostat is used on a voltage lower than the stated voltage, the anticipators will not work correctly. Therefore, the temperature will not be correctly maintained.

**Outdoor thermostat.**    The outdoor thermostat gives automatic changeover from heating to cooling or cooling to heating as the outdoor temperature changes. See Figure 25–26. These thermostats either may be bimetal actuated or use a remote bulb for operating the low-voltage SPDT mercury switch. This switch breaks one circuit and makes another on a temperature rise or fall in the outdoor air. The operating range of this control is from about 60°F to about 90°F. Outdoor thermostats are used on systems where accurate temperature control is necessary. The wiring connections for this control are shown in Figure 25–27.

*Operation.*    In operation, when the outdoor air temperature rises to the controller set point, the mercury bulb will dump, making one circuit and breaking another. This switches

**Figure 25–26**
Outdoor thermostat. (Courtesy of Johnson Controls, Inc., Control Products Division.)

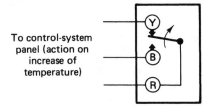

To control-system panel (action on increase of temperature)

**Figure 25–27**
Outdoor thermostat wiring connections.

the system to the cooling mode. The switch remains in this position until the outdoor air temperature drops to the controller set point. Then the mercury bulb dumps in the other direction, stopping the cooling unit and starting the heating unit.

**Fan-coil thermostat.**   Fan-coil thermostats are usually line-voltage controls used to control heating, cooling, or heating and cooling systems. See Figure 25–28. They are used to control fan motors, relays, or water valves on fan-coil units. Fan-speed selectors and temperature selectors are used so that the most comfortable conditions can be chosen. They have anticipators and bimetal-actuated snap-acting switches for controlling the equipment. Sequenced models also have bimetal sensing elements with a dead-band (neither heat nor cooling is provided).

*Operation.*   Fan-coil thermostats operate the same as any other heating, cooling, or heating and cooling thermostat, except that the equipment cannot be operated for a certain number of degrees on the thermostat between the heating and cooling selector switches.

**Outdoor heat pump thermostat.**   Outdoor heat pump thermostats are used to keep the auxiliary heat strips from coming on during mild weather to conserve energy. See Figure 25–29. These thermostats are NC SPST switches that are usually mounted in the outdoor unit. The sensing element is located where it will sense the outdoor ambient temperature. The control system to the auxiliary heat strips is broken when the contacts open on a rise in outdoor air temperature. An adjustable temperature range from 0°F to 50°F is furnished on most models, which saves energy by keeping the auxiliary heat strips off when the heat pump will handle the load.

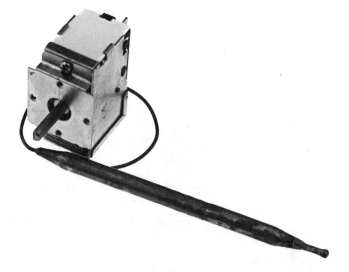

**Figure 25–28**
Fan-coil thermostat.
(Courtesy of Honeywell, Inc.)

**Figure 25–29**
Outdoor heat pump thermostat. (Courtesy of Ranco North America, Division of Ranco, Inc.)

*Operation.*    During mild weather operation, the outdoor heat pump thermostat contacts are closed until the outdoor temperature has risen to the control cut-out setting. The auxiliary heat strips may be energized as wanted when the contacts are closed. When the cut-out temperature is reached, the contacts open and heat strip operation is not possible. The outdoor ambient temperature must drop to the thermostat cut-in setting and close the contacts before the auxiliary heat can be automatically used again. There is an emergency heat switch on the indoor thermostat so that if the heat pump will not heat, the user can manually switch to the auxiliary heat strips for heat.

**Outdoor reset control.**    Outdoor reset controls are used to keep a proper balance between the temperature of the heating medium and the outdoor air temperature. They automatically lower or raise the temperature of the heating medium (water, steam, or warm air) control point as the outdoor air temperature changes. One sensing bulb is mounted to sense the heating medium and another bulb is used to sense the outdoor air temperature. See Figure 25–30. Outdoor reset controls use a set of SPDT contacts that change position on a change in the outdoor air temperature. Outdoor reset controls may be used on line-voltage, low-voltage (24 volts), or millivoltage systems. This control is not used to replace the safety high-limit control. The differential is adjustable within the limits of the control.

*Operation.*    Because both the outdoor air temperature and the temperature of the heating medium are measured, a combination of these two temperatures controls the operation of

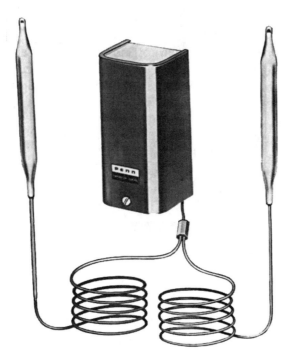

**Figure 25–30**
Outdoor reset control. (Courtesy of Johnson Controls, Inc., Control Products Division.)

this device. When the outdoor air temperature is high and the heating system capacity demand is low, the temperature of the heating medium is also low, reducing the amount of energy used. When the outdoor air temperature is low, the temperature of the heating medium is raised to compensate for the additional heat loss from the building.

As the temperature of the outdoor air begins to fall, the outdoor sensing bulb signals the controller that more heat is needed. The temperature of the heating medium is raised just enough to make up for the additional heat loss. Likewise, as the outdoor air temperature rises, the outdoor sensing bulb signals the controller that the temperature of the heating medium can be lower. This lowers the heating cost for the building while keeping the indoor air temperature at the level wanted.

**Discharge air averaging thermostat.**     The discharge air averaging thermostat is placed in the discharge air plenum of an air-conditioning system. It is used to cycle the equipment and keep an average discharge air temperature. Temperature averaging is a method used to save energy and, therefore, operating expenses. They are generally SPDT switches that can be used to cycle both the heating and cooling equipment to keep the wanted average temperature. To wire this control into the electrical system, see Figure 25–31.

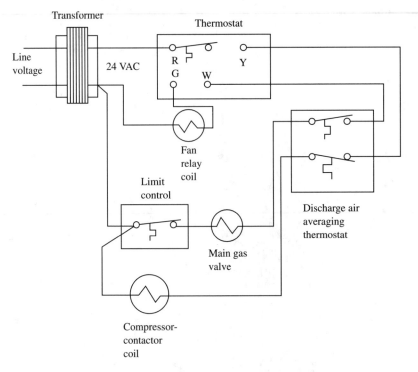

**Figure 25–31**
Typical discharge air averaging thermostat wiring diagram.

# **Summary**

- Temperature controls are temperature-sensing devices used to keep a wanted temperature by switching on or off the equipment in response to space demands.
- Temperature controls are generally divided into two different types: refrigeration temperature control and air-conditioning temperature control. This includes both cooling and heating.
- Refrigeration temperature controls are used to signal the equipment when to operate in response to the needs of the space being cooled, whether it is a commercial refrigeration case or a domestic refrigerator.
- The most popular methods of refrigeration temperature control are: (1) space and return-air sensing, (2) evaporator sensing, (3) product sensing, and (4) ice thickness sensing.
- The contacts on refrigeration temperature sensing controls are NO.
- Some refrigeration temperature controls have a second set of NO contacts that close when the temperature rises a few degrees above the cut-in point for normal operation. These NO contacts, when closed, energize an alarm circuit to notify the user that the space temperature is higher than wanted.
- Product-temperature sensing is a popular control method used on liquid chillers and on soft ice cream, slush ice, and yogurt machines.
- Evaporator sensing to control temperature is popular on medium- and low-temperature, self-contained refrigeration systems using a capillary tube as the flow control device.
- Ice thickness sensing for temperature control is popular on ice bank chilled water installations.
- Air-conditioning thermostats are used for sensing the space temperature and signaling the heating or cooling equipment either to run or to stop running in response to the space temperature.
- Three types of electric thermostats used in heating and cooling systems today are the bimetal type, the bellows type, and the proportioning (modulating) type.
- Two types of heat anticipators (resistors) used in room thermostats are heating anticipators and cooling anticipators.
- The heating anticipator is placed in the thermostat circuit in electrical series with the control circuit so that all current flowing in the control circuit must pass through the anticipator.
- The cooling anticipator is a 0.25-watt carbon type, nonadjustable resistor.
- In a cooling system, the cooling anticipator adds heat to the thermostat bimetal during the OFF cycle.
- The anticipator in a cooling system is wired in electrical parallel with the thermostat contacts.

- Staging thermostats are used on systems that have more than one-stage heating and more than one-stage cooling or any combination of heating and cooling stages.
- Modulating control systems are built around the Wheatstone bridge principle. Both the controller and the controlled component—usually a modulating motor operating a damper or water valve—use this principle.
- Fan-coil thermostats are usually line-voltage controls used to operate heating, cooling, or heating and cooling systems.
- Outdoor heat pump thermostats are used to prevent operation of the auxiliary heat strips during mild weather conditions.
- Outdoor reset controls are used to keep a correct balance between the temperature of the heating medium and the outdoor air temperature.
- The purpose of discharge air averaging thermostats is to cycle the equipment to keep an average discharge air temperature.

# Service Calls

*Service Call 1.*    A customer complains that a walk-in cooler is not operating correctly. A check of the system reveals that the compressor is not running. There is voltage to the line side of the contactor, but it is not closed, and no voltage is found at the contactor coil terminals. The technician checks the circuit back to the thermostat and finds that the thermostat contacts will not close. When the thermostat bulb is warmed above the space temperature the contacts still do not close—the bulb has lost its charge. The tech replaces the thermostat and puts the system back in operation. The running voltage, amperage, and refrigerant pressures are now in the normal operating range of the unit. The technician is satisfied that the system is repaired.

*Service Call 2.*    A customer complains that her home gets too warm before the unit goes off and gets too cool before it comes on. The system is still running. A check of the ON and OFF temperatures of the thermostat and the differential shows all are within the normal 3°F. The heat anticipator is currently set at 0.9 ampere so the unit is turned on. The current flow in the control circuit is at 0.4 ampere. The heat anticipator setting is too high allowing the unit to run too long before turning it off. The system will remain off too long before coming back on. The technician sets the heat anticipator to 0.4 ampere and checks the unit—it is operating correctly. The technician is satisfied that the system is repaired.

# Student Troubleshooting Problem

A customer complains that a reach-in refrigerator is freezing everything in it. The system is an air-cooled unit. Everything is running and cooling. It uses a temperature control with its sensing bulb attached to a return bend of the evaporator coil. The technician checks the thermostat and its contacts open and close at about 3°F differential at the case temperature. What could be the problem?

# Questions

1. Name the two types of temperature controls?
2. What is the purpose of a refrigeration temperature control?
3. What causes the contacts of a refrigeration temperature control to open or close on a change in temperature?
4. What type of temperature sensing is popular on walk-in coolers?
5. What is the normal position of the operating contacts in a refrigeration temperature control?
6. When two sets of contacts, one NO and the other NC, are used in a refrigeration thermostat, what is the purpose of the second set?
7. Where should the refrigeration thermostat sensing bulb be located?
8. Where should a refrigeration thermostat never be mounted?
9. How do product-temperature sensing thermostats give close temperature control of the product?
10. When the sensing bulb of a thermostat is installed in a well, what should be done?
11. What type of temperature control has a narrow differential?
12. On what type of system is evaporator temperature sensing used?
13. What type of temperature sensing uses an OFF-cycle defrost?
14. Where should the thermostat sensing bulb be located on evaporator temperature sensing systems?
15. What type of thermostat has its sensing element filled with water?
16. What is the most popular type sensing element used in room thermostats?
17. Name the three types of room thermostats.
18. What causes the snap action of the contacts in a room thermostat?
19. What causes burnt and pitted contacts in the thermostat?
20. How often should room thermostat contacts be filed?
21. What is used to reduce system temperature lag and overshoot?
22. What must be done when installing a new room thermostat?
23. At what value should a cooling anticipator be adjusted?
24. Which will provide a narrow room temperature differential, an anticipated or a nonanticipated thermostat?
25. What causes the contacts to open and close in a bellows-operated thermostat?
26. On what type systems are staging thermostats used?
27. What is the purpose of equipment staging?
28. On what principle are modulating control systems built?
29. Can a universal thermostat be used on any voltage?
30. What is used to cause the system to automatically change from one mode to another?

31. On what type thermostat is a dead-band used?
32. What is used to maintain a proper balance between the temperature of the heating medium and the outdoor air temperature?
33. What is used to cycle the equipment to keep an average discharge air temperature?

# Unit 26: HUMIDISTATS
# Introduction

Humidistats are used to control humidification equipment on air-conditioning systems during the heating cycle. Adding humidity to some buildings is wanted because the air gets drier when heated. Today, many different types of humidity-sensitive materials are available. For the most part they are organic and include such materials as nylon, wood, human hair, and sometimes animal membranes. In addition, other materials are used that change in electrical resistance with a change in humidity. Only a few of these are used in humidity controllers.

**Controls and sensors.**   The following describes the most commonly used types of controllers and the sensors used in them.

Mechanical sensors use the change in length of the sensing element in direct relation to a change in relative humidity. The sensors used in modern controllers are commonly made from human hair or some type of synthetic polymer. Sensors are normally attached to a linkage in the controller, and in turn control the mechanical, electrical, or pneumatic switching element in a valve or motor.

Normally, they are used to control a set point that the user or operator can select. Also available are controllers that automatically change the set point in response to the outdoor air temperature, which helps to reduce condensation on the windows and building walls.

When air passes over the sensing element, it senses any small changes in humidity. The electrical resistance between the two conductors varies in response to the humidity present. In most applications the sensor is placed in one leg of a Wheatstone bridge, where the output signal can be used to control the equipment or give a readout. Generally, a signal amplification is needed for correct operation.

There are several ways to wire these controls into a system; however, they should be wired so that the humidifier will run only when the indoor fan is running. This will keep moisture from collecting on the heat exchanger or the electric heating elements during the OFF cycle. One suggested wiring diagram is shown in Figure 26–1. This diagram provides a safety feature—the water solenoid cannot be energized unless both sources of electricity are supplied to it.

Most humidistats may be used on either line-voltage or low-voltage control systems. Usually a moisture-sensitive nylon ribbon is used to move the contacts open on a rise and close on a fall in humidity. Positive ON-OFF settings are sometimes provided for manual operation. The switch is an SPST, snap-acting type. See Figure 26–2.

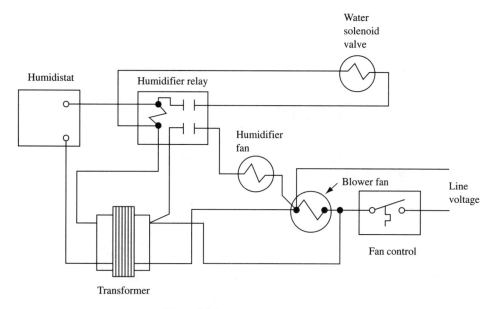

**Figure 26–1**
Typical humidifier wiring diagram.

**Figure 26–2**
Humidity controller. (Courtesy of
Ranco North America, Division of
Ranco, Inc.)

# Summary

- Adding humidity to some buildings is wanted because the air gets drier when heated.
- Many different types of humidity-sensitive materials are available. For the most part they are organic and include such materials as nylon, wood, human hair, and occasionally animal membranes.

- Some materials change in electrical resistance with a change in humidity.
- Mechanical sensors depend on a change in the length or size of the sensing element in direct relation to a change in relative humidity.
- Humidistat contacts open on a rise and close on a fall in relative humidity.

# Service Call

A customer complains that the inside of his windows are sweating. A check of the relative humidity shows it is too high for the outdoor temperature. The tech sets the humidistat to a lower RH percentage so that less moisture will be added to the building, and instructs the user to set the humidistat according to the outdoor air temperature. The technician is satisfied that the system is repaired.

# Student Troubleshooting Problem

A customer complains that the joints inside the building are cracking open. The system does have a humidifier and it is working. The humidity is 10 percent RH. What could be the problem? How can it be solved?

# Questions

1. When are humidistats used?
2. Of what are humidity sensors made?
3. What helps to reduce condensation inside a building?
4. When using an electric humidity sensor, where is the sensor placed?
5. When should a humidifier operate?

# Unit 27: AIRSTATS AND ENTHALPY CONTROLLERS

## Introduction

The airstat controller is used as a fan safety cutoff. It stops the fan when the return air plenum temperature rises to a point that indicates the possibility of a fire inside the building. The controller locks out to keep the fan from running and thus agitating the fire. The switch may be either bimetal actuated or a fusible link operating a spring-loaded SPST set of contacts. When this control is triggered, it either must be manually reset or the fusible link must be replaced.

To wire this control into the electrical system, see the diagram in Figure 27–1.

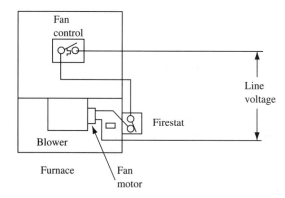

**Figure 27–1**
Firestat wiring diagram.

**Figure 27–2**
Enthalpy controller. (Courtesy of Ranco
North America, Division of Ranco, Inc.)

**Enthalpy controller.**    The enthalpy controller senses total heat (temperature and relative humidity) of the air. It is used to control the amount of outside air brought into the building in response to the total heat content or enthalpy of the outside air. A humidity sensor generally is a thin nylon ribbon and the temperature sensor is a liquid-filled bulb. See Figure 27–2. The enthalpy controller has an SPDT switch with terminals 1 and 2 that make on a rise in enthalpy above the control set point; terminals 2 and 3 make on a fall in enthalpy below the set point. These controllers can be mounted in any position in the outside air duct. Enthalpy controllers usually have a humidity range of up to 100 percent and a maximum operating temperature of 150°F.

# Summary

- The airstat controller is used as a fan safety cutoff. The airstat locks out to keep the fan from running and thus agitating the fire.
- The switch may be either bimetal actuated or a fusible link controlling a spring-loaded SPST set of contacts.
- The enthalpy controller senses total heat (temperature and humidity) of the air. It is used to control the amount of outside air being brought into the building in response to the total heat content, or enthalpy, of the outside air.

# Service Call

A customer complains that the indoor fan motor will not run. A check of the system reveals that the airstat is open, thus keeping the fan from running. When the airstat is reset, the unit starts operating satisfactorily. The customer and technician discuss what had happened in the house. The customer had placed an electric heater close to the return air grill during the cold night before. The technician explained to the customer that the electric heater had raised the return air temperature high enough to cause the airstat to sense a fire and thus stopped the fan. The technician advised the customer to not set the heater where it would affect the return air temperature. The technician is satisfied that the system is repaired.

# Student Troubleshooting Problem

A customer complains that the air in his large building was too stuffy and uncomfortable. A check of the system shows that an enthalpy controller was used to lower the operating cost by using outside air to help the conditioning unit. It is raining outside and the outside air dampers are completely open. The enthalpy controller has voltage at its terminals. It is demanding that the outside air dampers be closed. What could be the problem?

# Questions

1. What is the purpose of an airstat?
2. What type of contacts are used in airstats?
3. What is enthalpy?
4. What type of contacts are used in enthalpy controllers?

# Unit 28: GAS VALVES
## Introduction

The automatic control of gas to the main burners of a gas heating unit is done by using electrically powered mainline gas valves. Safe operation of these valves is guaranteed with pilot safety controls with a thermocouple. A thermostat is used for automatic operation.

**Definitions.**    The following are the accepted definitions of gas valves.

*Gas Valves.*  Gas valves are NC solenoid valves used for controlling gas to the combustion area of the equipment.

*Pilot Safety Controls.*  Pilot safety controls provide safe, automatic shutoff of gas valves during pilot flame failure in gas-fired appliances.

# Gas Valves

The main gas valve opens or closes on demand from the thermostat. Gas valves are designed to provide many operations. Its components are a gas-pressure regulator, a pilot safety, a main gas cock, and the main gas valve. These components are all contained in a single valve body that is generally called the combination gas control. See Figure 28–1. The purpose of the main gas valve is to admit gas to the main burners on demand from the thermostat. These valves are usually of the globe type. The main parts include the valve body, valve seat, valve disc, and valve stem. Gas is allowed to flow to the equipment when the valve disc is lifted off the valve seat. Gas valves are classified by the type of power used to open and close them. They are classified as follows: a solenoid valve, magnetic diaphragm valve, heat motor valve, and redundant gas valve.

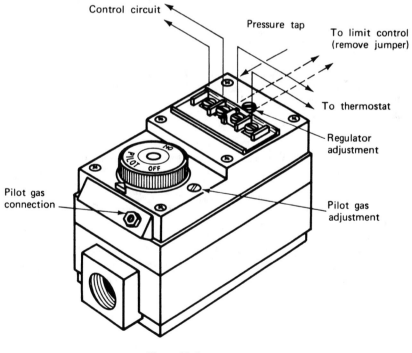

**Figure 28–1**
Combination gas control.

**Solenoid valve.**     The solenoid valve is generally considered an electrically operated stop valve. It is normally closed and is the most popular type used on heating systems.

*Operation.*     When the electrical circuit is made to the solenoid coil, the valve stem is drawn into the coil by magnetic force. The stem and the valve disc are fastened together so that the valve disc is pulled from the valve seat. When the valve stem is drawn into the solenoid coil, the valve opens. When the circuit breaks, the magnetic field collapses and the valve disc returns to the valve seat to close the valve. The noise made by the closing of the solenoid valve is lessened by the addition of a rubber or plastic seat. Valves of this type are known as soft-seat solenoids. A spring is added above the valve to slow the opening and to help stop the clacking noise made when the valve opens.

An improvement on this valve is the oil-filled solenoid. The oil reduces friction and almost completely eliminates the operating noise.

**Magnetic diaphragm valve.**     The magnetic diaphragm valve uses the inlet gas pressure as the major opening and closing force.

*Operation.*     On demand from the thermostat, the control circuit energizes the valve coil. When the valve coil is energized, a lever is magnetically pulled, closing the bleed gas passage to the pilot burner. The gas pressure above the diaphragm is vented through the bleed orifice to the pilot burner where it is burned. The gas pressure below the diaphragm is now greater than that above the diaphragm. The weight and disc are lifted off the seat, opening the gas valve.

When the control circuit is broken, the bleed gas passage is closed, allowing the gas pressure above and below the diaphragm to equalize. The weight of the valve disc then closes the valve.

**Heat motor valve.**     The heat motor valve uses the heat given off by an electrical heating coil to heat an expandable rod. The rod increases in length as it gets warmer. Lengthening of this heated rod gives the force necessary to move the valve mechanism.

*Operation.*     When the thermostat demands heat, the control circuit to the resistance coil is made and the coil is heated. The heat given off by the coil causes the expandable rod to expand in length. When a certain length is reached, the valve snaps open.

When the thermostat is satisfied, the control circuit is opened, and the coil temperature cools. When the rod has cooled enough, it shrinks in length and the valve snaps closed.

Note: Most of these valves have a certain amount of time delay before they will open or close to provide a smoother ignition and outage of the flame.

**Redundant gas valve.**     The purpose of redundant gas valves is, other than admitting gas to the main burners, safety. Two separately operated valves must open before gas can flow to the main burner. If one valve should fail to close, the other closes and stops the flow of gas. See Figure 28–2.

This type valve is used on gas-fired heaters and boilers, with or without intermittent pilot ignition, in place of the regular gas valve.

*Operation.*     Usually, any problems with this type valve are electrical rather than mechanical. Two electrically operated shutoff valves are in series with each other. No gas

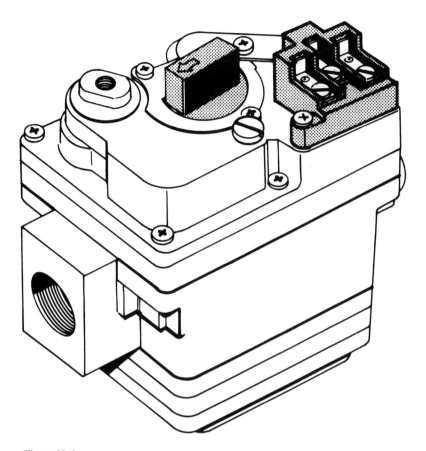

**Figure 28–2**
Redundant gas valve. (A Reston Publication reprinted by permission of Prentice-Hall, Englewood Cliffs, N.J.)

flows until both valves are open. Only one valve needs to close to stop the flow of gas, however. This is provided as a safety feature, because it is almost impossible for both valves to stick in the open position at the same time.

Several different types of redundant gas valves are used. One type uses an instant-acting solenoid valve at the valve outlet to control the flow of gas. A time-delay valve is used at the inlet of the valve. This time delay usually allows about 10 seconds before opening the gas valve.

The type of redundant gas valve used on furnaces with standing pilots is similar to the standard type gas valve, except that an internal heat motor valve is used. Thus, two 24-volt coils are used in these valves. See Figure 28–3.

In this diagram one coil is shown as a solenoid valve and the other as a resistor. They are in electrical parallel. Both are in electrical series with the limit control and thermostat. The resistor shows the heat motor that operates the second, or redundant, gas valve.

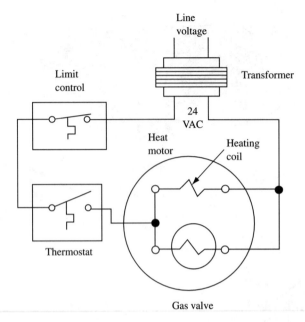

**Figure 28–3**

Two coils in one valve.

A third internal valve is a 100-percent safety shutoff valve that is energized by the electricity generated by a thermocouple (millivolts). If the pilot flame is too small to heat the thermocouple, the 100-percent safety shutoff will close, stopping the flow of gas to the main burner.

If either the limit control or the thermostat contacts opens the control circuit, the solenoid valve closes instantly. However, the heat motor valve takes a few seconds to cool down and close the second valve.

**Automatic spark-ignition gas valves.**    Automatic spark-ignition gas valves are equipped with a redundant pilot solenoid, main gas regulator, and the necessary wiring connections for the make and model being used. They use electrically operated solenoid valves controlled by a room thermostat or other temperature control. See Figure 28–4.

*Operation.*    The main gas valve is controlled by the proper flame sensor that plugs into the valve. Both the pilot and the main gas valve are controlled through the proper relight timer control, which deenergizes both valve coils if a pilot flame is not sensed within the lockout period. These valves should be installed only on systems equipped with the correct ignition systems as recommended by the manufacturer.

**High-low-off fire gas valves.**    The high-low-off main gas valves give all manual and automatic control functions needed for operating gas-fired heating equipment. They have an internally vented, diaphragm type main control valve and a separate thermomagnetic safety valve with pilot gas adjustment and a pilot filter. Most models have a pressure regulator for use with natural gas. See Figure 28–5.

(a)

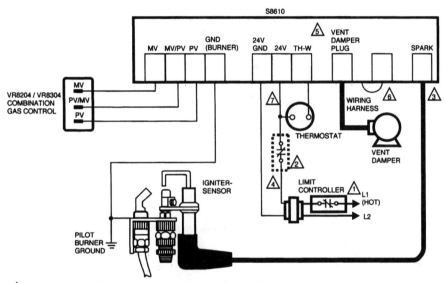

(b)

⚠1 POWER SUPPLY. PROVIDE DISCONNECT MEANS AND OVERLOAD PROTECTION AS REQUIRED.

⚠2 ALTERNATE LIMIT CONTROLLER LOCATION.

⚠3 MAXIMUM CABLE LENGTH 3 ft. (0.9 m).

⚠4 CONTROLS IN 24V CIRCUIT MUST NOT BE IN GROUND LEG TO TRANSFORMER.

⚠5 REMOVE PLUG ONLY IF USING VENT DAMPER. FUSE BLOWS ON STARTUP WHEN PLUG IS REMOVED AND VENT DAMPER WIRING HARNESS IS INSTALLED; THEN MODULE WILL OPERATE ONLY WHEN VENT DAMPER IS CONNECTED.

⚠6 REMOVE JUMER AND CONNECT SENSE TERMINAL ON TWO ROD APPLICATION ONLY.

⚠7 IF THE VENT DAMPER IS CONNECTED, WIRE 24V TERMINAL AS SHOWN. CONNECT VENT DAMPER CABLE IN PLACE OF PLUG SHIPPED WITH THE S8610U.

M8150

**Figure 28–4**
Automatic spark-ignition system main gas valves. (Courtesy of Honeywell, Inc.)

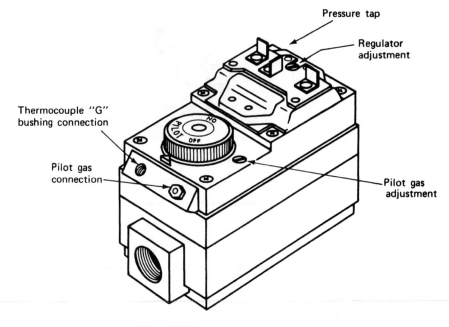

**Figure 28–5**
High-low-off combination gas valve.

The main gas valve diaphragm opens and closes in response to a thermostat and a limit control. The main line diaphragm valve also has a pressure regulator on natural gas models; however, models used on LP gas do not have a pressure regulator.

*Operation.*    When the electrical circuit is made between the C and W terminals of the valve, it automatically opens to the preset low-fire position. Then, when the circuit is made between terminals C and $W_2$, the valve opens to the high-fire position. The low-fire to high-fire shift is done by a heat motor in the valve operator. Time must be allowed for this heating action to take place. See Figure 28–6.

This type valve is used when the full Btu rating of the appliance is not needed but some heating is wanted, such as in the spring and fall months during mild weather.

# Summary

- Automatic control of gas to the main burners of a gas-fired heating unit is done by using electrically powered mainline gas valves.
- Its components are a gas-pressure regulator, pilot safety, a main gas cock, and the main gas valve.
- The purpose of the main gas valve is to let gas flow to the main burners on demand from the thermostat.
- Main parts include the valve body, the valve seat, the valve disc, and the valve stem.

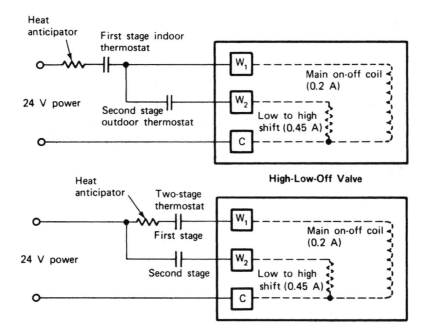

**Figure 28–6**
Schematic diagram for high-low-off combination gas valve.

- In the magnetic diaphragm valve, the gas pressure is the major opening and closing force.
- The heat motor valve uses the heat given off by an electrical resistance coil to move an expandable rod to open the valve.
- The purpose of redundant gas valves is, other than letting gas flow to the main burners, safety.
- Two separately operated valves that must open before gas can flow to the main burners are used in a redundant gas valve.
- Automatic spark ignition gas valves have a redundant pilot solenoid, main gas regulator, and the necessary wiring connections for the make and model being used.
- High-low-off main gas valves give all manual and automatic control functions needed for operation of the gas-fired heating equipment.

# Service Call

A customer complains that her residential heating unit is not heating. The home is found cool and the heating unit not operating. Electricity is at the unit. Voltage is shown from the secondary side of the transformer. The terminals of the gas valve have voltage. A continuity check of the valve coil shows it has infinite resistance. The gas valve coil is bad. The tech replaces the gas valve and turns the unit back on. The heating unit starts heating the building. A complete check of the unit shows it is operating as it should. The technician is satisfied that the system is repaired.

# Student Troubleshooting Problem

A customer complains that a residential heating unit will not shut off. A check of the system shows that the gas valve is in the open position. The temperature is above the thermostat setting. There is no voltage at the gas valve terminals. What could be the problem?

# Questions

1. What is a gas valve?
2. What two devices control the operation of the gas valve?
3. When is gas allowed to flow to the equipment?
4. What opens a magnetic diaphragm gas valve?
5. In what valve is an expandable rod used for the opening force?
6. What is the advantage of time delay in opening and closing gas valves?
7. Why are redundant gas valves used?
8. Can any redundant gas valve be used on any system?
9. When a high-low-off gas valve is being used in the high-fire position, what terminals are energized?
10. When are high-low-off gas valves used?

# Unit 29: PILOT SAFETY DEVICES
# Introduction

The safe and automatic control of gas to the main burners of a heating unit is provided by controls called pilot safety devices. Pilot safety devices are available in many forms. The most popular has been the thermocouple; however, the electronic ignition systems are gaining in popularity. These will be discussed in a later unit about electronic ignition systems.

**Thermocouple.**    A thermocouple is a commonly used source of electric current in which heat is changed into electric energy. In heating systems used today, the thermocouple is a widely used safety control. It is normally used with other pilot safety controls. The basic operation of the thermocouple must therefore be understood by the service technician.

*Operation.*    When two pieces of dissimilar metals, usually iron and copper, are connected and heated, a small dc electric current is generated. This current is caused by the interaction of the two metals and the resulting movement of electrons. See Figure 29–1.

Any current produced between the hot and cold junctions of the thermocouple increases as the temperature rises, up to a given current rating. Above this level, little if any rise in current is possible regardless of the amount of heat applied. Thermocouples should not be overheated. Excessive temperatures will result in a burned-out junction and a ruined thermocouple.

Thermocouples are normally used as safety devices and will hold an electromagnetic relay in the pulled-in position after the armature is manually pushed into the magnetic field. If for some reason the current is lowered or stopped, the armature will be pulled out of the coil by a spring.

Thermocouples are rated at 30 millivolts when the hot end is inserted into the pilot flame from 0.38 to 0.50 inch. See Figure 29–2.

Thermocouples are connected directly to the individual controls they operate. They do not depend on any outside voltage to do their job and should not be connected to any electrical control circuit with voltage different from their output. See Figure 29–3. A voltage other than the millivoltage of the thermocouple will ruin the control.

**Sail switch.**    Sail switches are used as safety controls on forced combustion air, gas-fired furnaces to detect when enough air is moving for correct combustion. They are designed to detect airflow or the lack of airflow through the furnace combustion chamber. These switches are also sometimes used to sense airflow in air distribution systems. Sail

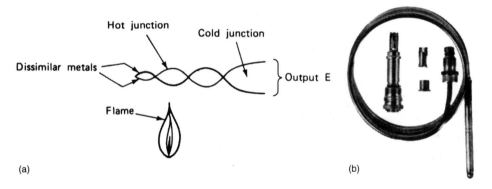

(a)                                                                                          (b)

**Figure 29–1**
(a) Thermocouple schematic, (b) thermocouple. (Courtesy of White-Rodgers Division, Emerson Electric Co.)

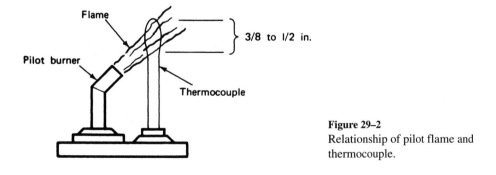

**Figure 29–2**
Relationship of pilot flame and thermocouple.

switches respond only to the velocity of air movement. They are used as safety devices on forced draft gas burners to keep the main gas valve from opening until enough combustion air is drawn into the combustion chamber. Some equipment manufacturers use them on electric heating systems to prove a minimum airflow before the strip heaters can be energized. Use of a sail switch allows auxiliary equipment such as electronic air filters to be wired separately from the blower motor. See Figure 29–4.

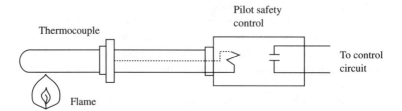

**Figure 29–3**
Thermocouple regulated control.

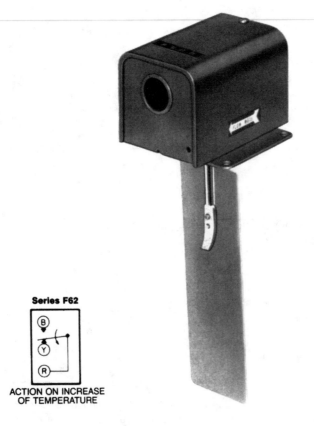

**Series F62**

ACTION ON INCREASE
OF TEMPERATURE

**Figure 29–4**
Sail switch. (Courtesy of Johnson Controls, Inc., Control Products Division.)

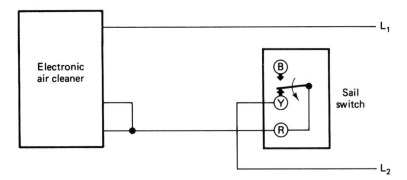

**Figure 29–5**
Typical sail switch schematic for an electronic air cleaner.

These switches are made with a sail mounted on an SPDT, micro, snap-acting switch. Some have a maximum operating temperature of about 180°F. Therefore, when replacing a sail switch, make sure to use one for the correct working temperature. Some are positional mounted so use care to install the replacement correctly. A typical wiring diagram for an electronic air cleaner is shown in Figure 29–5.

*Operation.*    The thermostat usually controls operation of the blower motor, whether it is a combustion blower or a circulating air blower, which in turn moves the air through the system as needed. The operating contacts of a sail switch are normally open. As the fan starts blowing air through the area being measured, the sail moves. When enough air is flowing through the area, the sail moves far enough to close the switch contacts. At this time the controlled equipment is energized and the system starts running.

If for any reason the airflow is reduced or is not enough to keep the contacts closed, the equipment will automatically shut down.

When the thermostat is fully satisfied, the fan motor is deenergized, stopping the flow of air through the monitored part of the system. As the flow is reduced, the sail switch contacts move to the open position to deenergize all the equipment it is controlling.

# Summary

- A thermocouple is a commonly used source of electric current in which heat is changed into electric energy.
- A thermocouple is made of two dissimilar metals, usually iron and copper. When they are heated a small dc electric current is developed.
- The current produced between the hot and cold junctions of the thermocouple increases as the temperature of the hot junction rises to a given current rating.

- Thermocouples are rated at 30 millivolts when the hot end is inserted into the pilot flame from 0.38 to 0.50 inch.
- Sail switches are used as safety controls on gas-fired furnaces to detect when enough air is flowing for correct combustion.
- Sail switches respond to the velocity of air movement.
- When replacing a sail switch make sure that one for the correct working temperature is being used.
- The operating contacts of a sail switch are normally open.

# Service Call

A customer complains that his heating unit is not heating. A check of the system reveals that the pilot is off and the furnace is cool. The pilot cannot be relit. The technician removes the thermocouple from the gas valve and connects a millivolt meter to the leads using the correct polarity. Again the technician lights the pilot and checks the millivolts, and only 14 millivolts are generated by the thermocouple so the tech replaces it. With the pilot relit and the furnace back in operation, a check shows that everything is operating correctly. The technician is satisfied that the system is repaired.

# Student Troubleshooting Problem

A customer complains that a rooftop heating unit is not heating. A check of the system shows that the combustion air blower is running, but nothing else. Power is supplied to the unit and the thermostat contacts are closed. A check of the control circuit shows that the sail switch contacts are not closed. What could be the problem? What must be done before making the repairs?

# Questions

1. How does a thermocouple generate electricity?
2. What is the output of a thermocouple?
3. What will happen when a thermocouple is overheated?
4. Will a thermocouple pull in a relay armature?
5. How are thermocouples connected into a circuit?
6. Why are sail switches used on combustion chambers?
7. What causes sail switches to operate?
8. What type switches are used in sail switches?
9. What must be done when replacing a sail switch?
10. What is the normal position of the contacts in a sail switch?

# Unit 30: FAN CONTROLS
## Introduction

In the days of hand-fired coal furnaces and boilers, the person who did the shoveling waited around long enough to make sure that the fire was burning properly. That is just one of the many jobs done by automatic controls. Automatic controls are on the job twenty-four hours a day to operate heating and cooling systems safely and economically and to give comfortable living conditions. One such control is the fan control.

The fan control is a device used to start and stop the circulating air fan in response to the temperature inside the furnace heat exchanger. A fan control has SPST NO contacts operated by temperature. The control is mounted to sense the air temperature within the circulating air passages of the furnace heat exchanger.

## Fan Controls

Forced-air circulating systems have controls known as heat watchers in addition to the burning controls. These additional controls cause the circulating air fan to run only when enough heat is available to heat the space.

Two types of fan controls are temperature actuated (see Figure 30–1) and electrically operated (see Figure 30–2).

**Temperature-actuated fan control.**  The temperature-actuated fan control uses either a bimetal strip, a bimetal disc, a helical bimetal, or a pneumatically operated sensing element. The sensing element is inserted into the discharge air plenum or directly into the circulating air passage of the furnace heat exchanger. In either case these elements

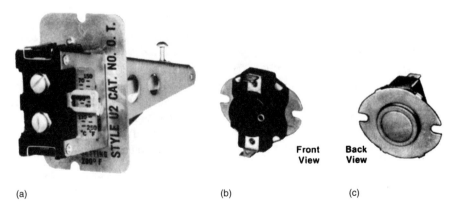

(a)                                          (b)                           (c)

**Figure 30–1**
Heat-actuated fan controls. (a) Blade type and (b) disc type. (Courtesy of White-Rodgers Division, Emerson Electric Co.)

**Figure 30–2**
Electrically operated fan control.
(Courtesy of Honeywell, Inc.)

are mounted in a position so that they sense the temperature of the discharge air from the furnace. See Figure 30–1. They are available with sensing elements of different lengths so that the temperature can be sensed in the exact place the manufacturer has learned to be the hot spot of the unit. When replacing a fan control be sure to use one with the same length element as the one being replaced. They are adjusted by a sliding lever, or a rotating dial, on a scale that is either printed or etched on the control element scale.

*Operation.*    In operation, when the thermostat calls for heat, the main burner gas is lighted, and the furnace heat exchanger begins to warm up. As the temperature of the furnace heat exchanger rises, the sensing element begins to move the dial toward the ON setting. When the ON setting of the control is reached, usually between 125°F and 150°F, the snap action of the switch closes the NO contacts and completes the line-voltage circuit to the fan motor. See Figure 30–3.

When either the thermostat is satisfied or the limit control NO contacts open, the main gas valve closes, stopping the flow of gas to the main burners. As the furnace cools, the air being blown into the conditioned space also cools and causes the bimetal sensing element to return to its OFF position. As the sensing element returns to its OFF position, the fan switch contacts open to stop the fan motor. The OFF temperature of the fan control is usually about 100°F.

**Electrically operated fan control.**    The electrically operated fan control (fan timer control) (see Figure 30–2) gives a timed fan on and off for forced warm-air furnaces. The heater coil is wired in electrical parallel to the low-voltage coil of the main gas valve or the heating relay. See Figure 30–4. This control is used on counterflow and horizontal furnaces because of the heat buildup in the furnace after the main gas valve is closed. The heater coil uses 24 volts ac. The switch is an NO SPST, heater-actuated bimetal control. About one minute after the thermostat calls for heat the fan is started; and stopped about two minutes after the thermostat is satisfied. This type of control causes the fan to operate even when there is no flame in the furnace because it is electrically heat actuated rather than flame heat actuated.

*Operation.*    When the thermostat demands heat, the heaters inside the fan control and the gas valve, or heating relay, are energized at the same time. After the time to start has passed, the fan control contacts close to start the fan motor. In operation, the main gas valve has also opened, or the heating relay has energized the heat strips. The furnace has warmed up enough to warm the building. When the thermostat is satisfied, the main gas valve, or heating relay, and the fan control are deenergized simultaneously. The flame is extin-

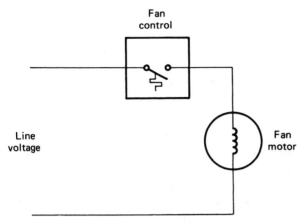

Fan
control

Line
voltage

Fan
motor

**Figure 30–3**
Schematic of fan motor wiring.

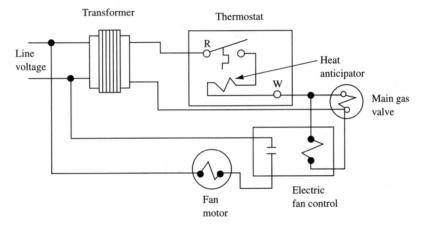

**Figure 30–4**
Schematic diagram of an electrically operated fan control.

guished immediately, or the heating relay is deenergized. The fan continues to run and cool the furnace down. After about two minutes the fan control bimetal has cooled enough to open its contacts and deenergize the fan motor. The furnace is ready for the next operating cycle.

# Summary

- The fan control is a device used to start and stop the circulating air fan in response to the temperature inside the furnace heat exchanger.
- A fan control has NO SPST contacts actuated by temperature.
- A fan control is mounted to sense the air temperature within the circulating air passages of the furnace heat exchanger.

- Fan controls allow the air circulating fan to operate only when enough heat is available to heat the conditioned space.
- Two types of fan controls are temperature actuated and electrically operated.
- Temperature-actuated fan controls use either a bimetal strip, a bimetal disc, a helical bimetal, or a pneumatically operated sensing element.
- Fan controls are available with sensing elements of different lengths so that the temperature can be sensed in the exact place the manufacturer has learned to be the hot spot of the unit.
- Fan controls are adjusted by a sliding lever, or a rotating dial, on a scale that is either printed or etched on the control dial.
- The electrically operated fan control gives a timed fan operation for forced warm-air furnaces when the heater coil is wired in electrical parallel to the low-voltage coil of the main gas valve, or the heating relay.
- The heater coil uses 24 volts ac. The switch is an NO SPST, heater-actuated bimetal control.
- The fan usually starts about one minute after the thermostat demands heat, and is stopped about two minutes after the thermostat is satisfied.

# Service Call

A customer complains that a residential heating unit fan will not stop running. A check of the system reveals that the fan is running because the temperature-actuated fan control contacts are closed. The furnace is cool. The tech attempts to adjust the fan control so that it will stop the fan motor, but it cannot be adjusted. The fan control contacts are stuck closed. The tech then replaces the fan control and adjusts it to bring the fan on at 135°F and stop the fan at 100°F. The technician then puts the heating unit in operation and gives it a complete check. The fan comes on and goes off at the fan control settings. The technician is satisfied that the system is repaired.

# Student Troubleshooting Problem

A customer complains that a horizontal furnace does not heat the building. The system's fan is controlled by an electrically operated fan control. The control contacts have 120 volts. Across the control heater connections is 24 volts. The technician removes the control wiring from the control and checks the continuity of the heater, which shows infinite resistance. What could be causing the problem?

# Questions

1. When do fan controls energize the circulating fan motor?
2. Name the two types of fan controls.
3. Where are temperature-actuated fan controls mounted?

4. When replacing a fan control what must be done?
5. What is the normal fan ON setting of a fan control?
6. What types of contact poles do fan controls have?
7. When is a fan control set at 100°F.
8. What causes an electrically operated fan control to operate?
9. What type of contact poles are used in electrically operated fan controls?
10. What are the usual time delays on an electrically operated fan control?
11. What is a disadvantage of the electrically operated fan control?

# Unit 31: LIMIT CONTROLS
# Introduction

Limit controls are used on all types of warm-air furnaces to keep the discharge air plenum temperatures from getting too high and possibly causing a fire. They have either NC SPST switch contacts or SPDT switch contacts, depending on their specific use. The contacts on the SPST type open on a rise in temperature and close on a fall in temperature. SPDT contact types open the NC contacts on a rise in temperature and at the same time close the NO contacts. NO contacts are used to energize the fan motor on horizontal and counterflow furnaces when the temperature inside the furnace gets too high. The snap-action switch may be actuated by either a bimetal strip, a bimetal disc, a helical bimetal, or a pneumatically operated sensing element. The sensing element is placed so that it will sense the air temperature inside the furnace heat exchanger. Sensing elements are available in different lengths so that the temperature in the hottest point in the furnace can be measured. Be sure to use the correct length so the furnace will operate correctly. See Figure 31–1.

**Adjustment.**    Only some limit control models are adjustable. The adjustable models have a range from about 180°F to 250°F with a fixed differential of 25°F. Adjustment is made by moving a lever along the printed scale on the control dial. They usually have a stop to prevent setting the OFF temperature higher than 200°F. The ON temperature setting is 25°F lower than the OFF temperature setting. Only the cutout is adjusted. The ON temperature setting follows the OFF temperature setting.

**Figure 31–1**
Limit control. (Courtesy of Honeywell, Inc.)

    ***Operation.***    In operation, when the OFF setting is reached, the electrical circuit to the main gas valve is broken. This can be done either by opening the 24-volt ac control circuit directly (see Figure 31–2) or by opening the line voltage to the primary side of the transformer. See Figure 31–3. This break allows the main gas valve to close, stopping the gas flow to the main burners. Both the furnace and the sensing element cools. When the ON setting is reached, the electrical circuit is made to the main gas valve and operation is continued.

    When the SPDT type is used, the operation is a little different. When the NC contacts open, the NO contacts close. At this time the fan motor is started to blow the hot air from the furnace. When the furnace has cooled down, the contacts switch, stopping the fan motor and energizing the main gas valve.

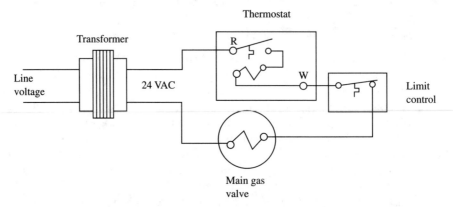

**Figure 31–2**
Limit control in 24 VAC control circuit.

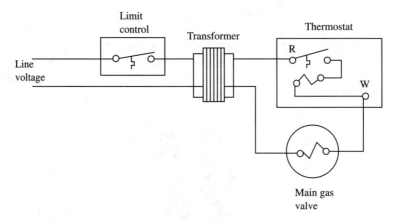

**Figure 31–3**
Limit control in line voltage to primary side of transformer.

This is a safety control and operates only when the temperature inside the furnace is too high. The SPST types are used for pilot duty only and will not handle a heavy current flow. However, the SPDT types have contacts rated heavy enough to carry the fan motor current.

**High-limit controller.**   High-limit controllers are safety controls used to break either the line-voltage or the low-voltage circuit and cause the main gas valve to close if the return air rises to the cut-off temperature. They give a lockout of the main burners in case of fan motor failure. A manual reset is used to keep the equipment from operating when there is an unsafe condition. A bimetal strip inserted directly into the air stream actuates an NC SPST switch. See Figure 31–4.

**Air switches.**   Air switches are used to control the fan in warm-air applications to stop reverse air circulation and high filter temperatures in counterflow and horizontal furnaces. Two-speed fan operation is possible through an SPDT switch. The bimetal is inserted directly into the air stream. See Figure 31–5. Air switches are designed for pilot duty only, with a maximum current flow rating of 50 volt-amps at 24 volts ac.

**Figure 31–4**
High-limit controller. (Courtesy of
Honeywell, Inc.)

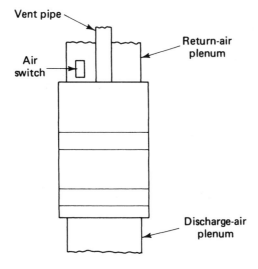

**Figure 31–5**
Air switch location.

# Summary

- The limit control is used on all types of warm-air furnaces to prevent excessive temperatures and a possible fire.
- They have either NC SPST switch contacts or SPDT switch contacts, depending on their use.
- The SPDT contacts open the NC contacts on a rise in temperature and close the NO contacts at the same time. NO contacts are used to start the fan motor on horizontal and counterflow furnaces when the temperature is too high inside the furnace.
- Limit controls may be actuated by a bimetal strip, a bimetal disc, a helical bimetal, or a pneumatically operated sensing element.
- The sensing element is placed so that it will sense the air temperature inside the furnace heat exchanger.
- Limit control sensing elements are available in different lengths. Be sure to use the correct length so that the equipment will operate correctly.
- Adjustable models have a range from about 180°F to 250°F with a fixed differential of 25°F. However, they usually have a stop to prevent setting them above 200°F. Only the cut-out temperature is set. The differential is nonadjustable.
- High-limit controllers are safety switches used only when too much heat is inside the furnace.
- High-limit controllers are used to break either the line-voltage or the low-voltage circuit and cause the main gas valve to close if the return air reaches the control cut-off temperature.
- High-limit controllers give a lockout of the main burners in case of fan motor failure.
- Air switches are used to control the fan in warm-air applications to stop reverse air circulation and high filter temperatures in counterflow and horizontal furnaces.

# Service Call

A customer complains that her heating unit is not operating. A check of the system shows that the upflow furnace is off on the limit control. The tech starts the fan to cool down the furnace. When it has cooled down, the limit control automatically resets. The main burners come on. The discharge air temperature is about 190°F, which is much too high—it should be about 145°F. The filter is clogged with dirt so the tech installs a new one. The discharge air temperature drops to about 150°F. The technician is satisfied that the system is repaired.

# Student Troubleshooting Problem

A customer complains that his heating unit is not operating. The system is a horizontal gas unit equipped with a high-limit controller. The unit is off on the high-limit controller. When the controller is reset, the unit starts running. What could be causing the problem? What could be done to keep it from happening again?

# Questions

1. What is the purpose of limit controls?
2. What is the purpose of the NO contacts on a limit control?
3. What is the range of adjustment of limit controls?
4. What is the highest temperature at which a limit control should be set?
5. Name two ways of wiring a limit control into the circuit.
6. At what temperature does a limit control with SPDT contacts operate?
7. What is a high-limit controller?
8. What is the purpose of air switches?

# Unit 32: COMBINATION FAN AND LIMIT CONTROLS
## Introduction

Combination fan and limit controls give fan and limit protection on forced warm-air furnaces and heaters. They are made with a snap-acting, sealed switch. Contacts in these controls can be used on millivolt, 24-volt, 120-volt, or 240-volt circuits.

*Operation.*    The limit control breaks the control circuit on a rise in temperature to stop main burner operation if the plenum temperature reaches the OFF setting. This is the same use as that of the single-limit control discussed earlier.

On a temperature rise, the fan switch makes the electric circuit to the fan motor when the plenum temperature rises to the ON setting. The fan motor stops when the plenum temperature drops to the fan OFF setting. This is identical to the operation of the single fan control discussed in Unit 30.

Combination fan and limit controls combine the workings of the individual fan and limit controls into a single compact unit. The scale setting is simplified. Both the fan and limit control settings are on the same scale. On some models a summer fan switch is easily used without removing the cover. It also allows for the selection of continuous fan operation or automatic fan operation. See Figure 32–1.

The two switches in these controls are never wired in series, but rather always in parallel. The limit control may be either in the low- or the line-voltage circuit, depending on the wants of the equipment manufacturer. The fan control is always wired into the line-voltage circuit so it can directly control fan operation. For some suggested wiring diagrams, see Figures 32–2 and 32–3.

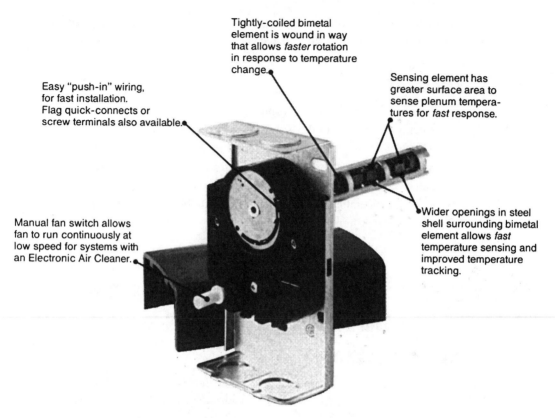

Tightly-coiled bimetal element is wound in way that allows *faster* rotation in response to temperature change.

Sensing element has greater surface area to sense plenum temperatures for *fast* response.

Easy "push-in" wiring, for fast installation. Flag quick-connects or screw terminals also available.

Manual fan switch allows fan to run continuously at low speed for systems with an Electronic Air Cleaner.

Wider openings in steel shell surrounding bimetal element allows *fast* temperature sensing and improved temperature tracking.

**Figure 32–1**
Combination fan and limit control. (Courtesy of Honeywell, Inc.)

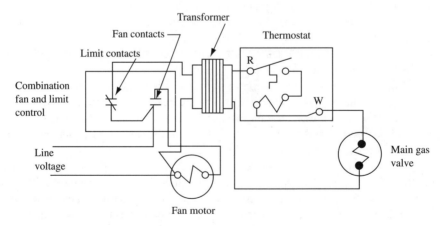

Transformer

Fan contacts

Limit contacts

Thermostat

R

W

Combination fan and limit control

Line voltage

Fan motor

Main gas valve

**Figure 32–2**
Typical control circuit using combination fan and limit control.

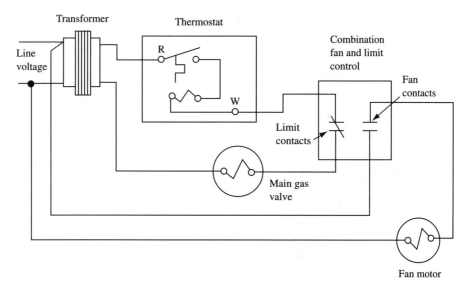

**Figure 32–3**
Typical control circuit with fan control in the line-voltage circuit and limit control
in low-voltage circuit.

# Summary

- Combination fan and limit controls give fan and limit control on forced warm-air furnaces and heaters.
- The limit control breaks the control circuit on a rise in temperature to stop main burner operation if the plenum temperature rises to the OFF setting.
- On a temperature rise, the fan switch makes the circuit to the fan motor when the plenum temperature rises to the ON setting.
- The combination fan and limit control combine the works of the individual fan and limit controls into a single unit.
- The two switches in these controls are never wired in series, but rather always in parallel.

# Service Call

A customer complains that the fire in a central heating unit is going on and off. The system is equipped with a combination fan and limit control. Also, the furnace temperature has reached the cut-off setting of the limit control and the blower is running. Most of the outlet grills in the house have also been closed, preventing adequate airflow through the unit. The technician informs the user that the grills must be left open to allow the correct airflow through the unit. After the grills are opened and operation of the complete unit checked, the unit operates correctly and the technician is satisfied that the system is repaired.

# Student Troubleshooting Problem

A customer complains that her residential heating unit is not heating the home and the flame goes on and off. A check of the system reveals that everything is running and the space temperature is below the thermostat setting. When the thermostat temperature lever is set above space temperature, voltage is found at the gas valve terminals. When the thermostat is set below space temperature no voltage is at the gas valve terminals. All the supply air vents are open and none of the return air vents are blocked. What could be the problem?

# Questions

1. What are the contact ratings of a combination fan and limit control?
2. What is the difference in the operation of a single fan control and the fan control in a combination fan and limit control?
3. What is the advantage of the combination fan and limit control in a single housing?
4. How may ventilation be given with a combination fan and limit control?
5. How are the switches wired in a combination fan and limit control?

# Unit 33: OIL BURNER CONTROLS
# Introduction

Approximately 65 million people use oil heating equipment in the United States. Because of the popularity of fuel oil, it is necessary to cover controls for their safe, automatic operation in a book of this type. Fuel oil burners use a few controls that are common to gas or electric heating equipment such as fan control, limit control, and thermostat. An oil burner is a pump and combustion air blower used to supply a mixture of atomized fuel oil and air, under pressure, to the combustion chamber where most of the combustion takes place.

**Protectorelay (stack-mounted) control.**　The stack-mounted protectorelay control is a combustion thermostat used to sense changes in the flue gas temperature. These changes in flue gas temperature cause the contacts of the stack control to open and close. Through the use of the thermal element, the contacts control operation of the safety switch in the relay unit, making sure that safe starting and correct burner operation are complete. Both recycling and nonrecycling protectorelays are available. These units are mounted directly on the flue pipe with the sensing element inside. They should be mounted as near as possible to the furnace or burner and at least 2 feet from the draft regulator toward the burner. See Figure 33–1.

*Operation.*　To start the burner, first be sure that the combustion chamber is free of oil. Then take the following steps:

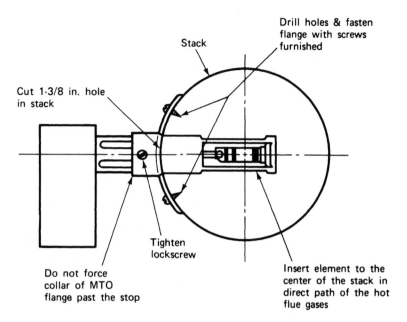

**Figure 33–1**
Typical stack-mounted control.

1. To put the contacts in step, pull the drive shaft lever outward 0.25 inch and release slowly.
2. Move the red reset lever to the right and release.
3. Open the hand valve in the oil supply line.
4. Set the thermostat to call for heat.
5. Close the electrical switch. The burner should start.

**Ignition timing.**    If the protectorelay drops out too soon after the burner starts, adjust the ignition timing lever toward *maximum.*

*Scavenger Timing.*    The following is a description of the *scavenger-timing* operation:

1. With the burner on, the drive shaft moves the protectorelay clutch finger outward. The stop arm halts the clutch finger, but the drive shaft moves a small amount further. This override is necessary for correct sequencing. *Note:* If the clutch finger does not reach the stop arm with the recycle lever at the minimum setting, the bimetal is not getting enough heat. The stack control must be moved.
2. Allow the burner to run a few minutes; then open and close the electric line switch. The burner should stop at once.
3. The burner should start again in about one minute.

4. If the burner starts too soon, carry on as follows:

   a. Open the line switch; wait five minutes for cooling.

   b. Move the cycle lever outward one notch. Close the line switch to start the burner and repeat steps 1, 2, and 3.

   c. Repeat steps a and b until the timing is correct.

To check the stack switch, use the following to prove the safety features:

1. Flame failure:

   a. Test for protectorelay recycling by shutting off the oil supply hand valve while the burner is running normally. Restore the oil supply after the burner shuts off. After a short scavenger period, the stack control restarts the burner.

   b. Test for safety shutoff after the flame failure by turning off the oil supply hand valve while the burner is running normally. When the burner shuts off, do not turn the oil supply back on at this time. The stack control will attempt to restart the system after a scavenger period; then in approximately 30 seconds, the safety switch will lock out. Reset the safety switch, and the burner will restart.

2. Ignition failure: Test for ignition failure by turning off the oil supply while the burner is off. Run through the starting procedure, omitting step 3. A lockout will occur. Reset the red safety switch.

3. Power failure: Turn off the electrical power supply while the burner is on. When the burner stops, turn the power back on and the burner will restart after a scavenger period.

For a typical wiring diagram of an oil burner system, see Figure 33–2.

**Burner-mounted combustion thermostat.**    This type combustion thermostat is mounted on the burner. It is a hermetically sealed switch. Its contacts open and close due to the temperature changes caused by the radiant heat of an oil burner flame. It is not a light-sensitive control and is not affected by normal soot deposits. Its small size allows easy mounting on the burner blast tube, where it will quickly sense the radiant heat of the flame pattern.

*Operation.*    This control is used to monitor the burner flame. During the burner starting period, the contacts must be in the closed position. The control contacts remain closed until enough radiant heat from the flame causes them to open.

When the proper flame is present, a rise in temperature of the sensing face opens the contacts within seconds and deenergizes the safety switch heater, allowing a normal start and run cycle. In case of flame loss during the burning cycle, the drop in temperature causes the contacts to close and deenergizes the safety circuit of the relay.

After a normal burner run, the combustion control contacts return to the closed, or starting, position. If an instant start is demanded by the thermostat, a forced purge will take place while the combustion contacts and ignition switch are being restored to the closed, or starting, position.

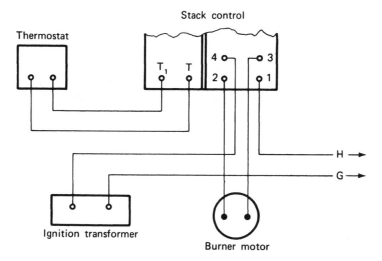

**Figure 33–2**
Typical wiring diagram.

**Kwik-sensor combination oil burner–hydronic control.** Kwik-sensor combination oil burner–hydronic controls use an immersion aquastat controller and an oil burner primary control to give high-limit and low-limit/circulator control for oil-fired hydronic (water or steam) heating systems. See Figure 33–3. They give intermittent (formerly called constant) ignition of the fuel oil. This type control is used with specific supporting controls. The manufacturer must be consulted for the correct types to be used. Some models mount directly on the burner, whereas other models mount externally. The external models are equipped with an armored capillary and a remote sensor. Most are capable of multiple-zone control with use of the correct valves.

**Magnetic valves.** Magnetic valves are used for the ON-OFF oil flow control to domestic oil burning equipment. See Figure 33–4. Magnetic valves are used to control the flow of fuel oil no heavier than number 2 at 125°F and with an ambient temperature range of 32°F to 115°F. When the coil is deenergized, the valve closes immediately.

*Operation.* When the thermostat demands heat, the burner motor starts. The magnetic valve is energized simultaneously. When the thermostat is satisfied, the valve closes instantly. This keeps oil from flowing into the firebox; this oil would be lighted at the start of the next ON cycle—probably causing the combustion chamber to overheat.

**Burner safety controls.** These burner safety controls are the same safety controls used on warm-air heating systems and steam or hot water boilers. They will stop the burner in case of high or low pressures or temperatures because of a lack of air or water flow. They are wired into the system in the same way. A typical wiring diagram is shown in Figure 33–5.

**Figure 33–3**
Kwik-sensor combination oil
burner–hydronic control. (Courtesy of
Honeywell, Inc.)

**Figure 33–4**
Magnetic valve. (Courtesy of Honeywell,
Inc.)

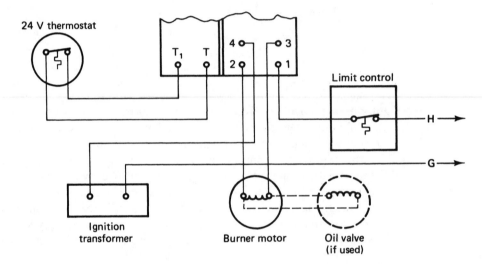

**Figure 33–5**
Wiring diagram with burner safety control.

# Summary

- An oil burner is a unit that supplies a mixture of atomized fuel oil and air, under pressure, to the combustion chamber where most of the combustion takes place.
- A stack-mounted protectorelay is simply a combustion thermostat used to sense changes in the temperature of the flue gases.
- Protectorelays should be mounted as close as possible to the burner and at least 2 feet from the draft regulator.
- The burner-mounted combustion thermostat is mounted on the burner. This thermostat is a hermetically sealed switch. Its contacts open and close in response to temperature changes caused by the radiant heat of an oil burner flame.

- This thermostat is not light sensitive and is not affected by normal soot deposits.
- Kwik-sensor combination oil burner controls use an immersion aquastat controller and an oil burner primary control to give high-limit and low-limit/circulator control for oil-fired hydronic heating systems.
- Kwik-sensor controls give intermittent ignition of the fuel oil. This type control is used with specific supporting controls, and the manufacturer must be consulted for the proper type to be used.
- Magnetic valves are used for the ON-OFF control of oil flow to domestic oil burning equipment.
- Oil burner safety controls are the same safety controls used on warm-air heating systems and steam or hot water boilers.

# Service Call

A customer complains that an oil burner is not heating properly. A check of the system reveals that the oil burner is off on the stack-mounted protectorelay. The tech resets the control and tries again to light the burner. It does not relight, so the tech adjusts the protectorelay, resets it again, and tries to relight it another time. It does not relight. The protectorelay is bad and is therefore replaced. Its operation is now good. The technician starts the unit and gives it a complete operational check. The technician is satisfied that the system is repaired.

# Student Troubleshooting Problem

A customer complains that a heating unit is not heating the residence. The system is an oil burner. The thermostat is found to be demanding heat. A check of the operating oil burner controls shows that the protectorelay is open. The relay is then reset, and the oil burner starts but no flame is lit. There is no oil flowing to the burner orifice and the burner shuts down on safety. The protectorelay is reset, and while the oil burner is trying to start, the voltage to the oil-line solenoid valve is 120 VAC. When one of the electric wires is disconnected from the solenoid valve coil, a click can be heard. There is also a click when the wire is reconnected. What could be the problem?

# Questions

1. What does an oil burner do?
2. What device is used to sense changes in flue gas temperature?
3. How are protectorelays installed?
4. Where should protectorelays be mounted?
5. What will happen if the protectorelay element does not get enough heat?
6. What will happen if the burner is started while the oil hand valve is closed?

7. What opens the contacts of a burner-mounted combustion thermostat?
8. Will too much light hitting the burner-mounted combustion thermostat cause it to stop?
9. What control gives intermittent ignition of the fuel oil?
10. What type of oil is the magnetic valve designed to control?
11. What type of valve is the magnetic valve?

# Unit 34: HYDRONIC HEATING CONTROLS

## Introduction

Heating a building with water or steam is common in large commercial systems and where energy conservation is important. Control of the boiler needs a special type and use of controls.

A boiler control may be defined as any control that gives safe, automatic, and economical operation of a boiler.

**Water level control.** The purpose of a water level control is to stop the burners if the water level gets low enough to cause damage to the boiler. The water level control is float operated to sense the water level inside the boiler. It will cut off the fuel to the main burner when a low water condition occurs. See Figure 34–1. The low water control is used on boilers with pressures up to about 30 psig. It is mounted on the boiler so the switch will open the control circuit to the main fuel valve when there is a low water level.

The low water cutoff is mounted on the boiler at normal operating water level. See Figure 34–2. Several locations and piping arrangements are possible for this type control. Because no normal water line is to be kept in a hot water boiler, any location of the control above the lowest allowable water level is correct. A steam boiler does, however, have a definite water level that must be kept. Follow the suggestions of the boiler manufacturer.

Construction of the hot water and steam boiler is basically the same. The main difference is the way the steam boiler is operated. Most things that cause low water in a steam boiler will also be true for a hot water boiler.

*Operation.* The water line in the boiler and the water line in the control drop at the same time and are at the same level. When the water level is lowered in the control float, the float drops. When the float drops it causes the control switch contact to open and break the control circuit to the fuel valve. See Figure 34–3 on page 221.

This is a basic safety control for boilers. It is a means of stopping the automatic firing unit if the water level drops below the minimum safe level.

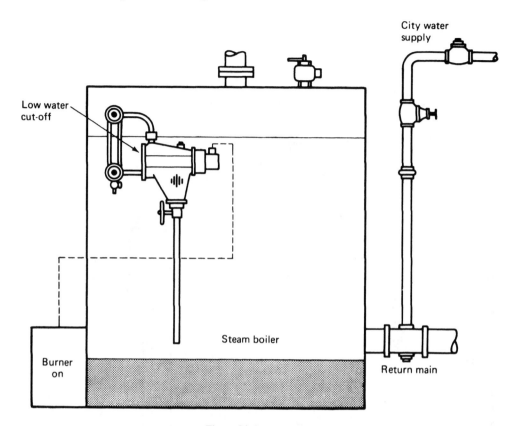

**Figure 34–1**
Float-operated low water cutoff.

**Combination feeder and low water cutoff.**   If the low water cutoff could be completely relied on to shut off the burner during low water conditions, there would be no problem. However, experience has shown that under some conditions, the low water cutoff cannot stop the burner.

The combination feeder and low water cutoff offers much more safety than the low water cutoff alone. It covers most installations and gives the most complete measure of safety possible. See Figure 34–4 on page 222.

This control provides:

1. The mechanical feeding of water to the boiler as fast as it is let out through the relief valve.
2. The electrical operation of stopping the burner during low water conditions.

This combination of water feeding and control of the electrical circuit offers the best protection. They are the best and are recommended for boilers.

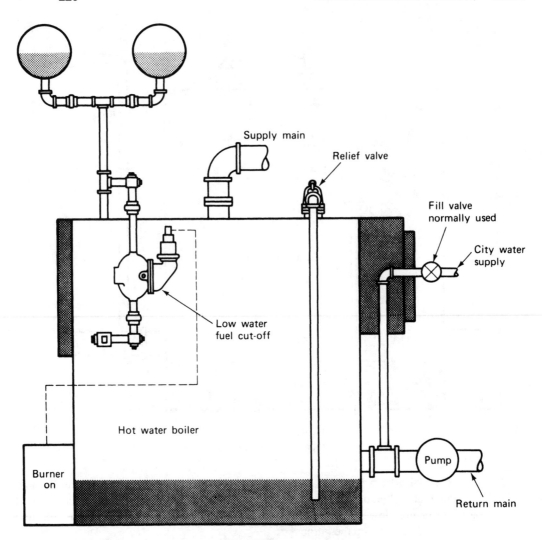

**Figure 34–2**
Installation of low water cutoff.

*Operation.*    In operation, the combination feeder and low water cutoff control lets water into the boiler to keep the needed water level. At the same time, it passes the electrical power to the control circuit that operates the burner in response to the needs of the boiler. If the water leaves the boiler faster than the feeder can let it in, the low water cutoff will stop the main burner before the boiler is overheated and possibly ruined.

*Switches.*    The low water cutoff and the combination feeder and low water cutoff controls have electrical contacts of various configurations. See Figure 34–5 on page 223.

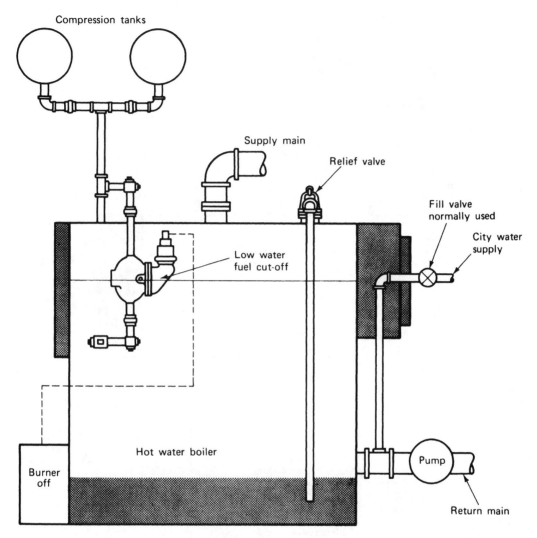

**Figure 34–3**
Low water line stops burner.

In Figure 34–5(b) the NC contacts complete the control circuit and cause the burner to operate on demand from the temperature or pressure control. Figure 34–5(c) shows that when there is a low water level, the control circuit is opened and the NO alarm contacts close, sending a danger signal to the operating engineer.

**Boiler temperature control (aquastat).**   In normal operation, the boiler temperature control regulates main burner operation by making or breaking the control circuit. The sensing element is placed in the boiler through openings furnished by the boiler

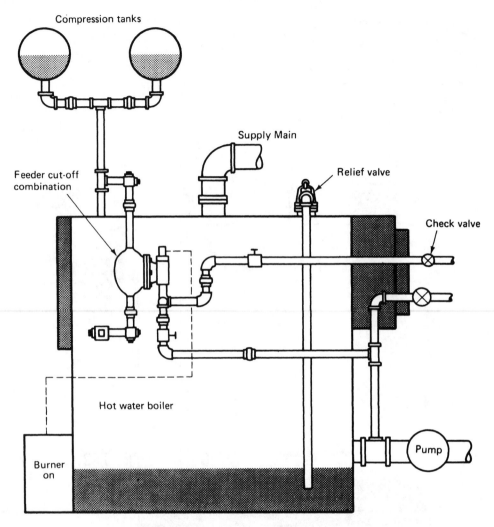

**Figure 34–4**
Combination feeder and low water cutoff.

manufacturer. See Figure 34–6. The switch may be either SPDT snap acting or a mercury tube. Both are triggered by a helically wound bimetal. Boiler temperature controls are adjustable so that the correct operating temperature can be chosen.

*Operation.* As the water temperature inside the boiler drops to the cut-in temperature of the control, the sensing element causes the switch contacts to close making the main burner valve control circuit. See Figure 34–7. As the water is heated to the aquastat cut-out temperature setting, the switch contacts open and break the control circuit, deenergizing the main burner fuel supply valve.

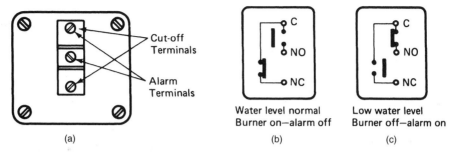

**Figure 34–5**
(a) Switch terminal locations, (b) water level normal, burner on–alarm off,
(c) low water level, burner off–alarm on.

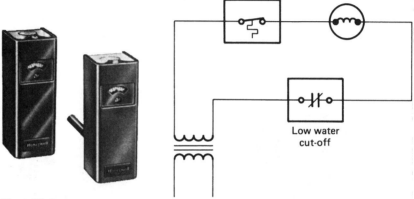

**Figure 34–6**
Boiler temperature control
(aquastat). (Courtesy of
Honeywell, Inc.)

**Figure 34–7**
Basic boiler wiring diagram with temperature
control.

**Boiler pressure control.**    The boiler pressure control is used on steam boilers. It is mounted above the boiler itself to sense the pressure at the most critical point in the system. See Figure 34–8. The line switch is operated by a diaphragm and bellows, which expand with a rise and contract with a fall in pressure. The contacts in these controls are wired in series in the control circuit. They either make or break the control circuit in response to the boiler pressure. See Figure 34–9. Boiler pressure controls are adjustable so the correct operating pressure for the system can be kept. See Figure 34–10.

*Operation.*    As the pressure inside the steam system drops to the cut-in setting of the boiler pressure control, the switch contacts close, making the control circuit to the main

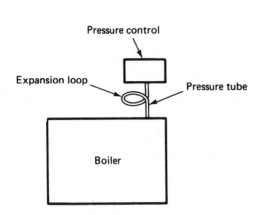

**Figure 34–8**
Boiler pressure control installation.

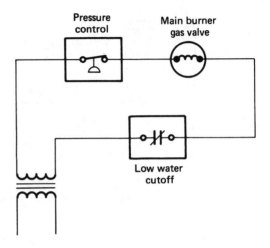

**Figure 34–9**
Basic boiler wiring diagram pressure control.

**Figure 34–10**
Boiler pressure control. (Courtesy of
Honeywell, Inc.)

**Figure 34–11**
High-limit control. (Courtesy of
Honeywell, Inc.)

burner fuel valve. As the burner runs, the pressure inside the system rises to the cut-out
setting of the pressure control. When the pressure control contacts open, the control circuit
is broken, thus deenergizing the main burner fuel valve.

**High-limit control.**    The high-limit control is a safety control. It is wired in series
in the control circuit to keep the boiler from overheating if either the temperature or pressure control fails to operate correctly. See Figure 34–11. High-limit controls may use either
SPST or SPDT switches. The SPST type only breaks the control circuit and deenergizes
the main burner fuel valve. The SPDT models may be used for several different things,
such as energizing a circulator pump or energizing an alarm circuit to alert the user that

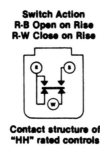

**Switch Action**
**R-B Open on Rise**
**R-W Close on Rise**

**Contact structure of**
**"HH" rated controls**

**Figure 34–12**
Contact configuration of high-limit
control. (Courtesy of White-Rodgers
Division, Emerson Electric Co.)

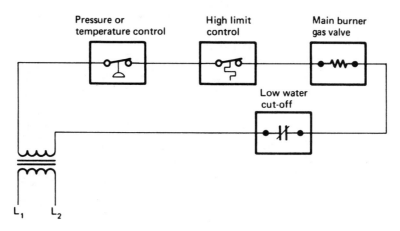

**Figure 34–13**
Basic boiler wiring diagram.

there is a problem. See Figure 34–12. The high-limit control is wired into the control circuit in series with the other controls. See Figure 34–13.

*Operation.*    If the boiler temperature or pressure rises above the setting of the pressure or temperature controller, the high-limit control will break the control circuit. This will
deenergize the main burner fuel valve to keep from overheating the boiler. A secondary
circuit, if used, will then be energized to do the wanted things, such as starting the circulator
pump or energizing an alarm circuit. As the boiler temperature or pressure drops to the cut-in
setting of the control, the contacts close making the control circuit. The alarm circuit is
deenergized and the boiler is again started for normal operation.

**High-low-off fire valves.**    The high-low-off combination diaphragm valves give
all the manual and automatic functions needed for normal operation of gas-fired heating
equipment. These valves have an internally vented diaphragm type, main control valve and
a separate thermomagnetic safety valve with a pilot gas adjustment and a pilot gas filter.
Most models include a pressure regulator for use with natural gas. See Figure 34–14.

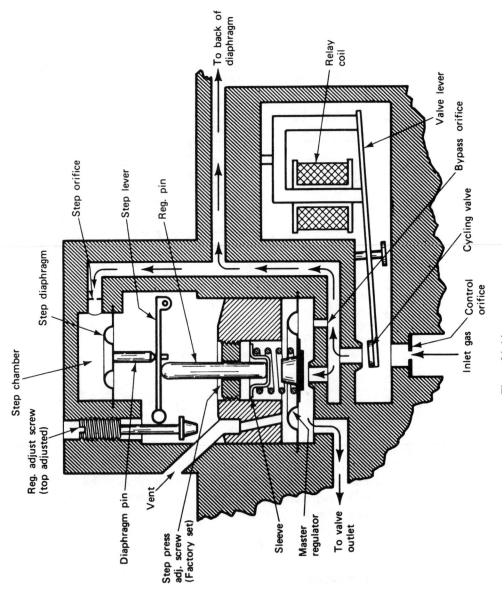

**Figure 34-14**
Cutaway view of high-low-off fire valve.

Reg. adjust screw
(top adjusted)

Step chamber

Step diaphragm

Step orifice

Step lever

Reg. pin

To back of
diaphragm

Relay
coil

Valve lever

Bypass orifice

Cycling valve

Control
orifice

Inlet gas

To valve
outlet

Master
regulator

Sleeve

Vent

Step press
adj. screw
(Factory set)

Diaphragm pin

The mainline gas valve diaphragm opens and closes in response to an aquastat or a boiler pressure control. The mainline gas valve diaphragm valve operates as a pressure regulator on natural gas models. The LP gas models are not equipped with pressure regulators.

When the electrical circuit is made between the C and $W_1$ terminals of the valve, the valve automatically opens to the low-fire position. Then, when the circuit is made between C and $W_2$ terminals, the valve opens to the high-fire position. The full shift from low-fire to high-fire is done by a heat motor in the valve operator. Time must be allowed for the motor to heat to operating temperature. See Figure 34–15.

This type of valve is used when the full Btu rating of the boiler is not needed but some heating is wanted.

**Liquid-level float switch.** Liquid-level float switches are used for several purposes. They are mounted so that they can sense the liquid level in a tank. See Figure 34–16. The

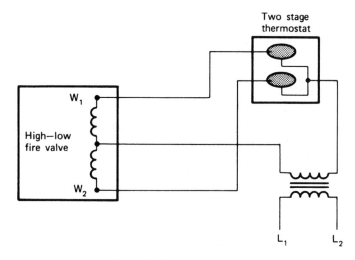

**Figure 34–15**
High-low fire wiring diagram.

**Figure 34–16**
Liquid-level float switch. (Courtesy of Johnson Controls, Inc., Control Products Division.)

switch may be either an SPST or SPDT type. They can be wired to open one circuit and close another when the liquid level rises above or falls below the wanted level. When the liquid level reaches a certain level, the float will open one set of contacts and close another, depending on the control design. Float switches are used for controlling the liquid level in steam boiler condensate tanks, wastewater tanks, and any other tank needing automatic filling and emptying. These controls are wired so that they will control the pump motor used to pump out the tank at the correct time or a fill solenoid valve that fills the tank with fluid when needed.

*Operation.*    When the float makes either set of contacts, the circuit to the wanted function is made and either the pump motor will start or a solenoid coil will be energized. The fluid is either removed from the tank or the tank is filled to the wanted level. The float then opens a set of contacts to stop the process. Sometimes the extra set of contacts may be used to energize an alarm circuit or an information light to show that the needed process is either complete or needs attention.

**Automatic changeover control.**    The automatic changeover control is used to give an automatic summer-winter changeover of the thermostat action in hydronic heating and cooling systems. See Figure 34–17. Automatic changeover controls are available in either direct clamp-on type or the bulb clamp-on type. A SPDT switching action is used. The switch opens red to black at 85°F and red to brown at 70°F.

*Operation.*    When the boiler main burner lights and the water is heated to 89°F at the point where the control is attached, the switch closes the red to yellow circuit. The red to yellow circuit then changes the thermostat to the heating mode. Automatic changeover controls will operate in the heating mode until the boiler is shut down and the chiller is started. When the water temperature at the point where the control is attached drops to 62°F, the switch opens the circuit between the red and yellow wires and makes the circuit between the red and blue wires or the red and black wires. The thermostat is changed over to the cooling mode and gives the wanted cooling control of the space. See Figure 34–18.

**Fan center.**    Fan centers provide low-voltage control of line-voltage fans, pump motors, and auxiliary circuits in forced warm-air or hydronic heating, cooling, or heating and cooling systems. See Figure 34–19. These controls have external wiring terminals to

**Figure 34–17**
Automatic changeover switch.
(Courtesy of Johnson Controls, Inc.,
Controls Group.)

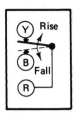

**Figure 34–18**
Switching action of automatic changeover control.

**Figure 34–19**
Fan center. (Courtesy of White-Rodgers Division, Emerson Electric Co.)

help in making the electrical connections. An enclosed relay is used to control the needed equipment. A transformer supplies low voltage for the control circuit. Two sets of contacts are used—they are a main pole and an auxiliary pole. The voltage and amperage used on these poles must not be greater than their rating or they will be ruined. The wiring should conform to the diagram accompanying the control.

# Summary

- A boiler control is defined as any control that provides safe, automatic, and economical operation of a boiler.
- The purpose of a water level control is to stop the burners in case of low water to prevent damage to the boiler.
- The low water cutoff is mounted on the boiler at the wanted water level.
- Because no normal water line is to be kept in a hot water boiler, any location of the control above the lowest allowable water level is satisfactory.
- A steam boiler does, however, have a specific water level that must be kept, and the recommendations of the boiler manufacturer should be followed.
- The water line in the boiler and the water line in the control drop at the same time and are at the same level.
- The combination feeder and low water cutoff offer much more safety than the low water cutoff alone.
- Boiler temperature controls regulate main burner operation making or breaking the control circuit.
- The high-limit control is a safety control wired into the control circuit to keep the boiler from overheating in case the temperature or pressure control fails to function correctly.
- High-low-off combination diaphragm valves give all the manual and automatic functions needed for the operation of gas-fired heating equipment.

- Liquid-level float switches are used with several purposes in mind. They are mounted so that they can sense the liquid level in a tank. They are used for monitoring steam boiler condensate tanks, wastewater tanks, and any other type tank needing automatic filling and emptying.
- The automatic changeover control is used to give an automatic summer-winter changeover of the thermostat operation in hydronic heating and cooling systems.
- Fan centers provide low-voltage control of line-voltage fans, pump motors, and auxiliary circuits in forced warm-air or hydronic heating, cooling, or heating and cooling systems.

# Service Calls

*Service Call 1.* A customer complains that his boiler is not heating as it should. The system has a steam boiler that uses a water level control. The boiler is low on water and is overheating to the point that the high-limit control is taking it off line. The technician blew the sediment from the water level control, which was a difficult job as there was much sediment, but the technician was successful. The technician filled the boiler with water and put the unit back in operation. Then the tech checked the operation of the water level control by closing off the water inlet hand valve and blowing down the control to remove some water from the boiler. The low water control stopped the boiler when the low water level was reached. The boiler was again filled with water, put back in operation, and given a complete operational check and found operating correctly. The technician instructed the owner that the water level control should be blown down about every two or three months to remove any sediment. The technician is satisfied that the system is repaired.

*Service Call 2.* A customer complains that the boiler in an office building keeps cycling on and off. A check of the system reveals that a hot water boiler is cycling on the high-limit control. The system has a bad temperature control. The hand valve in the makeup water line was closed. The technician removed enough water from the system to allow replacement of the control, and adjusted it to the correct temperature for the job. The hand valve in the makeup water line is opened. The tech again fills the system with water, checks the operation of the complete boiler, and finds it to be operating correctly. The technician is satisfied that the system is repaired.

# Student Troubleshooting Problem

A customer complains that a hydronic heating and cooling system has been switched over to cooling but no cool air is blowing from the grills. A check of the system reveals that the boiler is off and the chiller is running. The water leaving the chiller is 45°F. The thermostat is still in the heating mode. The temperature of the water line where the automatic changeover control is fastened is 47°F. The connections to the changeover control are still in the heating mode. What could be the problem? What must be done to repair it?

# Questions

1. What control stops the boiler when there is a low water condition?
2. What is the maximum pressure that a water level control is designed to work with?
3. Where must a water level control be installed on a hot water boiler?
4. What is the water level in a low water cutoff?
5. How fast will a water feeder allow water to enter the boiler?
6. What happens when the water leaves the boiler faster than the feeder can fill it?
7. How are the contacts operated in a steam boiler pressure control?
8. What protects the boiler in case the temperature or pressure control fails?
9. When is a regulator used with a gas valve?
10. On a high-low-off gas valve, what happens when terminals C and $W_1$ are energized?
11. What do liquid-level float switches do?
12. At what temperature does the automatic changeover control switch over to the cooling mode?
13. What is the purpose of the fan center?
14. What will happen if too much current passes through the fan relay?

# Unit 35: HYDRONIC COOLING CONTROLS

## Introduction

Cooling a structure with water is common in large commercial systems and where energy conservation is important. Control of the chiller needs a special type and use of controls. Chiller controls may be defined as any control that gives safe, automatic, and economical operation of a chiller.

**Chiller protector.**   Chiller protector controls are used specifically to protect chillers from freezing and perhaps damaging the tubes. See Figure 35–1. When the temperature drops to the control cut-out setting, its contacts open the control circuit to stop the compressor. The switch is an SPST type that the contacts open on a fall in temperature. While the compressor is off, any ice that may be on the chiller tubes melts so that the unit will operate correctly during the next ON cycle. The cut-out point of chiller protectors is usually about 38°F to 40°F. The sensing element is usually mounted in the water line leaving the chiller. This is so the water temperature is sensed after it has been cooled. Chiller protectors may be either automatic reset or manual reset. These controls are safety controls and should never be taken out of the circuit. If they are tripping, the problem must be found and corrected.

**Figure 35–1**
Chiller protector. (Courtesy of Johnson
Controls, Inc., Controls Group.)

*Operation.*    When the chiller is operating and the water is being cooled, the system
will operate as it should. If for some reason the water temperature reaches the cut-out temperature of the chiller protector, the control contacts open to break the compressor starter
control circuit. The compressor stops and the chiller temperature starts rising to the control
cut-in point. If the automatic reset type is used, the compressor will again start operating.
If the control is a manual reset type, the operator must reset the control before the compressor can be restarted.

**Freezestat/low-limit control.**    The *freezestat/low-limit control* is used to protect heating and cooling coils or other similar units from freeze damage. See Figure 35–2.
This control responds to the lowest temperature sensed along its entire bulb length. It has
an SPST switch that opens its contacts on a fall in temperature. They may be either manual or automatic reset types, depending on the model used. Freeze protection controls
should be set with a cut-out temperature that will keep the coil or protected part from freezing, usually above 32°F.

*Operation.*    While the compressor is running and the air leaving the coil is being
cooled, the unit will operate as wanted. However, if something happens to reduce the airflow or another control fails to operate as needed, the freezestat/low-limit control contacts will open, breaking the control circuit to stop the compressor. The compressor stays
off until the control either automatically resets or is manually reset. When the cut-in
setting of the control is reached, the compressor again starts and the system runs as designed, unless something is wrong with either the indoor air delivery or the refrigerant
charge in the system.

**Flow switch.**    Flow switches are used in hydronic systems to protect the chiller
from damage in case the pump fails to run or fails to move the correct water flow through
the system. See Figure 35–3. The switch has SPDT contacts, which are used to make one
circuit and break another. See Figure 35–4.

On an increase in water flow, the contacts make from red to yellow. On a decrease in
water flow, the contacts make from red to blue. This switch can be used to either stop the
compressor or energize an alarm circuit to start another pump, or do any combination of
things that will give the needed unit operation.

**Figure 35–2**
Freezestat/low-limit control. (Courtesy of Robertshaw Controls Company, Uni-Line Division.)

**Figure 35–3**
Flow switch. (Courtesy of Johnson Controls, Inc., Control Products Division.)

**Figure 35–4**
Flow switch switching action. (Courtesy of Johnson Controls, Inc., Control Products Division.)

*Operation.*    While the correct amount of water is flowing through the system, the control contacts are made from red to yellow. The compressor may operate as needed. However, if the pump should fail, the contacts will change red to blue, making the circuit to give the wanted operation. When the water flow is again started, the flow switch contacts change to make from red to yellow, and the unit operates normally.

**Fan center.**    Fan centers provide low-voltage control of line-voltage fans, pump motors, and auxiliary circuits in forced warm-air or hydronic heating, cooling, or heating and cooling systems. See Figure 35–5.

These controls have external terminals that are useful in making electrical connections. An enclosed relay is used to control the needed equipment. A transformer is used to give the low voltage for the control circuit. Two sets of contacts in the fan center are a main pole and an auxiliary pole. The voltage and amperage passing through each of these poles must not be greater than their rating or they will be damaged. The wiring should match the diagram that comes with the control.

**Figure 35–5**
Fan center. (Courtesy of Honeywell, Inc.)

# Summary

- Chiller controls may be defined as any control that gives safe, automatic, and economical operation of a chiller.
- Chiller protector controls are used specifically to protect chillers from freezing and perhaps damaging the tubes.
- These are safety controls and should never be taken out of the circuit.
- The freezestat/low-limit control is used to protect heating and cooling coils or other similar components from freeze damage.
- These controls should be set with a cut-out temperature that will prevent freezing of the coil or protected part, usually above 32°F.
- If something happens to reduce the airflow or another control fails to operate as designed, the freezestat/low-limit control contacts will open, breaking the control circuit and stopping the compressor.
- Flow switches are used in hydronic systems to protect the chiller from damage in case the pump fails to run or fails to pump enough water through the system.
- Fan centers give low-voltage control of line-voltage fans, pump motors, and auxiliary circuits in forced warm-air or hydronic heating, cooling, or heating-cooling systems.

# Service Call

A customer complains that her unit is not cooling in a large building. The system is cooled with a chiller, but the compressor is not running. The system is discovered off on a manual reset freezestat. The technician resets the control and the system starts running. The tech then checks the refrigerant pressures and finds them lower than normal. The liquid-line sight glass shows bubbles. The system is low on refrigerant. A leak test of the unit shows where the liquid line rubbed against the unit frame. The tech pumps down the refrigeration system and repairs the leak. The repaired part of the system is evacuated. With the unit back in operation and the necessary refrigerant charged into the system, operation of the complete system is normal and the technician is satisfied that the system is repaired.

# Student Troubleshooting Problem

A customer complains that a chilled water system is not cooling. A check of the system reveals that the flow switch has the system shut down. The alarm circuit is not connected to anything. A check of the voltage from the R to the Y terminals indicates none; a check from the R to the B terminals indicates voltage. This shows that not enough water is flowing through the system. The water flow through the pump is lower than needed. What could be the problem? What can be done to correct it?

# Questions

1. What happens when a chiller temperature falls too low?
2. What is the normal cutout of a chiller temperature control?
3. Where is the chiller water temperature measured?
4. What could possibly happen if the chiller got too cold?
5. Where does the freezestat/low-limit control sense the temperature?
6. How does the flow switch protect a chiller?
7. How are fan centers wired into the circuit?

# Unit 36: DISCHARGE PRESSURE CONTROLS (COOLING TOWERS)

## Introduction

Head or discharge pressure controls are necessary on refrigeration and air-conditioning systems. Two reasons for this follow: (1) When the head pressure rises above the normal operating pressure, the economy and performance of the system are reduced. (2) When the head pressure drops below the normal operating pressure, not enough pressure drop is available across the flow control device for proper distribution and evaporation of the refrigerant in the evaporator. A head-pressure control is any part that keeps the needed head pressure by controlling the coolant to the condenser.

**Cooling tower fan control.**    Tower fan controls are used to cycle the tower fan to keep the water temperature high enough to cause the needed head pressure for the refrigeration system. See Figure 36–1. These are single-stage temperature controls that may have either SPST or SPDT contacts. The contacts for the fan close on a temperature rise. The extra contacts may be used to make an alarm circuit or to start some type of emergency equipment to keep from overheating the tower water. The thermostat is wired into the tower fan circuit, as shown in Figure 36–2.

**Figure 36–1**
Cooling tower fan thermostat. (Courtesy of Johnson Controls, Inc., Control Products Division.)

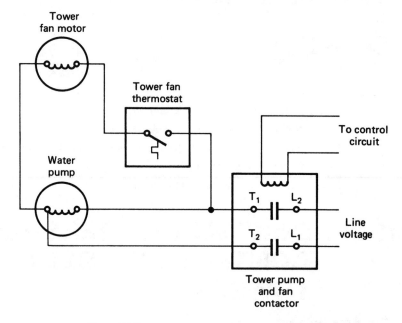

**Figure 36–2**
Wiring diagram for cooling tower fan thermostat.

*Operation.*    When the air around the tower drops to a chosen temperature, the thermostat will sense that the tower water is getting too cold. The contacts will open to stop the fan. The fan will stay off until the tower water temperature has risen to the thermostat cut-in setting. The thermostat contacts will then close, starting the tower fan to cool the water. This cycle will continue until the outdoor air temperature has risen to the temperature needing the tower fan to run continuously.

**Temperature-actuated modulating water valves.**    Temperature-actuated modulating water valves are used to control the flow of water through a water-cooled condenser to keep the needed head pressure. Available in several different temperature ranges, the correct valve must be used. The sensing bulb is mounted directly in the water stream and acts in response to the water temperature. These valves open on a rise and close on a fall in the condenser supply water temperature.

*Operation.*    When the condenser sits idle for a time, the water temperature drops to a point at which the valve seat becomes closed. No water flows through the condenser. When the indoor thermostat demands cooling, the compressor starts circulating the refrigerant. As the compressor continues to run, the refrigerant temperature rises high enough that more water is needed to keep the head pressure up. When the refrigerant temperature rises, the water temperature also rises. The control bulb senses this rise and causes the valve needle to move off its seat so that more water can flow through the condenser. The water valve modulates to keep this head pressure until the unit is stopped. Then the water valve slowly closes, stopping the flow of water through the condenser. When the condenser has cooled sufficiently, the valve closes and stops the flow of water.

**Pressure-actuated water-regulating valve.**    Pressure-actuated water-regulating valves are also modulating type valves. They are used with a particular refrigerant type. The correct valve for the application must be used. These valves act in direct response to the head pressure. They are connected to the high-pressure refrigerant line at the outlet of the receiver tank. A rise in refrigerant pressure will cause the valve to open. A fall in refrigerant pressure will cause the valve to close.

*Operation.*    When the system has been off for a time, the head pressure falls enough to cause the valve seat to close. No water is flowing through the condenser. When the indoor thermostat demands cooling, the compressor starts circulating the refrigerant, causing its pressure and temperature to rise. When the pressure rises, the valve senses this rise and starts opening. When the correct amount of water is flowing through the condenser to keep the needed head pressure, the valve begins to throttle, reducing the flow of water through the condenser. The valve modulates to keep this head pressure until the compressor stops running. Then the head pressure starts to fall and the valve starts to close off, reducing the flow of water. When the head pressure has fallen enough, the valve closes off completely.

**Three-way water-regulating valves.**    Three-way water-regulating valves are used with cooling towers. They are connected so that part of the water is diverted back to the tower as the head pressure drops. See Figure 36–3. These valves are connected to the high-pressure line, usually at the outlet of the receiver tank. They open and close as the refrigerant pressure rises and falls. These valves are used with specific types of refrigerant. The correct valve for the application must be used.

*Operation.*    When the compressor has been off and the discharge pressure drops, the valve is open to the bypass port and closed to the condenser port, allowing all the water to return to the tower. When the indoor thermostat demands cooling, the compressor starts. The refrigerant pressure rises high enough to cause the valve to modulate toward the open to the condenser port position. When the needed operating discharge pressure is reached, the valve modulates, allowing part of the water to go through the condenser and the rest to go back to the tower. After the compressor stops, the valve modulates to open the bypass port and close the condenser port. When the pressure has fallen enough, the valve completely closes off the condenser port and opens the bypass port.

From
Tower → ← To
Condenser

By-Pass →

(a)

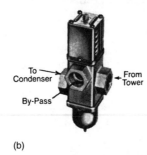

To →
Condenser ← From
Tower

By-Pass →

(b)

**Figure 36–3**
Three-way water-regulating valves.
(Courtesy of Johnson Controls, Inc.,
Control Products Division.)

**Figure 36–4**
Motor-actuated three-way mixing valve.
(Courtesy of Honeywell, Inc.)

**Motor-actuated three-way water-regulating valve.** Motor-actuated three-way water-regulating valves are used with a modulating motor, a matching thermostat, and the proper linkage. See Figure 36–4. They are piped into the system so that a part of the water may be sent back to the cooling tower. The pressure control has a potentiometer to match the one in the valve motor. Its pressure connection is fastened to the outlet of the receiver tank to sense the head pressure. They are used to keep the needed head pressure by sending some of the water back to the tower rather than sending it all through the condenser.

*Operation.*    When the system is at rest, the head pressure is equal to the ambient temperature. The valve is fully open to the bypass port. When the compressor starts, the head pressure begins to rise. With this rise, the modulating pressure control sends a signal to the valve motor. The valve motor responds by opening the valve to the condenser port, sending more water to the condenser. As the unit continues to operate, the head pressure rises slowly and the valve opens further, sending more water to the condenser. When the balance point between the water flow and the head pressure is reached, the valve stays in this position until a change in the head pressure is sensed by the pressure control. The valve then modulates to allow the correct amount of water to flow through the condenser. When the compressor is stopped, the head pressure begins to fall. The valve begins to modulate toward the open to the bypass port position.

# Summary

- A head pressure control is any device that keeps the needed head pressure by controlling the coolant sent to the condenser.
- Tower fan controls are used to cycle the tower fan to keep the water temperature high enough to cause the needed head pressure for the refrigeration system.
- Temperature-actuated modulating water valves are used to control the flow of water through a water-cooled condenser to keep the needed head pressure.
- The bulb is mounted directly in the water stream and acts in response to the water temperature.
- Pressure-actuated water-regulating valves are also modulating type valves. They are used with a particular refrigerant type. The correct one for the application must be used.
- Valves act in direct response to the system head pressure. They are connected to the high-pressure refrigerant line at the outlet of the receiver tank and open on a rise in refrigerant pressure.
- Three-way water-regulating valves are used with cooling towers. They are connected so that part of the water is sent back to the tower as the head pressure drops.
- Motor-actuated three-way water-regulating valves are used with a modulating motor, a matching thermostat, and the proper linkage.

# Service Call

A customer complains that her refrigeration system is not cooling the walk-in box. The system is a water-cooled unit. The box temperature is too high. The system is charged with HCFC-22 and the discharge pressure is 180 psig. It should be about 225 psig. The water regulating valve is set for too low discharge pressure. The technician adjusts the valve to cause the correct pressure, checks the operation of the complete system, and finds it to be operating correctly. The technician is satisfied that the system is repaired.

# Student Troubleshooting Problem

A customer complains that a large air-conditioning system is not cooling correctly. A check of the system reveals that the building temperature is higher than normal. The condenser is water cooled with a cooling tower. A three-way pressure-regulated water valve is used to control the head pressure. The head pressure is 255 psig for HCFC-22. The valve is bypassing too much water to the tower. What could be the problem? How could the problem be corrected?

# Questions

1. How does a tower fan control keep the needed head pressure?
2. How is the tower water thermostat wired into the circuit?
3. How do temperature-actuated water valves control the head pressure?
4. Where is the sensing bulb of a temperature-actuated water valve located?
5. How does a condenser water valve keep a constant discharge pressure?
6. To what do pressure-actuated water valves respond?
7. How are three-way water valves connected into the condenser water system?
8. How are motor-actuated three-way water valves controlled?
9. At what point is a water-regulating valve piped into the system?

# Unit 37: AIR-COOLED CONDENSER DISCHARGE PRESSURE CONTROLS

## Introduction

Head or discharge pressure controls are necessary for the safe and correct operation of refrigeration and air-conditioning systems. Two reasons for this follow: (1) When the head pressure rises above the normal operating pressure, the economy and performance of the system drops. (2) When the head pressure drops below the normal operating pressure, there is not enough pressure drop across the flow control device to cause the proper distribution and evaporation of the refrigerant in the evaporator. A head-pressure control is any control that keeps the needed head pressure controlling the amount of coolant to the condenser.

**Two-speed condenser fan controls.**    Some air-cooled condensing units have two-speed fan motors. Two-speed condenser fan motors are for controlling the head pressure. The speeds are controlled by a two-speed condenser fan thermostat or pressure control. See Figure 37–1. This thermostat is fastened to one return bend in the condenser coil. The pressure control is connected in the discharge line. See Figure 37–2. The exact location is usually recommended by the equipment manufacturer. Those recommendations should be followed for best results. The thermostat changes the fan motor speed from high to low speed and back to high speed as the refrigerant temperature changes. It is a nonadjustable switch used to change the fan motor speed at temperatures recommended by the equipment manufacturer.

*Operation.*    When the room thermostat demands cooling, the compressor and the condenser fan are energized simultaneously by the unit contactor or starter. The fan operates in the low speed until the refrigerant pressure and temperature rise to the fan speed changeover setting of the discharge thermostat. The fan continues to operate in the high-speed mode until the refrigerant temperature drops to the low-speed changeover

**Figure 37–1**
Two-speed condenser fan pressure control. (Courtesy of ICM Corporation.)

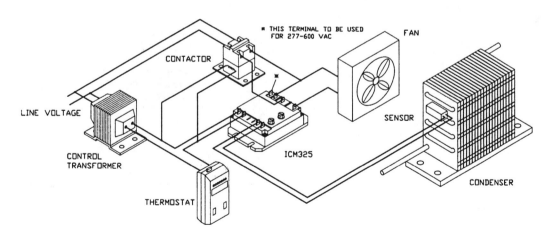

**Figure 37–2**
Location of condenser fan thermostat. (Courtesy of ICM Corporation.)

temperature. Then the fan operates in the low-speed mode. When the room thermostat is satisfied, the contactor or starter contacts open and both the compressor and the condenser fan stop. The condenser fan thermostat cools to the outside temperature and changes to the low-speed position for the next start period.

A typical wiring diagram of a two-speed condenser fan thermostat is shown in Figure 37–3.

**Low-ambient kit.**    The purpose of the low-ambient kit is to cycle the condenser fan to keep up the needed head pressure. See Figure 37–4. When outdoor temperatures are

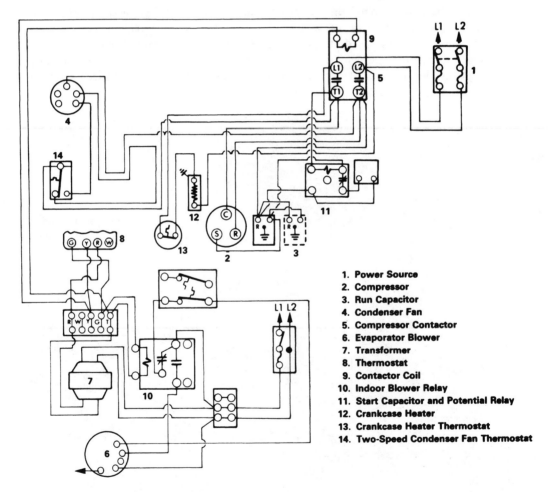

1. **Power Source**
2. **Compressor**
3. **Run Capacitor**
4. **Condenser Fan**
5. **Compressor Contactor**
6. **Evaporator Blower**
7. **Transformer**
8. **Thermostat**
9. **Contactor Coil**
10. **Indoor Blower Relay**
11. **Start Capacitor and Potential Relay**
12. **Crankcase Heater**
13. **Crankcase Heater Thermostat**
14. **Two-Speed Condenser Fan Thermostat**

**Figure 37–3**
Two-speed condenser fan wiring diagram. (Courtesy of Lennox Industries, Inc.)

low, the condenser fan should not run because it will drop the head pressure below the needed pressure for correct operation. A high-pressure control furnished with the kit is wired in series with the condenser fan motor. See Figure 37–5.

The pressure switch cycles the fan motor while the compressor continues to run. Usually when the outdoor ambient temperature drops to the point that the head pressure falls to 140 psig for HCFC-22, the pressure switch contacts open to stop the condenser fan motor. When the head pressure builds back up to the cut-in setting of the pressure control, the condenser fan is started to cool the condenser.

*Operation.*    When the room thermostat demands cooling, the compressor and condenser fan circuits are made at the same time. The condenser fan stays off until the head pressure has risen to the cut-in setting of the pressure control in the low-ambient kit. When the cut-in pressure is reached, the condenser fan is started. The fan continues to run until the pressure has dropped to the pressure control cut-out setting. At this point the contacts open to stop the condenser fan motor. The head pressure again builds up to the control cut-in setting and the condenser fan is again started. This cycle repeats itself until the room thermostat is satisfied.

Some problems with this type of head pressure control follow: (1) If the cut-in and cut-out pressures are set too close the fan will cycle rapidly and its life will be shortened. (2) When the cut-in and cut-out settings are set too far apart, the head pressure will have wide changes that will sometimes upset the refrigeration process.

**Condenser damper modulation.**    Airflow control dampers installed on the condenser air inlet are another method of controlling the head pressure on air-cooled

**Figure 37–4**
Typical low-ambient kit. (Courtesy of Lennox Industries, Inc.)

refrigeration and air-conditioning equipment. The air volume through a section of the con-denser is changed by changing the positions of the dampers. Both face and bypass dampers are used to keep from overloading the fan motor. See Figure 37–6.

The dampers are connected so that when the face dampers are open, the bypass dampers are closed, or when either one is partially open or closed, the other is partially open or closed the same amount. For full cooling, the face dampers are fully open and the bypass dampers are fully closed. See Figure 37–7. These dampers are usually controlled by a pres-sure switch that responds to the system head pressure.

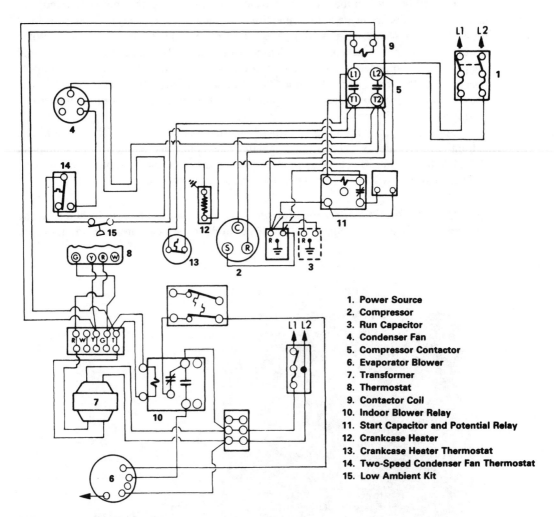

1. **Power Source**
2. **Compressor**
3. **Run Capacitor**
4. **Condenser Fan**
5. **Compressor Contactor**
6. **Evaporator Blower**
7. **Transformer**
8. **Thermostat**
9. **Contactor Coil**
10. **Indoor Blower Relay**
11. **Start Capacitor and Potential Relay**
12. **Crankcase Heater**
13. **Crankcase Heater Thermostat**
14. **Two-Speed Condenser Fan Thermostat**
15. **Low Ambient Kit**

**Figure 37–5**
Typical low-ambient kit wiring diagram. (Courtesy of Lennox Industries, Inc.)

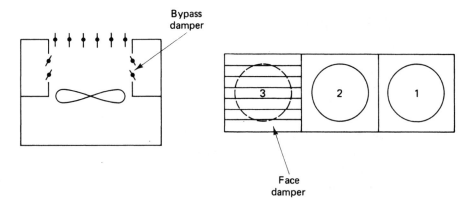

**Figure 37–6**
Damper control location.

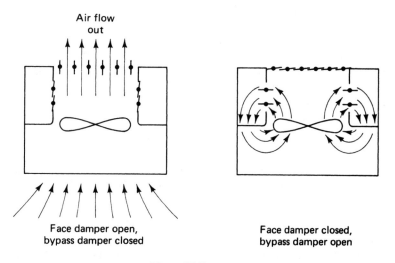

**Figure 37–7**
Damper operation.

***Operation.***    When the room thermostat demands cooling, the compressor and condenser fan start at the same time. However, the head pressure has dropped low enough to cause the face dampers to be fully closed and the bypass dampers to be fully open. See Figure 37–7. As the compressor continues to run, the head pressure rises, especially if the outdoor temperature is high enough that the condenser needs more cooling. When the head pressure has risen enough, the dampers gradually turn to let more air go through the condenser. If the outdoor air temperature falls and less condenser cooling is needed, the dampers modulate to close the face damper and open the bypass damper.

# Summary

- A head pressure control is any control that keeps the needed head pressure by controlling the coolant delivered to the condenser.
- Some air-cooled condensing units have two-speed fan motors used for controlling the head pressure.
- It is a nonadjustable switch that changes the fan motor speed at temperatures recommended by the equipment manufacturer.
- The purpose of the low-ambient kit is to cycle the condenser fan to keep the needed head pressure.
- Airflow control dampers installed on the condenser are another method of controlling the head pressure on air-cooled refrigeration and air-conditioning equipment.
- For full cooling, the face dampers are fully open and the bypass dampers are fully closed.

# Service Call

A customer complains that an air-cooled air-conditioning unit is not cooling. The system's condensing unit has a two-speed condenser fan control. The discharge pressure is 375 psig with HCFC-22 refrigerant. The fan thermostat is in the low-speed mode. The pressure control is bypassed and the condenser fan starts running in high speed. The head pressure drops to 250 psig. The fan pressure control is bad. The technician replaces the control and puts the system back in operation. The complete system is now operating correctly and the technician is satisfied that the system is repaired.

# Student Troubleshooting Problem

A customer complains that an air-cooled air-conditioning unit is not cooling the building. The system's condenser is equipped with face and bypass dampers for head pressure control. The outdoor temperature is 95°F. The dampers are only about half open. The head pressure is 170 psig for HCFC-22. A check of the fan pressure control shows that it is operating properly. What could be the problem?

# Questions

1. What is the purpose of two-speed motors on air-cooled condensers?
2. How are two-speed condenser fan motors controlled?
3. What causes the fan thermostat to change motor speeds?
4. When using a condenser fan speed control, in what speed does the condenser fan start?
5. How is the low-ambient kit wired into the circuit?
6. What controls the low-ambient switch operation?

7. In what positions are the contacts of a low-ambient control?

8. When using a low-ambient kit, when is the fan motor started?

9. Why are both face and bypass dampers used for head pressure control on air-cooled condensers?

10. How are head pressure control dampers usually controlled?

# Unit 38: MODULATING MOTORS
# Introduction

The ON-OFF type control is probably the least expensive type to use. When the thermostat demands, it turns the system on; when the need is satisfied, it turns the system off.

A significant improvement over the ON-OFF control is made by dividing the conditioning load into several separate elements and keeping enough of them on continuously to give an even flow of conditioned air to the area. One method of doing this is to use modulating motors to turn on and off the number of conditioning elements needed to keep an even temperature inside the space.

**Modulating motor.**    A modulating motor is one that will change position when the correct signal from the controller is received. Modulating motors are available for a variety of functions, such as modulating spring return, reversing two position, reversing proportion, and two-position spring return.

*Modulating Spring-Return Motor.*    Modulating spring-return motors are used to operate dampers and valves. They have a helical spring that returns the motor shaft to the normally deenergized position. See Figure 38–1. Depending on system needs, the motor will go either to the full open or the full closed position on a break in power. They have a fixed 90-degree stroke. The full 90-degree stroke is made in 34 seconds. These motors are used with either 120-volt or 24-volt control systems. The motor may be wired for either ON-OFF or floating operation to extend its use. Be sure to check the motor manufacturer's recommendations.

**Figure 38–1**
Modulating spring return. (Courtesy of Honeywell, Inc.)

*Reversing Two-Position and Proportional Motors.*    Reversing two-position and proportional motors are used to operate valves and dampers. A selection of torque output, depending on the timing needed, allows the correct motor choice for the job. See Figure 38–2. The controllers used with these motors may be two-position SPDT or modulating controls. For operation in both directions, these motors need 24-volt electrical power. If the power is interrupted, the motor stays in that position until the power is restored. The dampers or valves operated by these motors may either be fully closed, fully open, or modulating, depending on the needs and the type of controller used.

*Two-Position Spring-Return Motor.*    Two-position spring-return motors are used where returning the controlled part to the normally deenergized position on a power failure or interruption is needed. It takes 60 seconds for this motor to make a full 180-degree cycle and 24 volts of electrical power for the motor to run. See Figure 38–3.

*Operation.*    The modulating motor circuit works to position the controlled part (usually a damper motor or motorized valve) at any point between fully open and fully closed (except two-position motors); this will balance the air delivery to that needed by the controller mechanism. See Figure 38–4.

The power unit is a low-voltage capacitor motor that turns the motor shaft with a gear train. Limit switches are used to limit the rotation to the motor rating. The gear train and all other moving parts are immersed in oil to eliminate the need for routine lubrication and to guarantee long, quiet service.

**Figure 38–2**
Reversing two-position modulating
motor. (Courtesy of Honeywell, Inc.)

**Figure 38–3**
Two-position spring-return modulating
motor. (Courtesy of Honeywell, Inc.)

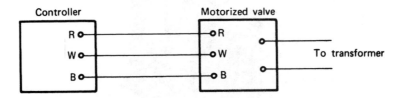

**Figure 38–4**
Modulating motor circuit-field wiring.

The power unit is started, stopped, and reversed by the single-pole double-throw (SPDT) contacts of the balancing relay. See Figure 38–5. The balancing relay consists of two solenoid coils with parallel axes, into which are placed the legs of a U-shaped armature. The armature is pivoted at the center so that it can be tilted by changing the magnetic strength of the two coils.

A contact arm is fastened to the armature so that it will touch one or the other of two stationary contacts as the armature moves back and forth on its pivot. When the relay is in balance, the contact arm floats between the two contacts, touching neither of them.

A balancing potentiometer is included in the motor. The potentiometer is electrically identical to the one in the controller. The finger is moved by the motor shaft so that it travels along a coil and makes contact wherever it touches. See Figure 38–6.

Figure 38–5 shows how a balancing relay is made. As the relay is used in the modutrol circuit, the flow of current passing through the coils is governed by the relative positions or the controller potentiometer and the motor-balancing potentiometer. See Figure 38–7. Thus, when equal amounts of current are flowing through both coils, the contact

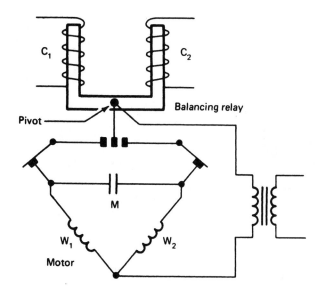

**Figure 38–5**
Diagram of balancing relay and motor circuit.

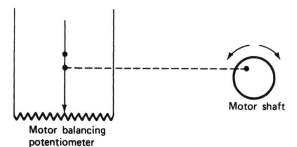

**Figure 38–6**
Schematic diagram of a balancing potentiometer.

blade is in the center of the space between the two contacts. The motor is not running. When the finger of the controller potentiometer is moved, more current flows through one coil than the other. The relay is unbalanced. The relay armature is then turned so that the blade touches one contact causing the motor to run in the correct direction.

The contact made by the balancing relay can only be broken if current flowing through $C_1$ is equal to the amount flowing through $C_2$. This movement is caused by the motor-balancing potentiometer linked to the motor shaft. As the motor turns, it moves the contact finger of the motor-balancing potentiometer toward the position to equalize the resistances in both legs of the circuit. There is a definite position of the finger for each motor shaft position through its complete rated degrees of rotation.

Figure 38–7 shows the current flowing from the transformer, through the potentiometer contact finger, and down through both legs of the circuit. In the position shown, the thermostat potentiometer contact finger and the motor-balancing potentiometer contact finger divide their respective coils so that the resistance on both sides of the circuit are equal. Coils

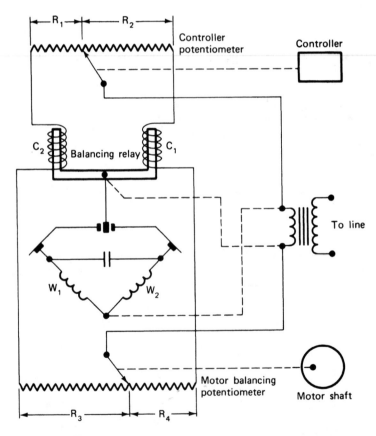

**Figure 38–7**
Diagram of control circuit in balanced condition.

$C_1$ and $C_2$ of the balancing relay are balanced. The contact blade is floating between the two stationary contacts. No current is flowing to the motor, and the motor is not turning.

Modulating motors may be used for the following applications:

1. *Airflow diversion*: Where a parallel airflow pattern is used, a diverting damper is needed to direct the airflow through either the heating unit or the cooling unit.

2. *Airflow changeover:* Where a resistance damper is used to decrease the airflow on the heating cycle.

3. *Ventilation control:* Where an outside air duct is used for bringing outdoor air into the system during the cooling season but not during the heating season.

4. *Zoning:* Where both heating and cooling may be wanted in different areas at the same time.

5. *Valve operation:* Where steam or water may need to be directed in different amounts, such as steam or hot water heating coils or condenser cooling water to a condensing unit.

# Summary

- A great improvement over the ON-OFF control can be done by dividing the conditioning load into several separate elements and keeping enough of them on all the time to give an even flow of conditioned air to the space.

- A modulating motor is a motor that will change position when the correct signal is received.

- Modulating motors are available for many different functions, such as modulating spring return, reversing two position, reversing proportion, and two-position spring return.

- A modulating spring-return motor is generally used to operate dampers and valves.

- Depending on system needs, the motor will go either to the fully open or the fully closed position on a power failure.

- Reversing two-position and proportional motors are used to operate valves and dampers. A selection of torque output, depending on the timing needed, allows correct selection of the motor for the job.

- The controllers used with these motors may be two-position SPDT or modulating controls.

- Two-position spring-return motors are used where returning the controlled device to the normally deenergized position on a power failure or interruption is needed.

- The modulating motor circuit operates to position the controlled device (usually a damper motor or motorized valve) at any point between fully open and fully closed (except two-position motors); this will balance the delivery to that needed by the controller.

- The power unit is a low-voltage capacitor motor that turns the motor shaft with a gear train. Limit switches are used to limit the rotation to the rating of the motor.

- The power unit is started, stopped, and reversed by the single-pole double-throw (SPDT) contacts of the balancing relay.
- The balancing relay consists of two solenoid coils with parallel axes, into which are inserted the legs of a U-shaped armature. The armature is pivoted at the center so it can be tilted by changing the magnetic strength of the two coils.
- When equal amounts of current are flowing through both coils, the contact blade is in the center of the space between the two contacts and the motor is not turning.

# Service Call

A customer complains that an air-conditioning system in a large building is not cooling. The system's unit has a water-cooled condenser and the head pressure is controlled by a modulating motor connected to a three-way valve. The valve is mostly open to the condenser. The head pressure is 180 psig for refrigerant HCFC-22, which is too low. The motor has a burnt potentiometer. The technician therefore replaces the potentiometer with the correct replacement and puts the unit back into operation. The discharge pressure rises to 215 psig. The unit starts cooling. The unit is now operating correctly according to the tech's complete check of the operation. The technician is satisfied that the system is repaired.

# Student Troubleshooting Problem

A customer complains that the building's air-conditioning unit is keeping the office building too cool. The system's air handling unit is equipped with face and bypass dampers. The dampers are almost completely open to the cooling coil. The voltage at the modulating motor terminals is 24 VAC. The technician removes the wiring from the terminals and checks the resistance of the motor winding. It has infinite resistance. What could be the problem?

# Questions

1. What is a modulating motor?
2. What type motor is used to operate valves and dampers?
3. What voltage does the modulating spring-return motor need?
4. What type controller can be used with reversing, two-position, and proportional motors?
5. What voltage does the reversing, two-position, and proportional motor need?
6. What happens when the power supply is interrupted to a reversing, two-position, or proportional motor?
7. What type of motor needs 60 seconds to make a full 180-degree cycle?
8. What is the purpose of the modulating motor being immersed in oil?
9. What operates the power unit on a modulating motor?
10. What happens when the balancing relay is balanced?

# Unit 39: STEP CONTROLLERS
# Introduction

An improvement over the ON-OFF control is made by dividing the conditioning load into several separate elements. Enough of them are kept on continuously to provide an even flow of conditioned air to the space. This is done by using step controllers. A step controller is a series of switches (steps) operated by a modulating motor.

**Step controllers.**    Step control is much better than ON-OFF control because it meets load needs at all times. See Figure 39–1. Temperature differences within the space are much greater with ON-OFF control under identical conditions. However, the modulating effect of step control depends on the number of steps used. Ideally, enough steps should be used to hold the air temperature passing through the coil to a rise or fall of 5°F per step.

A step controller consists of a series of switches operated by an actuator. The actuator seeks a switch position matching the controller needs, thereby closing or opening the correct number of switches as needed. See Figure 39–2. The switches can directly start the conditioning elements or control these elements through actuators.

The actuators usually have a modulating type motor and operate from a modulating type thermostat. Some are the proportioning type and are controlled by a staged (step) controller. Many types have a reversible motor-driven cam and step switch assembly with limit switches, feedback potentiometer, recycle relay, multiple tapped transformer, and terminal

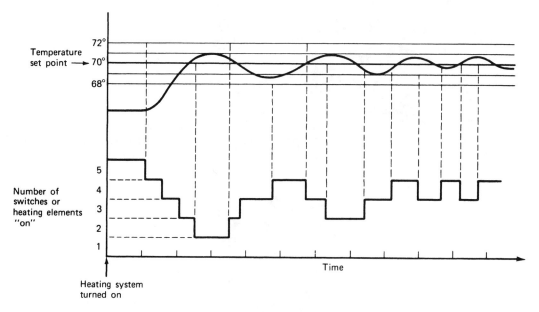

**Figure 39–1**
Temperature variation with step control.

strips for connecting the control circuit. Some models have up to ten adjustable switches. These switches may be mounted in any position.

Step controllers may be used for the control of any electric heating unit with elements that can be divided into separate circuits. Also, they may be used to start or stop refrigeration compressors connected in tandem to produce the necessary capacity. Compressor unloaders are also controlled with this type control.

*Operation.* When the thermostat is satisfied, the step controller motor is positioned so that all the switch contacts are open and no equipment is running. When the thermostat demands operation, the step controller motor begins to move to close the switch contacts in sequence. As the motor turns farther, more of the switches are closed until all of them are made and all of the equipment is running. As the thermostat becomes more satisfied, the motor begins to turn in the opposite direction to open the switches as the motor reaches a chosen degree of rotation. When the thermostat is completely satisfied, the step controller motor returns to the at-rest position and stays there until another demand is made by the thermostat.

**Auxiliary (end) switches.** Auxiliary switches are mounted on the shaft of modulating motors. See Figure 39–3. The motor operates the switches at chosen motor output

**Figure 39–2**
Step controller. (Courtesy of Paragon Electric Co., Inc.)

**Figure 39–3**
Mounted auxiliary switches. (Courtesy of Johnson Controls, Inc., Control Products Division.)

shaft positions. Typical applications include capacity control on multistage refrigeration systems, multispeed fan controls, staging of multiple fuel oil nozzles on boilers, and other capacity change needs. End switches use either mercury bulb contacts or microswitches.

*Operation.*    When the system is off and the motor is positioned so that all switch contacts are open, nothing will be running. As the thermostat demands operation, the modutrol motor starts turning to the full demand position. As it moves, more of the switches are closed to start more of the equipment to handle the needs of the building or space being conditioned. When the motor has moved to the full demand position, all the switches are closed and all the equipment is running. As less equipment is needed, the motor starts turning in the opposite direction. At each of the chosen degrees of shaft rotation, a switch is opened to stop a part of the equipment. This reduces the cost of operation while keeping the space design conditions. When the thermostat is fully satisfied, the motor returns to the full OFF position and stays there until the thermostat again demands operation.

# Summary

- A step controller has a series of switches (steps) operated by a motor.
- Step control is much better than ON-OFF control because it continuously meets load requirements.
- Ideally, enough steps should be used to hold the air temperature passing through the coil to a rise or fall of 5°F per step.
- The actuators on these units usually have a modulating type motor and operate from a modulating type thermostat.
- Many types have a reversible motor-driven cam and step switch assembly with limit switches, feedback potentiometer, recycle relay, multiple tapped transformer, and terminal strips for connecting the control circuit.
- When the thermostat is satisfied, the step controller motor is positioned so that all the switch contacts are open and no equipment is running. When the thermostat demands operation, the step controller motor begins to move to close the switch contacts in sequence.
- Auxiliary switches are mounted on the shaft of modulating motors.
- The motor operates the switches at chosen motor output shaft positions.

# Service Call

A customer complains that a refrigeration unit is freezing everything in a dairy case. The unit has a step controller to control the capacity. The step switches are all set to demand full cooling when the motor reaches the first step. The switches are set to keep a 5°F difference between each step. When the unit is put back in operation, the step controller starts moving toward the full ON position. When it reaches the first step the compressor is started. It then moves toward the second step and more load is placed on the compressor. When the thermostat is set below the case temperature, the motor starts turning toward the OFF position. When the thermostat is set above the space temperature, and the motor starts turning to the full ON position. With the thermostat now set at the wanted temperature, the technician is satisfied that the system is repaired.

# Student Troubleshooting Problem

A customer complains that the flowers in a florist case will almost wilt and then almost freeze. The system has a three-step controller to control the unit capacity. The refrigeration system is operating correctly. When the thermostat set point is raised and lowered, the step controller shows a 10°F differential between the steps. What could be the problem? How could the problem be solved?

# Questions

1. What are step controllers?
2. What is usually used to control step controllers?
3. What is used to control the capacity of a refrigeration compressor?
4. When a step controller first demands equipment operation, to what position does it go?
5. How often is the equipment capacity reduced with step controllers?
6. What is the purpose of staging equipment?

# Unit 40: TIMERS AND TIME CLOCKS

## Introduction

Timers and time clocks are controls used for the automatic and economical operation of air-conditioning, heating, and refrigeration equipment. They have a variety of contact poles and contact current ratings. They may be set to give almost any ON-OFF cycle arrangement wanted.

**Definitions.**    The following are the accepted definitions of timers and time clocks.

*Time Clock.*    The time clock is an electrically operated control used to govern the different cycling functions of equipment over a given time period. See Figure 40–1.

*Timer.*    The timer is similar to the time clock except that its operation is limited to less complicated systems. Generally, timers control the ON-OFF cycles of only one appliance, such as a window air-conditioning unit, with one or two ON-OFF cycles per 24 hours. Timers have a receptacle for plugging in the controlled appliance. See Figure 40–2.

**Figure 40–1**
Time clock. (Courtesy of Paragon
Electric Co., Inc.)

**Figure 40–2**
Timer. (Courtesy of Intermatic Incorporated.)

*Operation.*    To help in understanding how these controls operate, the technician should first become familiar with the components inside a time clock. See Figure 40–3. They are designed and built to operate correctly even under bad conditions.

The high-powered switching mechanism is made of channeled brass U-beam blades. They are operated by a cam to give instant and positive make and break action. See Figure 40–4. The contacts are self-cleaning and made of a special alloy to reduce pitting. They are rated to carry an inrush of current ten times their normal rating without arcing or sticking.

Time clocks are usually powered by heavy-duty industrial type motors. See Figure 40–5. Time clock motors are the synchronous type. They are quiet in operation and are self-lubricating. The motors never need service or attention and are practically immune to temperature and humidity conditions.

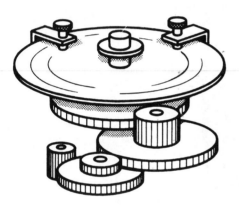

**Figure 40–3**
Gears.

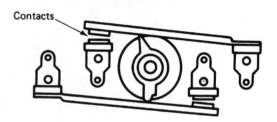

**Figure 40–4**
Switching mechanism.

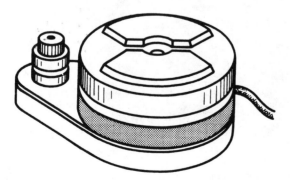

**Figure 40–5**
Industrial motor.

The terminal board is designed for fast and easy wiring with plenty of room for hands and wiring. See Figure 40–6. The control dial is divided accurately into units of time. See Figure 40–7. Most timer control dials are painted with contrasting colors to aid in the placement of the trippers. The trippers are adjustable to give the needed ON and OFF periods. These dials are designed for 24-hour operation. The trippers open and close the contacts. They are fastened to the rotating dial by thumb screws. The dial is set at the correct time by pushing the dial toward the back of the control and turning it until the current time matches the time indicator on the dial. Reading the manufacturer's instructions before attempting to set the dial is recommended to ensure correct procedure.

Seven-day clocks are used for skipping operations. The skipper dial allows operations to be skipped on selected days of the week. The time to be skipped is adjusted by placing screws into the correct holes in the dial for the days on which the operation is to be skipped. See Figure 40–8.

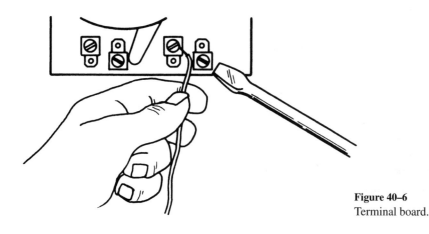

**Figure 40–6**
Terminal board.

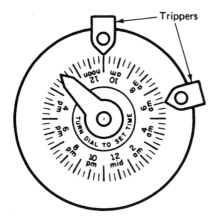

**Figure 40–7**
Control dial.

The hand trip allows equipment operation through the time switch without disturbing the dial settings. See Figure 40–1. The trip may be moved to the ON position to see if the system is operating properly or to start the equipment when an extra load or similar demand is placed on it.

In operation the time clock motor is wired in parallel with the switching contacts. As the motor turns the dial and trippers, the contacts are opened or closed, depending on the needed operation at that time. A simple diagram is shown in Figure 40–9. Most time clocks

**Figure 40–8**
Time clock showing skipper dial.
(Courtesy of Paragon Electric Co., Inc.)

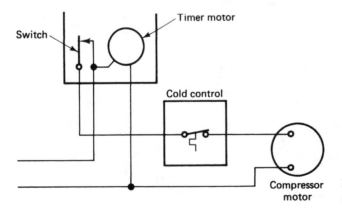

**Figure 40–9**
Simple wiring diagram.

have a spring-wound carryover mechanism. The carryover mechanism keeps the time clock on schedule for 36 hours during power failures. When the power comes back on, the carryover automatically rewinds itself, thus providing ideal control in areas where power outages are frequent.

A time clock provided with all the previously mentioned parts can be used to control almost any operation wanted. The following are some more common uses of time clocks.

1. Heating control
2. Ventilation control
3. Air-conditioning control
4. Defrost control

Timers give automatic control over many functions such as the cycling of window air-conditioning units, heating units, fans, and lights. The 24-hour dial is divided into half-hour increments with permanently attached trippers, allowing the same needed ON-OFF periods in every 24 hours. See Figure 40–2.

Following are some suggested uses for the timer.

1. Time-delay relays
2. Motor control (to 1 horsepower)
3. Pump control
4. Heating control
5. Domestic refrigeration control

**Defrost time clocks.**    The following describes some of the more popular types of defrost time clocks.

*Time-Initiated Time-Terminated Control.*    A time-initiated time-terminated control can be set up for six defrost periods per day. See Figure 40–10. At least four hours between each defrost period is needed. The clock is adjustable for times from 4 to 110 minutes in 2-minute increments. It has a switch contact rating of 30 amps per pole, with 2 horsepower maximum at 120 to 240 volts ac. It is available with four contact poles for electric heat, hot gas defrost, or a compressor shutdown defrost.

*Operation.*    The initiation time and termination of the defrost cycles of these controls are both set on the timer dial with timer pins. See Figure 40–11. The defrost cycle is started at a given time or times each day. It is also terminated at the same given time or times each day. The controls do not take into consideration whether the frost is completely removed or if the unit stays in defrost longer than necessary.

*Time-Initiated Temperature-Terminated Control.*    The defrost periods that the time-initiated temperature-terminated control is used are adjustable from one to six per day

**Figure 40–10**
Time-initiated time-terminated defrost timer. (Courtesy of Paragon Electric Co., Inc.)

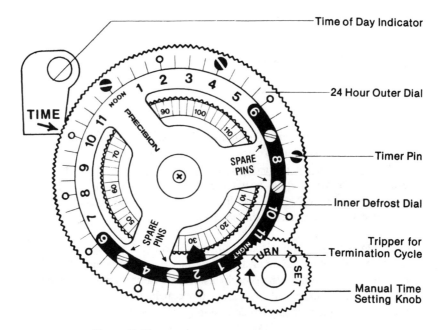

**Figure 40–11**
Timer dial. (Courtesy of Precision Multiple Controls.)

with a minimum of four hours between each defrost cycle. See Figure 40–12. The defrost periods are adjustable for times from 4 to 110 minutes each with 2-minute increments. The control has a switch contact rating of 30 amperes per pole with a 2 horsepower maximum, using 120 to 240 volts ac. There are three contact poles for an electric heat, hot gas defrost, or compressor shutdown defrost cycle.

This control is used with an external temperature or pressure sensor. An inner dial gives a backup defrost termination to protect the system against a part failure.

***Operation.***    At a chosen time each defrost period is started by the timer. When the temperature of the sensor reaches a chosen temperature, the control terminates the defrost cycle. Thus, the length of the defrost cycle varies according to the temperature sensor. A backup in the timer takes the system out of defrost after a chosen time has passed.

***Time-Initiated Pressure-Terminated Control.***    The time-initiated pressure-terminated control gives one to four defrost cycles per day or one to six in 24 hours. See Figure 40–12. The length of the defrost cycles is adjustable from 4 to 110 minutes in 2-minute increments with a minimum of four hours between each defrost cycle. The switch contact ratings are 30 amps per pole, 2 horsepower maximum rating at 120 to 240 volts ac. Three contact poles are available for a hot gas defrost, electric heat, or compressor shutdown for defrost. The control has an adjustable pressure, cut-in setting dial calibrated from 35 to 110 psig. Also, a mechanical timed backup termination is used, which protects the system against sensor failures.

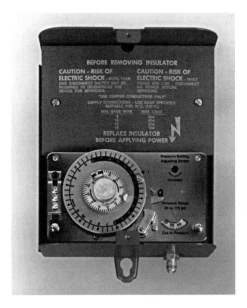

**Figure 40–12**
Time-initiated pressure-terminated defrost timer. (Courtesy of Paragon Electric Co., Inc.)

***Operation.*** System operation is begun by setting the trippers for each day that operation is wanted. However, when the day selected for the system not to operate comes around, the tripper will not start the system. When this 24-hour period passes, the system will again run according to the tripper placement.

***Seven-Day Four-Pole Time Switch with Carryover Mechanism.*** This seven-day time switch will give seven-day operation with a different program for each operating day. It uses a spring-wound carryover mechanism for use during power failures. See Figure 40–1. It continues to run up to a maximum of ten hours on reserve power. When the power comes back on, the carryover mechanism rewinds itself.

***Operation.*** This timer operates just the same as those previously described. The only exception is that with a power failure, the clock keeps the set points at the wanted times.

***Multicircuit Defrost Controls.*** A multicircuit defrost control is a multicircuit timing control that operates from one to twenty-four switches individually. See Figure 40–13. These are snap-acting switches that give positive switching of the circuits. These timers are used to control a series of operations with a varied time sequence such as a defrost cycle. They are field adjustable and are easily programmed.

With this control each switch is independently controlled by the combination action of two separate time shafts. This arrangement allows for extremely short ON cycles with extremely long OFF cycles, such as a two-minute ON cycle each day. It also permits the switches to be cycled separately at chosen times for independently adjusted ON cycles. For example, one switch can be adjusted to operate once each day with a two-hour ON cycle. Another switch can be set to operate eight times per day with a ten-minute ON cycle.

**Figure 40–13**
Multicircuit defrost control. (Courtesy of Paragon Electric Co., Inc.)

Each switch is opened and closed by an individually adjustable timer. This consists of a cam assembly with a drive gear ready to turn one revolution at a high speed then stop, giving one cycle to the switch. The starting of each cycle timer is determined by an individual timer turning at the slow speed of one revolution per 24 hours. Each program timer includes a molded dial with slots into which trippers can be placed.

*Operation.*    In operation, a tripper of the program timer serves to start the cycle timer, which then goes through one cycle and stops. Adjustment of the timer controls the length of the ON cycle. Placing the trippers in the program timer controls the ON cycle.

*The 24-Hour Time Switch, Skip-a-Day Series.*    The 24-hour time switch, skip-a-day clock is used to allow the operator to stop operation on Saturday, Sunday, or any

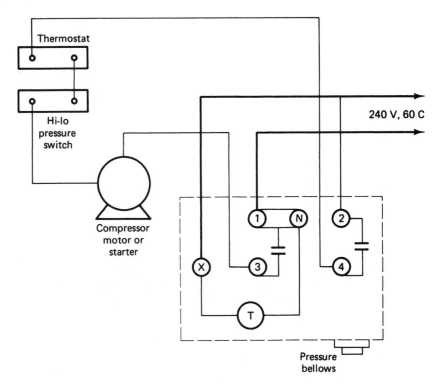

**Figure 40–14**
Wiring using 240 VAC single-phase compressor motor. Breaking both sides of 240 VAC line. (Courtesy of Paragon Electric Company, Inc.)

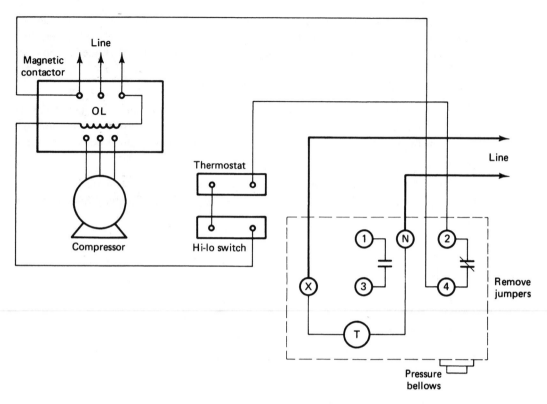

**Figure 40–15**
Wiring for compressor motor with magnetic starter. Clock motor independent of load circuit.
(Courtesy of Paragon Electric Co., Inc.)

other chosen day. See Figure 40–8. The timer has a 24-hour dial with the skip-a-day feature built in. It has either SPDT or DPST switching arrangements. The contacts are rated at 30 amps with two horsepower maximum using 120 to 240 volts ac. This timer also has a manual ON-OFF switch that allows for hand operation of the system without disturbing the clock settings.

*Operation.*     System operation is begun by setting the trippers for each day that operation is wanted. However, when the day chosen for the system not to operate comes around, the tripper will not start the system. When this 24-hour period has passed, the system will again operate as wanted according to the tripper placement.

**Wiring diagrams.**     The wiring diagrams in Figures 40–14 through 40–19 on pages 265–270 are only suggestions for wiring time clocks into the circuit for different system needs. Always use the manufacturer's wiring suggestions.

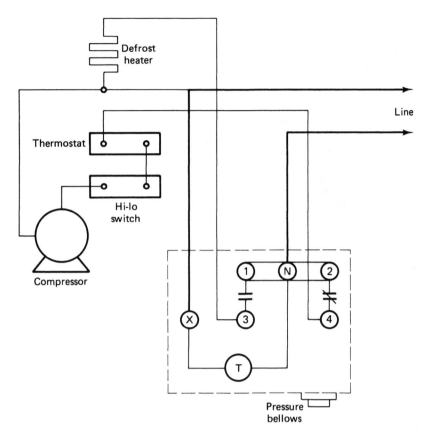

**Figure 40–16**
Wiring using single phase for electric heat system without magnetic starter.
(Courtesy of Paragon Electric Co., Inc.)

# Summary

- The time clock is an electrically operated control used to control the cycling of equipment over a given time.
- The timer is similar to the time clock except that it is limited to less complicated systems. Generally they control the ON-OFF cycles of only one appliance such as window air-conditioning units for operation with one or two ON-OFF cycles per 24 hours. They generally have a receptacle for plugging in the controlled appliance.
- The contacts are usually self-cleaning and made of special alloy to keep them from pitting. They are rated to carry an inrush of current ten times their normal rating without arcing or sticking.
- The control dial is divided accurately into units of time.

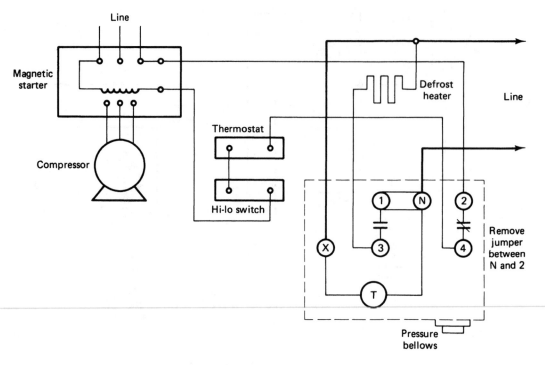

**Figure 40–17**
Wiring for compressor motor with magnetic starter and heater load. Clock motor in common heater circuit with independent compressor motor circuit. (Courtesy of Paragon Electric Co. Inc.)

- The trippers are adjustable to give the wanted ON and OFF periods. These dials are designed to cover 24 hours. The trippers open and close the contacts and are fastened to the rotating dial by thumb screws.
- The skipper dial allows operations to be skipped on chosen days of the week by placing screws into the correct holes in the dial. Seven-day clocks are needed for this feature.
- In operation the time clock motor is wired in parallel with the switching contacts. As the motor turns the dial and trippers, the contacts are opened or closed, depending on the need at that time.
- Timers provide automatic control over a great variety of functions such as cycling of window air-conditioning units, heating units, fans, and lights.
- A time-initiated time-terminated control can be set for six defrost periods per day.
- The defrost periods for which the time-initiated temperature-terminated control is used are adjustable from one to six per day with a minimum of four hours between each defrost cycle.

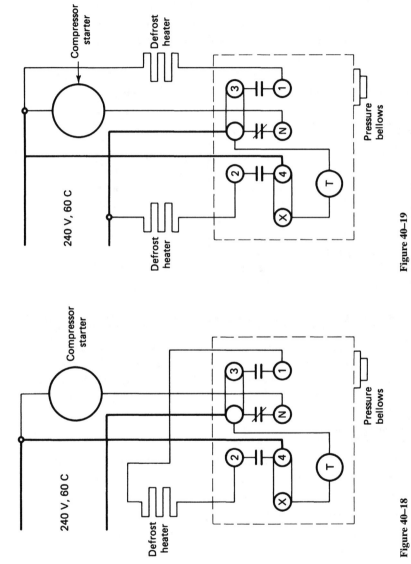

**Figure 40–18**
Wiring using 240 VAC single-phase line for electric heat system with split-load defrost heaters. (Courtesy of Paragon Electric Co., Inc.)

**Figure 40–19**
Wiring using 240 VAC single-phase line for electric heat system with split-load defrost heaters. (Courtesy of Paragon Electric Co., Inc.)

- The time-initiated pressure-terminated control gives one to four defrost cycles per day or one to six in 24 hours.
- The length of these defrost cycles is adjustable from 4 to 110 minutes in 2-minute increments with a minimum of four hours between each defrost cycle.
- Seven-day time switch permits seven-day operation with a different program for each operating day using a spring-wound carryover mechanism.
- A multicircuit defrost control is a multicircuit timing control that operates from one to twenty-four switches individually.
- The 24-hour time switch, skip-a-day clock is used to allow the operator to stop operation on Saturday, Sunday, or any other selected day.

# Service Call

A customer complains that a commercial refrigeration unit is getting too warm. A check of the system reveals that the compressor is not running and the refrigerated case is too warm. The defrost timer is in the defrost cycle and the timer motor is not running. The voltage to the timer motor is 240 VAC. The technician turns the power off, removes the wiring from the motor terminals, and performs a continuity test of the motor. The motor has infinite resistance. The timer motor winding is open. The tech then replaces the timer and sets the defrost times and the current time. When the power is turned back on, the unit starts running and cooling. The technician then checks the timer to make sure that the motor is running, and is satisfied that the system is repaired.

# Student Troubleshooting Problem

A customer complains that a commercial refrigerator cabinet is too warm. A check of the system shows that the cabinet is 40°F. The unit has an electric defrost system. Nothing is running. The electric defrost heaters are on. The defrost timer has 240 VAC at it's timer terminals. What could be the problem? How can the problem be repaired?

# Questions

1. What type of switching arrangement do timers and time clocks have?
2. How much inrush current can timer and time clock contacts handle?
3. What control can be set for up to six defrosts per day?
4. Which control has a four-hour minimum between defrost periods?
5. Which control has three contact arrangements for an electric heat, hot gas defrost, or compressor shutdown defrost?
6. Which control has backup defrost termination?
7. Which control has a variation in the length of the defrost cycle?
8. Which control allows a different program for each day?
9. What control will keep the set points when a power failure occurs?

# Unit 41: COMMERCIAL DEFROST SYSTEMS

## Introduction

Defrost systems are used on most commercial refrigeration systems because their evaporator coil temperature is usually below 32°F and frost and ice will collect on its surface. This frost or ice acts as an insulator and decreases the efficiency and performance of the unit; therefore, it must be periodically removed. To be efficient the frost or ice must be removed from the coil automatically. A definite set of controls for defrosting is needed. Commercial defrost systems are a variety of controls used to complete the automatic defrosting of evaporating coils.

*Operation.*    The simplest defrost system uses a low-pressure control to cycle the compressor. The low-pressure control is set low enough to keep the lowest temperature needed in the refrigerated space. It is set high enough so that most of the frost will melt during the OFF cycle. This control maintains compressor operation. The compressor is started only after the temperature and the corresponding refrigerant pressure have risen enough to cause most of the frost to melt.

This type of defrost control system is not very efficient because not all of the ice and frost is melted during each OFF cycle. When the low-pressure control is used for defrosting, the complete system must at times be shut down and manually defrosted. Low-pressure defrost systems are inconvenient and inefficient.

The most economical defrost system is one that automatically and completely removes all the frost on the evaporator during each defrost cycle and returns to the refrigeration cycle when all the frost is removed. This type of system is usually time initiated and temperature terminated. It allows a more even temperature during the busy time of the day. The defrosting is done during a period of least activity.

## Classifications of Commercial Defrost Systems

The three types of commercial defrost systems are hot gas defrost, electric defrost, and water defrost. Because it is the most popular, the hot gas defrost system is discussed first.

**Hot gas defrost system.**    The hot gas defrost is the fastest and most economical method used for automatic, positive defrosting of commercial refrigeration evaporators and heat pump outdoor coils during wintertime operation. Refrigerant vapor, superheated by compression, is circulated through the evaporator as a continuous supply of heat to melt the frost.

During the defrost cycle, the evaporator and condenser fans are turned off. Some means of reversing the flow of refrigerant is used to direct the hot gas into the evaporator coil. Reversing the refrigerant flow is done by a reversing (four-way) valve. The valve directs the hot refrigerant from the compressor discharge to the evaporator coil. In effect the evaporator and condenser coils change jobs.

**Types of hot gas defrost systems.**    The different types of hot gas defrost systems are named by the types of controls used to complete the defrost cycle. They are time-initiated time-terminated systems, time-initiated temperature-terminated systems, and time-initiated pressure-terminated systems. Following is a discussion of these different types of control systems.

*Time-Initiated Time-Terminated Systems.*    The time-initiated time-terminated systems have a time clock, discussed in Unit 40. The trippers on the time clock are set to give a defrost cycle at a chosen time, or times, each day, depending on the system needs. The time clock initiates a defrost cycle at the set time. The defrost cycle continues until the time clock terminates it. See Figure 41–1. The defrost intervals should be chosen by box usage so that defrosting is complete and the evaporator coil is free of frost just before heavy usage.

A good-quality time control should be used to initiate the defrost cycle. The drain line should be heated where it is exposed in the refrigerated space to the condensate from freezing during the defrost period. This may be done with an electric tape around the pipe or with the liquid line in contact with the drain line. The drain should be trapped outside the refrigerated space to keep condensate from freezing there during normal operation.

*Time-Initiated Temperature-Terminated Systems.*    The time-initiated temperature-terminated systems use a time clock, discussed in Unit 40. The ON trippers are set to start the defrost cycle, or cycles, at a chosen time each day. The system stays in the defrost mode until the temperature and corresponding pressure in the evaporator have risen to a point showing that all the frost has been melted from the evaporator coil. This temperature is sensed by a defrost termination control. At this time the defrost cycle is terminated and the system returns to normal operation. See Figure 41–2. The defrost intervals should be chosen by box usage so that defrosting is complete and the evaporator coil is free of frost just before heavy usage.

A good-quality time control should be used to initiate the defrost cycle. The drain line should be heated where it is exposed in the refrigerated space to keep from freezing the condensate water. This may be done with either an electric tape around the pipe or the liquid line in contact with the drain line. The drain should be trapped outside the refrigerated space.

*Time-Initiated Pressure-Terminated Systems.*    The time-initiated pressure-terminated defrost systems use a time clock, discussed in Unit 40. The ON trippers start the defrost cycle or cycles at a chosen time, or times, each day. The system remains in the defrost mode until the low-side refrigerant pressure has risen to the point that shows all the frost has been melted from the evaporator coil. At this point the defrost cycle is

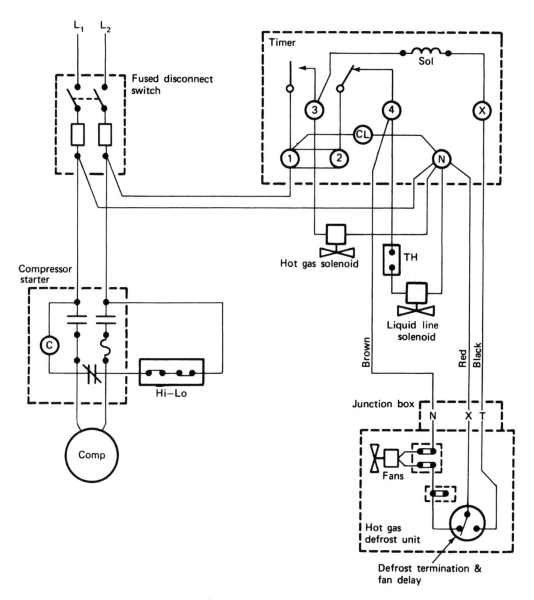

**Figure 41–1**
Typical hot gas defrost wiring diagram.

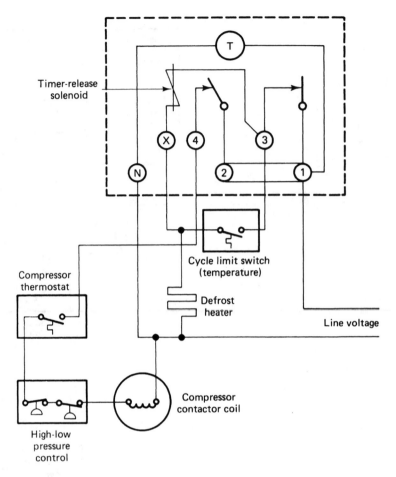

**Figure 41–2**
Typical wiring diagram for a time-initiated temperature-terminated defrost system.

terminated by the pressure control. The system returns to the normal cooling cycle. See Figure 41–3. The defrost intervals should be chosen so that defrosting is complete and the evaporator coil is free of frost just before periods of heavy use.

A good-quality time control should be used to initiate the defrost cycle. The drain line should be heated where it is exposed in the refrigerated space during the defrost period to keep from freezing the condensate. This may be done with either an electric tape around the pipe or the liquid line touching the drain line. The drain line should be trapped outside the refrigerated space to keep from freezing the condensate during normal operation.

**Electric defrost system.**   When the electric defrost system is used, the compressor is shut down and a liquid-line solenoid is closed (deenergized) during the defrost

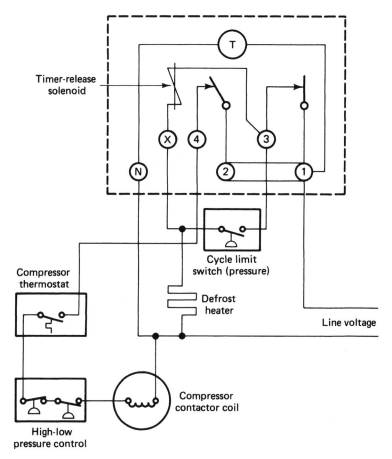

**Figure 41–3**
Typical wiring diagram of a timed-initiated pressure-terminated defrost system.

period. An electric heater attached to the evaporator coil is energized by the time clock to heat the evaporator coil and melt the frost. At the same time, the evaporator and condenser fans are stopped to keep from blowing warm air into the refrigerated space. As the coil temperature is raised to a chosen temperature, the defrost cycle is terminated and then the cooling cycle is started. See Figure 41–4 for wiring diagrams.

A good-quality time control should be used to initiate the defrost cycle in all defrost systems. Drain lines should be heated where exposed in the refrigerated space to keep from freezing the defrost water during the defrost period. This may be done with either an electric tape around the pipe or the liquid line touching the drain line. The drain line should be trapped outside the refrigerated space to keep from freezing the condensate during the cooling cycle.

The liquid-line solenoid should be good quality to keep liquid refrigerant from entering the low side of the refrigeration system during the defrost cycle regardless of the thermal expansion valve setting.

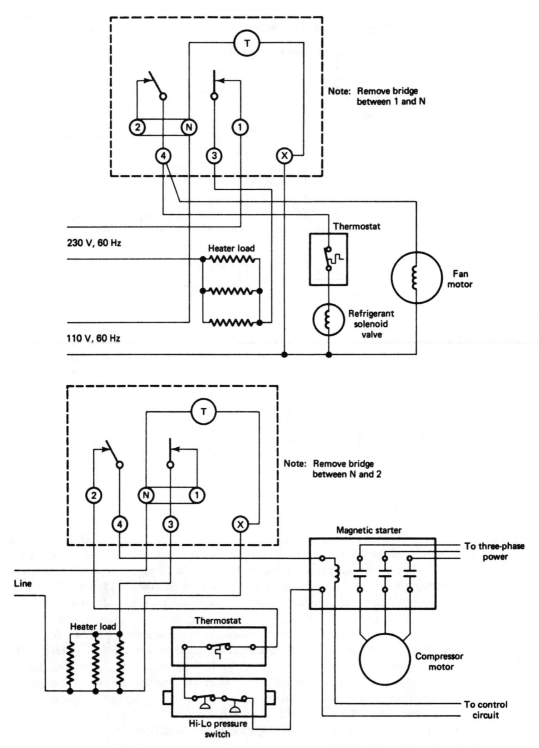

Note: Remove bridge
between 1 and N

230 V, 60 Hz

Heater load

Thermostat

Refrigerant
solenoid
valve

Fan
motor

110 V, 60 Hz

Note: Remove bridge
between N and 2

Magnetic starter

To three-phase
power

Line

Heater load

Thermostat

Compressor
motor

To control
circuit

Hi-Lo pressure
switch

**Figure 41–4**
Typical wiring diagrams for electric heat defrost system.

276

Following are typical examples for use of electric defrost systems.

1. Frozen food cabinets
2. Dairy cases
3. Vegetable cases
4. Beverage coolers
5. Ice cream display cases
6. Walk-in coolers

*Operation.*    These types of defrost controls use a time control similar to those discussed in Unit 40. When the defrost control senses that the evaporator needs defrosting, the time control stops the compressor, evaporator fan, and condenser fan and deenergizes the solenoid valve coil. It energizes the electric resistance heater to melt the frost from the evaporator coil. When the control senses that all the frost has been melted, the system is returned to its normal cooling mode.

# Summary

- Commercial defrost systems use a variety of controls to complete the automatic defrosting of evaporating coils.
- The simplest defrost system uses a low-pressure control to cycle the compressor. The low-pressure control is set low enough to keep the lowest temperature needed in the refrigerated space. It is set high enough so that most of the frost is melted during the OFF cycle. This control has control of compressor operation.
- The most economical defrost system is one that automatically and completely defrosts the evaporator during each defrost cycle and returns to the refrigeration cycle when all the frost is melted. This type of system is usually time initiated and temperature terminated.
- Three types of commercial defrost systems are hot gas defrost, electric defrost, and water defrost.
- The hot gas defrost is the fastest and most economical method used for automatic, positive defrosting of commercial refrigeration evaporators and heat pump systems during wintertime operation.
- During the defrost cycle, the evaporator and condenser fans are turned off. Some means of reversing the flow of refrigerant is used to direct the hot gas into the evaporator coil to melt the frost. The reversing process is done by a reversing (four-way) valve.
- The different types of hot gas defrost systems are designated by the types of controls used to cause the defrost cycle, such as time-initiated time-terminated systems, time-initiated temperature-terminated systems, and time-initiated pressure-terminated systems.

- The defrost intervals should be chosen by box usage so that defrosting is complete and the evaporator coil is free of frost just before heavy usage.
- The time-initiated pressure-terminated defrost systems use a time clock for this purpose.
- When the electric defrost system is used, the compressor is shut down during the defrost period. The solenoid valve coil is deenergized. The time clock energizes an electric heater to heat the evaporator coil and melt the frost.
- The liquid-line solenoid should be of good quality to keep liquid refrigerant from entering the low side of the system during the defrost cycle regardless of the thermal expansion valve setting.

# Service Call

A customer complains that a commercial ice cream cabinet is too warm, causing the ice cream to melt. A check of the system shows that the cabinet is too warm and the evaporator coil is completely iced over. The technician checks the defrost timer and finds that one of the defrost pins had fallen out of its hole and fell to the bottom of the control cabinet. The tech advances the timer to demand a defrost period to remove the frost from the evaporator. The tech then replaces the pin at the correct defrost time, but leaves the control in the defrost demand position until all the frost is melted from the evaporator. The clock is set to the current time and day. The timer motor is working. The technician is satisfied that the system is repaired.

# Student Troubleshooting Problem

A customer complains that the ice cream in her commercial ice cream cabinet is melting. The system's evaporator is covered with ice and frost. Everything is running but there is not much cooling. The technician checks the defrost timer and finds the trippers in the bottom of the timer enclosure. What must be done to correct the problem?

# Questions

1. What type of commercial defrost system is not very efficient?
2. What must be periodically done when a low-pressure control is used for a defrost control?
3. What is the most economical defrost system?
4. What is the fastest and most economical defrost system?
5. How is the flow of refrigerant reversed during a hot gas defrost cycle?
6. What is done in a commercial refrigeration cabinet to keep the defrost condensate from freezing in the drain line?
7. Why are the evaporator fans stopped during the defrost period?
8. On what type defrost system is both the evaporator and the condenser fans stopped during defrost?

# Unit 42: DOMESTIC REFRIGERATION DEFROST CONTROLS
## Introduction

Defrost systems are used on refrigeration systems because the evaporator operating temperature is below freezing. Defrosting is needed because a layer of frost will accumulate on the evaporator coil. This frost, or ice, acts as an insulator and will decrease the refrigerator efficiency and production. Defrost systems consist of a variety of controls used to complete the automatic defrosting of evaporator coils.

**Domestic refrigerator defrost controls.**    The automatic defrost controls used on domestic refrigerators and freezers are timing controls that operate when the unit is plugged into the electrical outlet. They have a set of NC contacts and a set of NO contacts. The NC contacts make the circuit to the compressor and all fans. The NO contacts control the defrost heaters. See Figure 42–1. These controls are set to defrost the unit at chosen times during a 24-hour period. During the defrost period the compressor and all fans are stopped and an electric heater is energized. In some units a condensate drain heater is used to keep the condensate from freezing in the drain line where it is exposed in the freezer compartment.

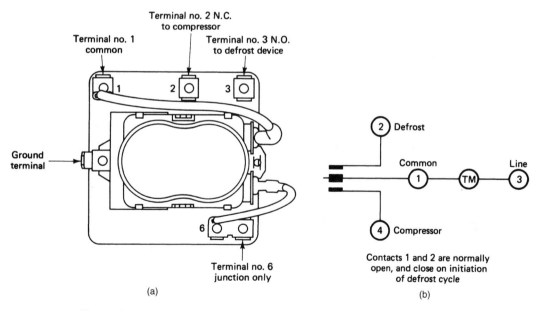

(a)                                                                  (b)

**Figure 42–1**
(a) Typical defrost timer connections; (b) internal diagram of a defrost timer. (Courtesy of Gem Products, Inc.)

*Operation.*    When the time clock reaches the chosen time for defrosting, a defrost cycle is initiated. At this time the NC contacts open and the compressor and all fans stop. The NO contacts close, energizing the electric defrost heater and drain heater, when used. In a short time, the contacts return to their normal cooling position. The defrost cycle continues until the evaporator temperature reaches about 50°F. The defrost limiter control then terminates the defrost cycle and the unit goes into the normal cooling mode.

**Defrost limiter.**    The defrost limiter used on domestic refrigerators and freezers is a temperature-sensing control that terminates the defrost cycle when the evaporator temperature reaches about 50°F. This control is fastened to the evaporator to sense its temperature. When the temperature is reached that shows all the frost has been melted from the coil, the defrost cycle is terminated. The defrost limiter has a set of NO contacts that close on a fall in temperature.

*Operation.*    When the timer contacts switch, showing that a defrost period is needed, the compressor and all fans are stopped and the electric heating elements are energized. The defrost limiter contacts are closed because of the lower temperature of the coil. The unit goes into defrost. The electric heating element begins to warm the evaporator coil and melt the frost. When the evaporator temperature reaches the cut-out setting of the limiter, its contacts open and the system goes into the normal cooling cycle.

# Summary

- Defrost systems are a variety of controls used to complete the automatic defrosting of evaporator coils.
- The automatic defrost controls used on domestic refrigerators and freezers are timing controls that operate when the unit is plugged into the electrical outlet. They have a set of NC and a set of NO contacts. The NC contacts make the circuit to the compressor and all fans. The NO contacts energize the defrost heaters.
- The defrost limiter used on domestic refrigerators and freezers is a temperature-sensing control that terminates the defrost cycle when the evaporator temperature reaches about 50°F.

# Service Call

A customer complains that a domestic refrigerator is too warm. A check of the system shows that the refrigerator is in the defrost mode and that the defrost control is working correctly. The defrost limiter is still in the defrost mode so it is considered bad. The tech replaces the limiter. When the refrigerator is plugged into the wall outlet, the compressor starts running and cooling. The temperature inside the refrigerator begins to fall. The system is now operating normally. The technician is satisfied that the system is repaired.

# Student Troubleshooting Problem

A customer complains that a domestic refrigerator is not keeping the food cold enough. A check of the system shows that the evaporator is covered with frost. The voltage to the unit is 120 VAC. There is no voltage to the compressor circuit. The defrost timer has 120 VAC to the power terminals. There is no power to the compressor circuit. What could be the problem?

# Questions

1. When do the defrost controls operate on a domestic refrigerator?
2. What is the contact arrangement of a domestic defrost control?
3. When are domestic defrost controls set to start defrosting?
4. What type of defrost system is used on domestic refrigerators?
5. At what temperature does the domestic refrigerator come out of defrost?
6. What terminates the domestic defrost cycle?
7. What type of contacts are used in a defrost limiter?

# Unit 43: HEAT PUMP DEFROST CONTROLS
# Introduction

When heat pump systems are operating in the heating mode, the outdoor coil acts like an evaporator. When the temperature of the outdoor coil drops below 32°F, frost and ice begin to collect on it. This ice and frost reduce the efficiency and the output of the system. It must be removed to keep the unit operating as efficiently as possible. Defrost systems are a variety of controls used to automatically defrost evaporating coils.

*Operation.*    The frost is melted because the controls make the reversing (four-way) valve change positions and direct the hot gaseous refrigerant to the outdoor coil. When this shift takes place, both the indoor and outdoor fans are stopped. The outdoor fan is stopped to help give a fast defrost. The indoor fan is stopped to keep from blowing cool air into the conditioned space.

The most popular methods of automatic defrost initiation and termination are (1) air-pressure differential across the outdoor coil (demand), (2) outdoor coil temperature, (3) time, and (4) time and temperature.

**Air-pressure differential (demand) defrost systems.**    Air-pressure differential defrost systems use a diaphragm type control to measure the air-pressure differential across the outdoor coil. The defrost cycle is initiated on demand rather than by using a time clock. See Figure 43–1. These controls use a slow make or break switch operated by a slack-bag diaphragm. Low-pressure taps permit operation on positive or negative pressure applications. At the calibration pressure, the switch closes, making the circuit to the defrost control. See Figure 43–2.

*Operation.*     As frost forms on the outdoor coil, the difference in the air pressure across the coil rises. The defrost control measures this pressure difference. As it rises to a chosen differential, the inflated slack-bag diaphragm closes a pilot-duty switch. This switch makes the initiation circuit to the defrost control. The defrost control then stops both the indoor and outdoor fans, switches the reversing valve to the defrost position, and makes the circuit so the indoor resistance heaters can operate when needed. See Figure 43–3.

**Outdoor coil temperature.**     In the outdoor coil temperature system, the defrost cycle is initiated by the defrost thermostat. See Figure 43–4. This is a double-bulb thermostat that senses a drop in unit efficiency and initiates the defrost cycle. Time, outdoor air temperature, wind, and atmospheric conditions do not affect this control.

**Figure 43–1**
Air-pressure differential defrost control.
(Courtesy of Ranco North America,
Division of Ranco, Inc.)

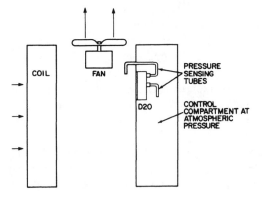

**Figure 43–2**
Typical installation of air-pressure sensor.
(Courtesy of Ranco North America,
Division of Ranco, Inc.)

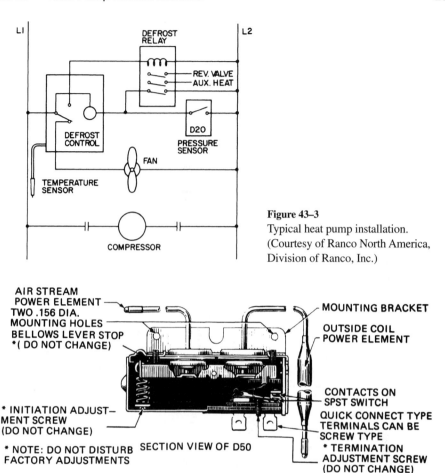

**Figure 43–3**
Typical heat pump installation.
(Courtesy of Ranco North America,
Division of Ranco, Inc.)

**Figure 43–4**
Defrost thermostat. (Courtesy of Ranco North America, Division of Ranco, Inc.)

The signal that starts the defrost cycle initiation is caused by a rise in the temperature difference between the outdoor air temperature around the outdoor coil and the temperature of the coil itself. The control then compares this temperature with the temperature difference of an ice-free coil at the same outdoor temperature.

*Operation.*   As the unit operates and the ice accumulation causes a great enough temperature difference, the defrost cycle is initiated. The outdoor fan stops, the reversing valve changes positions, and the auxiliary heating strips are ready to be energized when needed. See Figure 43–5. As the compressor continues to run, the outdoor coil is warmed and the ice is melted from it. The outdoor coil remains at 32°F until all the frost has been melted. The unit continues to operate in defrost until the temperature of the outdoor coil rises to the temperature needed to cause the reversing valve to change position. This tem-

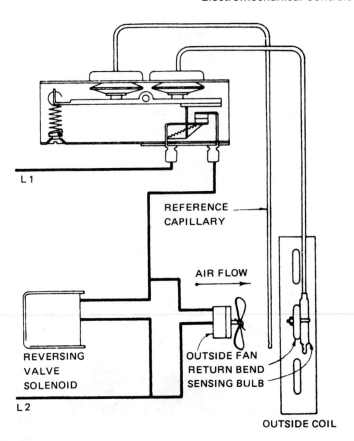

**Figure 43–5**
Schematic wiring diagram of thermostatic defrost control. (Courtesy of Ranco
North America, Division of Ranco, Inc.)

perature is usually about 60°F on the preset control. At this temperature the defrost cycle
is terminated. The system then automatically returns to the heating mode.

**Time.**    A time defrost control initiates a defrost cycle after a given amount of
compressor running time. The defrost control uses a timer motor with contacts to give the
necessary switching action. See Figure 43–6.

The time defrost system has several disadvantages. The unit may be switched into
defrost when it is not needed, it may wait too long before initiating a defrost cycle, it may
keep the unit in defrost too long, or it may not keep it in defrost long enough. These conditions will usually cause the unit to operate at less than peak efficiency.

*Operation.*    When the compressor has been running for a chosen time, usually 90
minutes, the defrost timer contacts close to change the system into the defrost mode. After

a chosen amount of time has passed, usually about 15 minutes, the timer contacts open. The defrost cycle is terminated and the unit changes to the heating mode.

**Time and temperature.**   The time and temperature method of heat pump defrost is very popular with manufacturers. This method uses a combination of clock timer and defrost thermostat for sensing when the defrost cycle should be initiated and terminated. See Figure 43–7. The thermostat contacts are normally open, and close on a fall in temperature. The sensing bulb is located on one outdoor coil tube near the refrigerant outlet. When the sensing bulb senses that the outdoor coil temperature has dropped to 32°F, the thermostat contacts close. These contacts are in electrical series with the NO contacts in the clock timer. See Figure 43–8. The clock motor is electrically in parallel to the outdoor fan motor and runs only when the compressor is running. See Figure 43–9.

*Operation.*   When a chosen time period has passed, usually 90 minutes (these timers are adjustable for either 30-, 60-, or 90-minute cycles), the timer motor closes its NO set of contacts. These contacts stay closed for only a few seconds during each cycle of the time clock. If the defrost thermostat contacts are closed at the same time, the defrost relay coil is energized and a defrost cycle is initiated. If the thermostat contacts

**Figure 43–6**
Defrost timer. (Courtesy of Lennox Industries, Inc.)

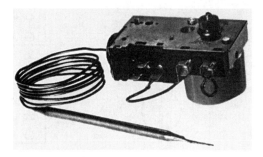

**Figure 43–7**
Defrost timer and thermostat. (Courtesy of Ranco North America, Division of Ranco, Inc.)

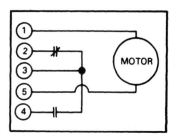

**Figure 43–8**
Defrost time clock timer schematic. (Courtesy of Lennox Industries, Inc.)

are not closed at this time—showing that a defrost cycle is not needed—the clock timer will start another cycle.

When the unit goes into a defrost cycle, the clock timer motor and the outdoor fan are deenergized. The defrost thermostat will keep the unit in defrost until the outdoor coil temperature has risen to about 65°F. At this temperature the thermostat contacts open, deenergizing the defrost relay and terminating the defrost cycle.

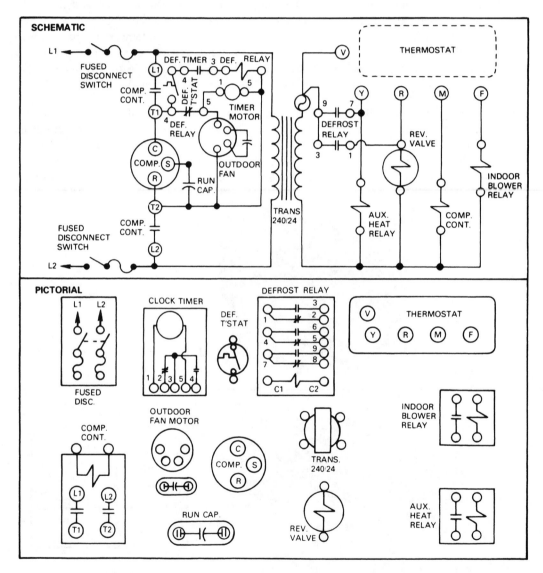

**Figure 43–9**
Wiring diagram of a heat pump. (Courtesy of Lennox Industries, Inc.)

# Summary

- Defrost systems use a variety of controls to complete the automatic defrost of evaporating coils.

- The frost is melted because the controls make the reversing (four-way) valve change positions and direct the hot gaseous refrigerant into the evaporating coil. When this shift occurs, both the indoor and outdoor fans are stopped.

- The most popular methods of automatic defrost initiation and defrost termination are (1) air-pressure differential across the outdoor coil, (2) outdoor coil temperature, (3) time, and (4) time and temperature.

- Air-pressure differential defrost systems use a diaphragm type control to measure the air-pressure differential across the outdoor coil. They initiate a defrost cycle on demand rather than by using a time clock.

- In the outdoor coil temperature system the defrost cycle is initiated by a defrost thermostat.

- The outdoor coil defrost system is a double-bulb thermostat that senses a decrease in unit efficiency and initiates the defrost cycle. Time, outdoor air temperature, wind, and atmospheric conditions do not affect this type of control.

- A time defrost control usually initiates a defrost cycle after a chosen amount of compressor running time. The defrost control uses a timer motor with contacts to give the necessary switching action.

- The time defrost system has several disadvantages. The unit may be switched into defrost when it is not needed, it may wait too long before initiating a defrost cycle, it may keep the unit in defrost too long, or it may not keep it in defrost long enough. These conditions will usually cause the unit to operate at less than peak efficiency.

- The time and temperature method of heat pump defrost is quite popular with manufacturers. This method uses a combination of clock timer and defrost thermostat for sensing when the defrost cycle should be initiated and terminated.

- When a chosen time has passed, usually 90 minutes (these timers are adjustable for either 30-, 60-, or 90-minute cycles), the timer motor closes its NO set of contacts. These contacts stay closed for only a few seconds during each cycle of the time clock.

# Service Call

A customer complains that a residential heating unit is not heating the home. The system is a heat pump. The defrost is controlled by a time-temperature defrost control. The outdoor coil is completely defrosted. The compressor is not running. A check of the high-pressure control shows that the unit is off there. It is a manual reset type. The tech resets the control and the compressor starts running; however, it is still in the defrost mode. The tech then places a jumper between terminals 2 and 3 on the defrost control. The reversing valve changes and the unit goes into the heating mode. A check of the power shows 240 VAC at the clock motor terminals, but the clock is not running. The defrost control is bad. The technician replaces the control, puts the unit back in operation, and completely checks the unit. It is now heating the home. The technician is satisfied that the system is repaired.

# Student Troubleshooting Problem

A customer complains that a heat pump unit is not keeping the residence cool enough and the electric bills are higher than normal. A check of the system shows that everything is running. The thermostat is set on 76°F and the return air temperature is 82°F. The voltage to the unit is 240 VAC. The amperage is lower than normal. The tech checks refrigerant pressures, and with HCFC-22 refrigerant the suction pressure is 50 psig. The discharge pressure is 225 psig for an air-cooled heat pump. The outdoor temperature is 95°F. What could be the problem? What can be done to solve the problem?

# Questions

1. Why are the indoor and outdoor fans stopped when a heat pump goes into defrost?
2. What type of time clock is used with the demand defrost system?
3. How are the contacts operated on a demand defrost system?
4. What does a demand defrost system measure?
5. What causes the defrost signal on a double-bulb thermostat defrost system?
6. At what temperature does the outdoor coil remain until all the ice is melted?
7. At what temperature does the double-bulb defrost system stop the defrost cycle?
8. With a timed defrost, how often does the unit go into defrost?
9. How long does a timed defrost system keep the unit in defrost?
10. On a time and temperature defrost system, when do the thermostat contacts close?
11. On a time and temperature system, when does the time clock run?
12. When a heat pump is defrosting, what happens when the outdoor coil reaches 65°F.
13. What type of defrost system uses the pressure drop across the coil and a thermostat to accomplish the defrost cycle?

# Unit 44: LOCKOUT RELAY
## Introduction

Lockout relays are used as motor protectors when conditions exist that could possibly cause damage to the motor. They are used to stop the motor when severe operating conditions are present.

**Lockout relay circuit.**    Lockout relays are used to break the electrical circuit if an overload occurs in the protected circuit. The coil of the lockout relay is wired in parallel to

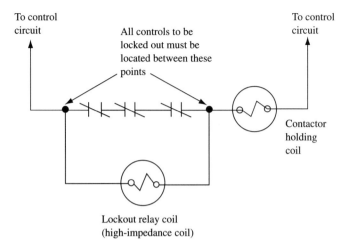

To control
circuit

To control
circuit

All controls to be
locked out must be
located between these
points

Contactor
holding
coil

Lockout relay coil
(high-impedance coil)

**Figure 44–1**
Lockout relay connections.

the contacts of all the other controls in the control circuit to the contactor or starter holding coil. Its contacts are placed in series with the other control contacts in that circuit. The relay has a very high impedance, high resistance coil that is energized when any set of contacts in the control circuit open. See Figure 44–1. The high resistance of the lockout relay coil drops the electrical power enough to keep the contactor holding coil from pulling in and closing its contacts.

*Operation.*    During normal operation, the contacts of the lockout relay are normally closed. Because the resistance of all the contacts in the control circuit and the contactor holding coil is less than that of the lockout relay coil, the coil will not pull in to open its contacts. When something happens to cause any set of contacts within the monitored part of the control circuit to open, the current will then pass through the relay coil causing it to pull in and open its contacts. The current will now continue to flow through the relay coil because its contacts are open and will not allow the current to pass through the control circuit.

When a lockout relay has stopped the unit, it can be restarted by manually turning off the electric power to the unit, waiting a few seconds, and then turning it back on. The unit should start if there is no longer a problem.

# Summary

- Lockout relays are used as motor protectors when conditions exist that could possibly damage the motor.
- Lockout relays are used to stop the motor when under severe operating conditions.
- The contacts of the lockout relay are placed in series with the other control contacts.

- The lockout relay uses a high-impedance coil.
- The contacts of the lockout relay are normally closed.
- The resistance of the contacts in the control circuit and the contactor holding coil is less than that of the relay coil. The contactor or starter holding coil will not pull in when all the contacts are closed and the relay contacts are open.

# Service Call

A customer complains that an air-conditioning system is not cooling. The system's air-cooled condensing unit is not operating. The technician finds that the system is off on the lockout relay, and resets the system. The tech checks the voltage, current, and pressures of the system to determine what caused the lockout. The discharge pressure is 395 psig for HCFC-22, which is higher than normal. The condenser coil is dirty, so the technician cleans it and puts the system back in operation. After the condenser has dried, a check of the pressures, voltage, and amperage to the unit shows they are normal. The technician is satisfied that the system is repaired.

# Student Troubleshooting Problem

A customer complains that a residential cooling system is not cooling as it should. The system's thermostat is set on 76°F. The return air is 85°F. The outdoor temperature is 95°F. The unit is air-cooled and uses HCFC-22 refrigerant. Nothing is running but the indoor fan. The voltage to the outdoor unit is 240 VAC. There are 24 VAC from the secondary side of the transformer. The unit has a lockout relay. Its contacts are open. The system has an overload relay and high- and low-pressure controls. What could cause the lockout relay to open its contacts?

# Questions

1. What do lockout relays do?
2. In what position is the lockout relay coil wired?
3. Why does the lockout relay not pull in when the circuit is energized?
4. How can a unit that is off on the lockout relay be restarted?

# 4 Electronic (Solid-State) Controls

## Introduction

Almost all air-conditioning and refrigeration systems manufactured in recent years have solid-state controls. The great improvements in these controls have led to a more accurate, more compact, and more varied control system. Solid-state controls provide better efficiency of the entire electronic system, such as variable-speed motor control, better motor protection, more accurate and dependable defrost controls on heat pump and refrigeration systems, better motor starting devices, and automatic ignition of gas-fired furnaces.

# Unit 45: ELECTRONIC THERMOSTATS
## Introduction

Electronic thermostats were introduced to conserve energy and to add comfort to the occupants of the space being conditioned. They also provide better storage conditions for the products needing refrigeration, because electronic thermostats are more responsive to temperature changes than the electromechanical types.

This type thermostat uses several different methods of controlling equipment operation. The clock-operated type was one of the earlier models, in which a clock controlled the ON and OFF settings of the thermostat. Current models use a microprocessor to control the equipment. Some have only one setback period per day and must be reset every day. Others can be set for several temperature changes per day and need only to be set one time. The model used will determine the flexibility of equipment operation.

**Saverstat.** The Saverstat® is a solid-state programmable thermostat. It is used with single-stage heating and cooling systems using a 24-volt control system. It does not need isolation relays for accurate operation. It is powered by a standard AA battery so that

**Figure 45–1**
Saverstat electronic thermostat.
(Courtesy of Maple Chase, a Coleman
Company.)

the microprocessor will get the cleanest possible power. The thermostat is programmed with only one button. See Figure 45–1.

The Saverstat uses no subbase—it is mounted directly on the wall. It also has a large memory so that spring and fall reprogramming is not needed. An automatic short-cycle protection is also part of the control. The Saverstat can be set for four time and temperature changes per weekday, and two changes per weekend day.

Programming is done by pressing the SET button and using the UP or DOWN change buttons to make set point adjustments.

# Summary

- Electronic thermostats were introduced to conserve energy and to add comfort to the occupants of the space being conditioned.
- Later models use a microprocessor to control equipment operation.
- Each model has its own degree of flexibility.
- The Saverstat is programmed with only one button.
- The Saverstat has a large memory so that spring and fall reprogramming is not needed.
- The Saverstat includes an automatic short-cycle protection.

# Service Call

A customer complains that the cooling equipment has not cooled at the correct times since the customer returned from vacation. The system is controlled by a Saverstat. The time is several hours behind the present time and the batteries are weak. The microprocessor is not getting the correct power. The technician replaces the batteries with the correct type and resets the thermostat to the correct time and wanted setback modes. The technician is now satisfied that the system is repaired.

# Student Troubleshooting Problem

A customer complains that sometimes when he stays up past his normal bedtime the house gets too cold. A check of the system reveals that the thermostat is programmed to go into the energy saving mode at 10:30 P.M. The customer does not want the thermostat reprogrammed for the occasional times when he stays up late. What can be done to solve the problem?

# Questions

1. Why are electronic thermostats becoming more popular?
2. How do electronic thermostats manage energy usage?
3. What method is used to program the Saverstat?

# Unit 46: SOLID-STATE FAN CONTROLS (CONDENSERS)

## Introduction

Solid-state condenser fan controls are used to control the speed of the condenser fan in order to maintain the correct discharge pressure during mild weather operation. The power for the fan comes through the contactor along with the power for the compressor. The control is installed between the contactor load contacts and the fan motor. Be sure to follow the *correct* manufacturer's wiring diagram when testing or replacing this control.

**Discharge pressure controls.**    Discharge pressure controls regulate the discharge pressure at low outdoor temperatures by changing the airflow through the condenser. See Figure 46–1. Keeping the head pressure up helps to ensure that enough pressure differential remains across the expansion device to reduce downtime and unsatisfactory operation of refrigeration and air-conditioning equipment. By keeping this pressure differential, the efficiency of the equipment is kept at a peak value.

**Figure 46–1**
Solid-state discharge pressure control.
(Courtesy of ICM Corporation.)

*Operation.*    When 24 volts ac is first applied to the unit, the output voltage of the head pressure control is in the hard-start mode for a chosen time. On some controls this time is factory set and on others it may be set in the field. It is usually adjustable from 0.1 to 5.0 seconds.

When the hard-start mode is completed, the output voltage of the head pressure control changes to the thermistor. Fan speed control will not happen until the liquid-line temperature drops below about 103°F. This temperature may vary depending on the make, model, and other parts used. When starting, the full voltage will be applied to the fan motor to avoid loss of condensing efficiency and to make sure that the fan starts turning in the correct rotation. As the temperature that is being sensed drops, the output voltage of the control also drops. The output voltage may drop to the chosen low temperature cutoff setting. When the low temperature cutoff setting is reached, the output voltage of the control will drop to zero. The low temperature cutoff setting may be factory set or adjusted in the field. Figure 46–2 is a typical ladder diagram showing this control.

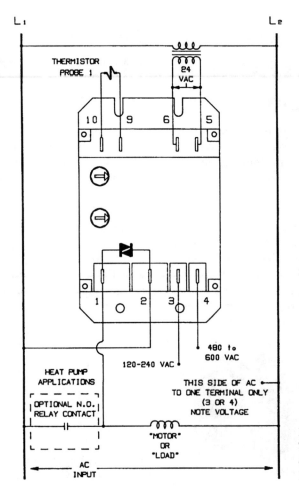

**Figure 46–2**

Typical solid-state discharge pressure control ladder diagram. (Courtesy of ICM Corporation.)

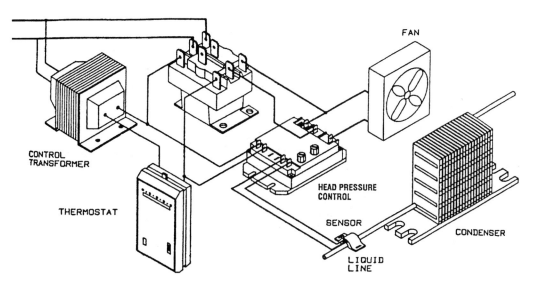

**Figure 46–3**
Typical solid-state, discharge pressure control installation schematic. (Courtesy of ICM Corporation.)

The fan motor will restart when the temperature rises above the low temperature cutoff setting by a chosen number of degrees. This is known as the system hysteresis (dead-band or differential). The system hysteresis is usually set at the factory before shipment.

When the 24-volt control system is broken, the control is deenergized. When the control circuit is again made, the output voltage is at the maximum for the hard-start time. The output voltage will be determined by the value of the thermistor input. Figure 46–3 is an example of how the control is installed on the system.

# Summary

- Solid-state condenser fan controls are used to control the condenser fan speed so that the correct discharge pressure can be kept during mild weather operation.
- Condenser fan controls operate to ensure the maintenance of pressure differential across the expansion device.
- When 24 volts ac are first applied to the unit, the output voltage of the head pressure control is in the hard-start mode for a chosen time.
- Fan speed control will not happen until the liquid-line temperature drops below about 103°F.
- When the low temperature cutoff setting is reached, the output voltage of the control will drop to zero.

# Service Call

A customer complains that an air-conditioning unit is not operating correctly. A check of the system reveals that the air-cooled condenser fan motor is running at full speed. The outdoor temperature is 65°F. The discharge pressure is 185 psig for HCFC-22. The fan should be running at a slower speed, but the fan speed control output voltage is 240 volts. It should be something lower than line voltage. The input voltage is also 240 volts. The fan speed control is bad so the tech replaces the control with the correct replacement and puts the system back in operation. After the hard-start mode has passed, the fan speed slows to keep the correct discharge pressure. The technician is satisfied that the system is repaired.

# Student Troubleshooting Problem

A customer complains that an older commercial air-conditioning system is not cooling as it should. A check of the system shows that the indoor temperature is 85°F. The outdoor temperature is 95°F. The system is air cooled and uses HCFC-22 refrigerant. Everything is running, but it is only dropping the temperature 14°F across the indoor coil. The suction pressure is 75 psig. The discharge pressure is 380 psig. The voltage is 240 VAC with the unit running. The amperage is higher than normal for this size unit. The unit has a head pressure control. The condenser is clean. What could be the problem?

# Questions

1. From where does the fan get its power?
2. How do condenser fan controls regulate the discharge pressure?
3. How do condenser fan controls help maintain high equipment efficiency?
4. About how long does the condenser fan control stay in the hard-start mode after first starting?
5. What happens when the liquid-line temperature drops below a certain temperature?
6. What is the system dead-band?

# Unit 47: SOLID-STATE GAS BURNER IGNITION CONTROLS

## Introduction

The burner is the device used for mixing the gas and air for correct combustion. The two types of ignition controls used in modern gas heating furnaces are a main burner and a pilot burner. The pilot burner is used to light the main burner gas and to give the path needed to operate the pilot safety devices. The main burners operate in either ON or OFF modes, relighting the burner each time the thermostat demands heat. Until recently, gas burners mostly used standing pilots; however, more equipment manufacturers are using intermittent pilot ignition to save energy. Standing pilots may be lighted manually or by an automatic pilot ignition control. In either case, intermittent pilots must be lit with each demand for heat. A standing pilot is lit once and then burns continuously. Intermittent pilots are used to conserve energy and for convenience to the user. There are several methods for relighting the pilot such as spark ignition and hot surface ignition.

**Automatic gas burner ignition system.**    The automatic gas burner ignition system automatically lights or relights the burner gas as needed for the job. Three general types of automatic gas burner ignition systems used today are the intermittent, hot surface, and direct spark or electronic ignition system.

*Intermittent Ignition Systems.*    The intermittent type is generally used for lighting a standing pilot. See Figure 47–1. This unit uses a purge time delay of about five minutes if the pilot goes out. This delay allows any collection of gas in the furnace to escape. Relighting the pilot, when needed, is automatic when the power to the unit is 24 volt ac.

*Operation.*    The following describes the operating cycle of an intermittent ignition system.

When the thermostat demands heat, the ignition module makes a safe start check to test all the internal components of the module for a flame-simulating condition. If any check finds a flame-simulating failure, no ignition will occur. If none is found, the module begins the safety lockout timing, powers the spark ignitor, and opens the pilot gas valve so gas can flow to the pilot burner. The pilot must light within the safety lockout timing of the module being used, otherwise the pilot gas valve closes.

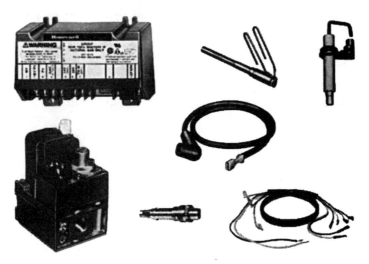

**Figure 47–1**
Intermittent ignition system. (Courtesy of Honeywell, Inc.)

Most modules used on power burners have a prepurge cycle before the sparking starts. During the prepurge cycle, the combustion blower runs to clear the combustion chamber of any unburned fuel. The module allows 30 to 45 seconds for the prepurge cycle. Usually a separate air-sensing switch ensures that the module starts timing only after the correct airflow is sensed.

Depending on the make and model being used, the module may go into a safety lock-out if the pilot gas does not light. On non-100-percent shutoff models, the ignitor continues to spark and the pilot gas (but not the main burner gas) flows as long as the thermostat demands heat. On 100-percent shutoff models, the module locks out, both main valves close, and the system must be manually reset. Usually the ignition spark also stops, although a few modules allow the spark to continue after the system has locked out.

The intermittent method of flame sensing works as follows. When the pilot lights, current flows from the sensor through the pilot flame to the pilot head and then to the ground. Because of the size difference between the sensor and the burner, the current flows in one direction. It becomes, therefore, a pulsating dc, or rectified, current. This current tells the module that a flame has been sensed. The ignition stops, either in response to the flame current or to the end of the timed ignition period. If a flame is sensed, the second main gas valve opens, allowing gas to flow to the main burner. The main burner lights and burns until the thermostat is satisfied and stops the flame-burning process.

As long as the rectified flame current stays above the minimum, the module keeps the valves in the gas control open. If the current drops below the minimum, or becomes un-steady, the main gas valve closes, stopping the flow of gas to the main burner. The module then makes another safe start check and restarts the ignition sequence.

The low-voltage electricity is supplied by a class 2 transformer and must supply a minimum of 24 volts ac, 60 hertz during the ignition and running cycles.

Repairs of this control cannot be made in the field. Faulty controls should be returned to the factory and new ones installed according to the manufacturer's specifications. A typical unit wiring diagram is shown in Figure 47–2.

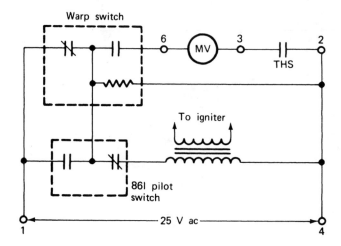

**Figure 47–2**
Schematic wiring diagram.

**Direct electronic ignitors.**    The major types of electronic spark ignitors are (1) those that light the pilot gas with a spark across electrodes, (2) those that light the main burner gas with a spark across electrodes, (3) those that light the main burner gas with a spark-plug type ignitor, and (4) hot surface ignitor. These are solid-state controls that are universal in use and should be given careful consideration when pilot or main burner lighting problems occur.

*Principles of Operation.*    These types of ignition systems are commonly called direct spark ignition (DSI) and an intermittent ignition device (IID). The IID system changes the standing pilot in that it uses solid-state electronic circuitry and a flame sensor to replace the thermocouple and the pilot safety control used on standing pilot type systems.

*Sensing Methods.*    When the standard thermocouple type system is used, heat is necessary for thermocouple operation. Heat is not needed with the IID system because flame conduction, or rectification, is used. To understand the principle of flame conduction and rectification, let us first come to know and understand the structure of a gas flame.

When gas is burning with the correct air-to-gas ratio, a blue flame having the following three zones will be visible. See Figure 47–3.

*Zone 1:* An *inner cone* that will not burn because there is too much fuel.
*Zone 2:* A blue envelope, known as the *intermediate cone,* around the inner, fuel-rich cone. This envelope consists of a mixture of vapor from the fuel-rich inner cone and the secondary, or surrounding, air. Zone 2 is where combustion takes place.
*Zone 3:* A blue envelope known as the *outer cone* which has too much air.

The second cone is most important here because this is the combustion zone, which is the area where the best flame sensing is possible.
A flame is simply a series of small, controlled explosions that cause the immediate area to become ionized. Ionization causes the area around the flame to become conductive. This conductivity is what is so important for correct flame conduction. See Figure 47–4.

*Operation.*    In operation, the flame can be thought of as a switch. The switch is between the pilot burner tip and the flame sensor. When the flame does not touch both the

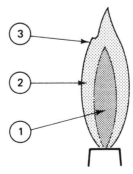

**Figure 47–3**
Gas flame structure. (Courtesy of Carrier Corp.)

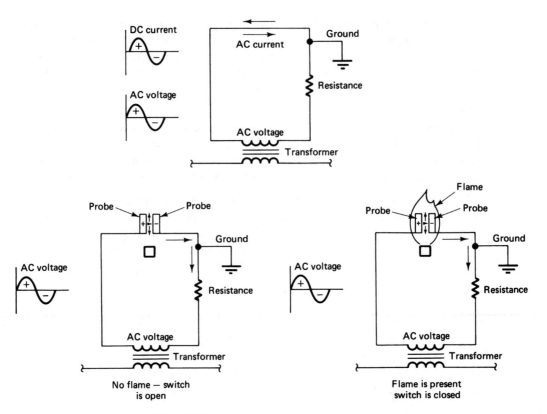

**Figure 47–4**
Electric current through a flame. (Courtesy of Carrier Corp.)

pilot tip and the sensor, the switch is open. When the flame touches both the pilot tip and the flame sensor, the switch is closed. In Figure 47–5, the flame is used to conduct an ac signal. Both probes have about the same amount of area exposed to the flame. The electric current flows through the flame. Unfortunately, this is not good enough to be used as a safety signal because it does not identify the current conducted by the flame. Therefore, a short circuit could be mistaken as a flame. For the flame to be correctly identified, an important difference known as flame rectification must be made. See Figure 47–5.

    ***Flame Rectification.***    In flame rectification, the flame and probes are used in a similar way but with one major difference. The area of one probe exposed to the flame is larger than the area of the other probe exposed to the flame.

    As shown in Figure 47–5, the flame is still used to conduct the ac signal. Both probes are touching the flame. The probe having the largest surface area attracts more electrons than the smaller one; therefore, it becomes the negative probe. The flow of electrons is from the positive probe to the negative probe. Also note that the ac voltage sine wave is not changed, but the negative portion of the current sine wave is no longer shown. The positive portion now represents a pulsating dc electric current. This is how flame rectification occurs.

To use this principle, a pilot and flame sensor have replaced the two probes. See Figure 47–6. After ignition of the pilot gas, a dc microamp current flow is conducted through the flame, from the flame to the sensor (the positive probe), and to the pilot tip—in this case, the negative probe—completing the circuit to ground. The sensing circuit uses this dc current flow to energize a relay that energizes the main gas valve.

***Sequence of Operation.*** The following general discussion describes the sequence of operation of electronic ignition systems.

1. When the temperature control demands heat, the spark transformer in the control circuit and the pilot valve are automatically energized at the same time.

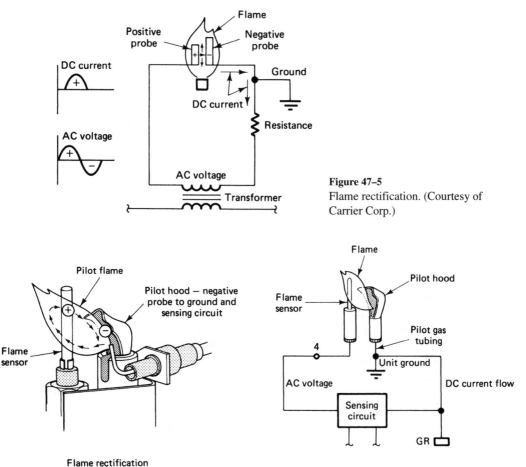

**Figure 47–5**
Flame rectification. (Courtesy of Carrier Corp.)

Flame rectification
pilot and probe

Rectification circuit

**Figure 47–6**
Pilot and flame sensor as used in flame rectification. (Courtesy of Carrier Corp.)

2. The spark ignites the pilot gas on each demand for heat from the thermostat.

3. The flame is proven by the flame sensor. The ignition control is then shut off and the spark stopped. At this time, the main gas valve is opened. (Some models allow the spark to continue for a short time after the main burner is lit.) On 100-percent lockout models, a shutdown of the entire system will occur if the pilot gas should not light within a fixed amount of time (usually only seconds).

4. The main burner gas lights and the system operates normally.

5. When the temperature control is satisfied, the main burner valve and the pilot valve are deenergized, stopping all gas flow to the burners.

*Application Guidelines.*     The following conditions have a direct effect on every IID application.

*Voltage:* The voltage supplied to electronic ignition controls should be within the following ranges.

1. 120-volt ac controls: The voltage should be from 102 to 132 volts ac.

2. 24-volt controls: The voltage should be from 21 to 26.5 volts ac.

All 24-volt ignition systems should have a transformer that will give enough power under maximum load conditions.

*Gas Pressure:* The gas pressure to the unit should be a minimum of 1-inch water column (WC) above the equipment manufacturer's recommended manifold pressure. Under no circumstances should the inlet gas pressure be lower than the equipment manufacturer's suggested inlet gas pressure.

The maximum inlet gas pressure for natural gas units should be a maximum of 10.5-inch WC. On LP gas units the inlet gas pressure should be a maximum of 14-inch WC.

*Temperature:* Electronic ignition control modules should not be exposed to temperatures above 150°F or below −40°F.

*Pilot Applications:* The applications of the pilot and sensor are the most important considerations of the IID application.

The pilot flame must touch the pilot burner tip and surround the flame sensor probe. A microammeter can help ensure that the correct amount of current is flowing through the pilot flame. The current needed depends on the type of system and the manufacturer's design. If the minimum signal is not kept at all times, the main burner will cycle rapidly or the main burner gas may not be ignited at all. On systems using flame rectification, the response time may be as little as 0.8 second from the time the flame is lost. Any deflection of the pilot flame away from the sensor tip or the pilot burner could result in rapid cycling of the main burner gas valve or keep the main burner from coming on.

Two other conditions that can cause ignition failure of the main burner gas or rapid cycling of the main burner are that the pilot flame is too small or that the gas pressure is too low for the pilot flame to surround the flame sensor. In either case, the pilot gas may ignite, but the main burner gas valve will not be energized. It is also possible for drafts or unusual air currents to blow the pilot flame away from the sensor. Blowing of the pilot flame may also be caused by the main burner ignition concussion or main burner flame rollout.

Another point to consider is the pilot flame condition. If the pilot flame is hard and blowing, the grounding area of the pilot is reduced so much that the necessary current is not being conducted, and there may be a shutdown of the system.

Correct placement of the flame sensor is critical. It should be placed so that it will make contact with the second, or combustion, area of the pilot flame. Passing the flame sensor through the inner cone of the pilot flame is not recommended. Therefore, a short flame sensor may give a much better signal than a long one. The final location (length) is best learned by using a microammeter.

**Hot surface ignitors (silicon carbide HSI).**    Several different types of hot surface ignition (HSI) systems are available. The one used will depend on the equipment manufacturer's design and the make and model of the ignitor.

***Description.***    The HSI system is used for the direct ignition of natural, liquefied petroleum (LP), manufactured, mixed, or LP gas-air mixtures. The main gas is lit by a hot surface ignitor. The flame is sensed either by a sensing probe (remote sensing) or by the hot surface itself (integral sensing). When the main burner flame is sensed, the ignitor is deenergized. The main gas valve remains open allowing gas to the main burner until the room thermostat is satisfied.

If burner gas ignition is not sensed within the allowed time trial for ignition, the main gas valve and the ignitor are broken. On three-trial models, a period of time known as the interpurge is programmed into the microprocessor to permit unburned gas to escape from the combustion chamber before another trial for ignition. The number of ignition trials (one or three) depends on the model of the unit. All hot surface ignition systems must use a redundant gas valve. The specifications for the G750 HSI control are listed in Table 47–1.

***Operating Mode Definitions.***    Following are the operating mode definitions for the Johnson Controls G750 HSI ignition system. Most other brands will have similar operating mode definitions; however, the specific make and model instructions should be used. Note: Some models may not use all the operating modes listed.

1. Prepurge: Initial time delay of the control before ignition of the main burner gas.
2. Warm-up time: A hot surface element (ignitor) is energized for a fixed period so that the temperature can rise higher than the gas ignition temperature.
3. Trial for ignition: The main gas and ignitor are energized for a short period. (*Note:* The ignitor will be turned off before the end of the trial for ignition period to allow sensing on the integral models.)
4. Interpurge: The time delay between the trials for ignition when both the gas and ignitor are deenergized to allow any unburned gas to escape before the next trial for ignition. (This step happens only if a correct ignition did not happen during the trial for ignition period.)

**TABLE 47–1**  SPECIFICATIONS.

| | | |
|---|---|---|
| Mounting | | Surface Mount—Any Position |
| **Electrical (Control)** | **Operating Voltage** | 24 Volts, 60 Hz |
| | **Operating Current** | 0.1 A, 24 VAC |
| | **Contact Rating (MV)** | 2.0 A Continuous / 5.0 A Inrush, 24 VAC |
| **Electrical (Igniter)** | **Operating Voltage** | 120 Volts, 60 Hz |
| | **Operating Current** | 6 A, 120 VAC |
| | **Contact Rating** | 120 Volts (Resistive) |
| **Ignition** | **Means** | Hot Surface (HSI) |
| | **No. of Trials** | Multiple (One or Three) |
| | **Trial Times** | 4, 6, 8, 10 or 15 Seconds |
| | **Igniter Warm-Up Times** | 15, 30 or 45 Seconds |
| | **Prepurge Times** | 0, 4, 15 or 30 Seconds |
| | **Interpurge Time** | 30 Seconds |
| **Flame Detection** | **Means** | Flame Rectification |
| | **Flame Failure Response Time** | 0.8 Seconds (Maximum) |
| | **Flame Output Voltage** | 24 Volts RMS |
| | **Flame Output Frequency** | 60 Hz |
| | **Flame Current Signal** | 0.2 Microamps (Minimum) |
| **Ambient Temp. Range** | | $-40°F$ ($-40°C$) to $160°F$ ($70°C$) |
| **Moisture Resistance** | | 95% RH @ 160°F (70°C) Non-Condensing |
| **Wiring Connections** | **Igniter** | 1/4″ Spade |
| | **Control** | 1/4″ Spade |
| **Case Material** | | Thermoplastic |
| **Type of Gas** | | Natural, Liquified Petroleum (LP), Manufactured, or Mixed |
| **Standards** | | ANSI Z21.20 C22.2, No. 199 |
| **Shipping Weight** | **Bulk Pack of 48** | 36 lbs. (19 kg) |

(Courtesy of Johnson Controls, Inc.)

5. Lockout: The main gas did not light on any of the allowed trials for ignition. The thermostat contacts must be opened for at least 30 seconds to reset the control.

6. Run: The main gas is on after a successful ignition attempt.

***Operation.***    The G750 HSI control is energized on demand for heat. See Figure 47–7. If the control uses a prepurge mode, the furnace prepurge fan or relay is also energized through the thermostat circuit. In the prepurge mode, the control will delay for the

**Figure 47–7**
G750 HSI control. (Courtesy of
Johnson Controls, Inc.)

time chosen (e.g., 30 seconds) before making the electric power circuit to the hot surface ignitor. If the prepurge is not chosen, the ignitor is energized within one second after the demand for heat by the thermostat. The ignitor warms and after the chosen heating time (e.g., 30 seconds), the main gas valve opens to allow gas to the main burner. A flame must be sensed within the trial for ignition period (e.g., 4 seconds) or the electric power to the main gas valve and the ignitor is switched off by the G750 control.

When the main burner gas ignites within the chosen time trial for ignition, the ignitor is deenergized and used to sense the main burner flame current (integral sensing), signaling the G750 to keep the main gas valves open when the thermostat demands heat. (Models equipped with the remote flame-sensing mode give a similar function by using an individual flame sensor probe.) The control will lock out if a flame is not sensed at the end of the trial for ignition period. On multitrial (three) models, the trial for ignition will be repeated for a total of three times after the 30-second interpurge period. The control will lock out if a main burner flame is not sensed at the end of the third trial for ignition.

To reset the G750, the thermostat must be turned off for a minimum of 30 seconds. If the main burner flame goes out (flame loss with the thermostat demanding heat), the control will automatically restart the ignition sequence. Wiring diagrams for the G750 HSI control are shown in Figures 47–8 through 47–11. These diagrams are only representative. Be sure to use the correct diagram for make and model being used.

**Spark plug and sensor.**    Following is a description of the spark plug and sensor main gas ignition systems as used on the Lennox G 14 series units.

The spark plug and sensor are on the lower left side of the combustion chamber. See Figure 47–12. Note that the sensor is the top plug and is longer than the ignition spark plug. These plugs have different thread diameters and cannot be interchanged.

The ignition spark plug is used with a primary control for igniting the initial gas and air mixture. See Figure 47–13 on page 309. The temperature in the combustion chamber keeps the spark plug free of oxides. It should not need any regular upkeep. Compression rings are used to form a seal between the plug and the combustion chamber.

For correct operation the spark plug gap must be set within certain specifications. See Figure 47–14 on page 309. Note that a ground-strap angle is used. When compared with other spark plug ignition systems this is unusual. A feeler gauge can be used for checking the gap.

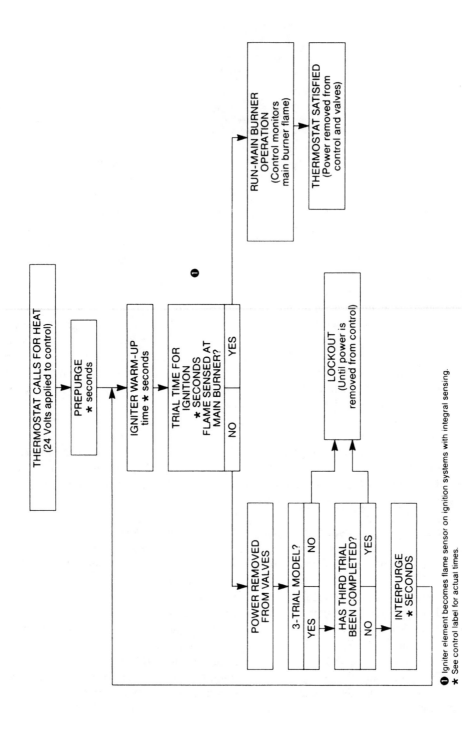

THERMOSTAT CALLS FOR HEAT
(24 Volts applied to control)

PREPURGE
★ seconds

IGNITER WARM-UP
time ★ seconds

TRIAL TIME FOR
IGNITION
★ SECONDS
FLAME SENSED AT
MAIN BURNER?

NO    YES

❶

RUN-MAIN BURNER
OPERATION
(Control monitors
main burner flame)

THERMOSTAT SATISFIED
(Power removed from
control and valves)

POWER REMOVED
FROM VALVES

3-TRIAL MODEL?

YES    NO

HAS THIRD TRIAL
BEEN COMPLETED?

NO    YES

INTERPURGE
★ SECONDS

LOCKOUT
(Until power is
removed from control)

❶ Igniter element becomes flame sensor on ignition systems with integral sensing.
★ See control label for actual times.

**Figure 47-8**
Sequence of operation. (Courtesy of Johnson Controls, Inc.)

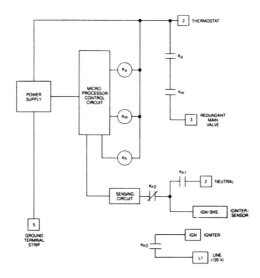

**Figure 47–9**

G750 wiring schematic (integral or remote sensing). (Courtesy of Johnson Controls, Inc.)

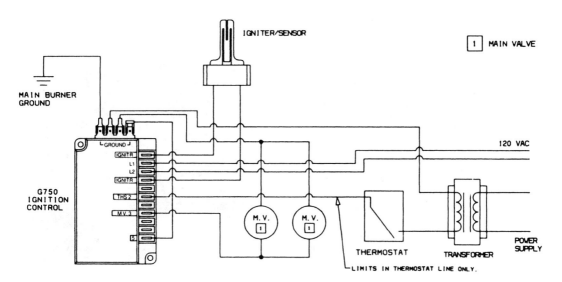

**Figure 47–10**

Connection diagram (integral models). (Courtesy of Johnson Controls, Inc.)

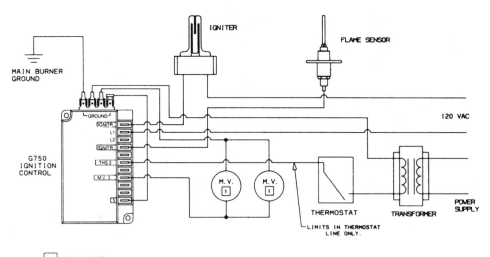

**Figure 47–11**
Connection diagram (remote sensing models). (Courtesy of Johnson Controls, Inc.)

**Figure 47–12**
Location of spark and sensor on Lennox G 14 series furnace. (Courtesy of Lennox Industries, Inc.)

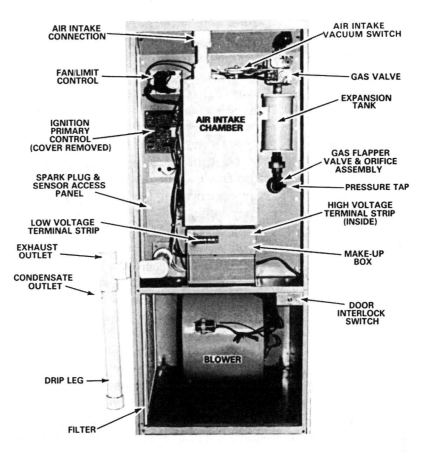

**Figure 47–13**
Location of primary control. (Courtesy of Lennox Industries, Inc.)

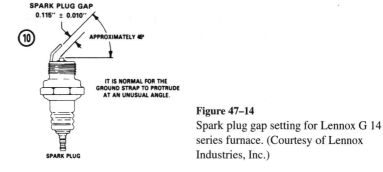

**Figure 47–14**
Spark plug gap setting for Lennox G 14 series furnace. (Courtesy of Lennox Industries, Inc.)

The sensor is also a type of spark plug that has a single center electrode (no ground strap) and compression rings to form a seal between it and the combustion chamber. It should not need any regular upkeep.

*Operation.*     Following is a description of the operating sequence for the Lennox spark-ignition system. See Figure 47–15.

1. The line voltage feeds through the blower access door interlock switch. The blower access panel must be in place before the circuit is made to the furnace components.
2. The transformer gives the 24 volts ac to the control circuit.
3. When the thermostat demands heat, the thermostat heating bulb contacts are closed.
4. The control circuit voltage passes from the W leg through the exhaust outlet pressure switch (Canadian Gas Association (CGA) only), the air intake vacuum switch (American Gas Association (AGA) and CGA units), and the limit control to energize the primary control.
5. The electric power then passes through the primary control to the purge blower. The blower is energized for about 30 seconds, giving a prepurge cycle.
6. At the end of the prepurge cycle, the purge blower continues to run. The gas valve, the fan control heater, and the ignition spark plug are all energized for about eight seconds.
7. The sensor senses, by flame rectification, whether the main burner gas is lit and deenergizes the spark plug and purge blower. The combustion process continues.
8. After about 30 to 45 seconds, the fan control contacts close to start the indoor fan motor on low speed.
9. When the thermostat is satisfied, the heating bulb contacts open. The primary control is deenergized, stopping the electrical power to the main gas valve and the fan control heater. At this time the purge blower is energized for a 30-second purge period. The indoor fan remains on.
10. When the recirculating air temperature reaches 90°F, the fan control contacts open, stopping the indoor fan motor.

**The 100-percent lockout module.**     The 100-percent lockout module is used with any non-100-percent shutoff ignition control. These modules should be used according to the manufacturer's recommendations. The purpose of this control is to give the lockout needed on these systems. The lockout function is intended to shut the system down completely if the pilot gas fails to ignite. To start a new ignition period the electric power must be opened for at least 30 seconds. See Figure 47–16.

**Spark ignitor.**     The spark ignitor is made of an inner electrode, a ceramic insulator, bracket, and ground strap. It is used to make a spark for direct ignition of the main burner gas. The gap between the electrode and the grounding strap should be set as recommended by the equipment manufacturer. The maximum temperature ratings, as designated by the unit manufacturer, should not be exceeded.

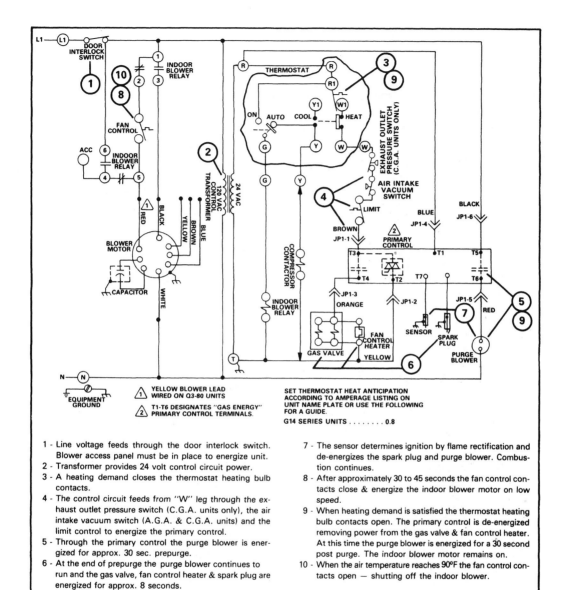

1 - Line voltage feeds through the door interlock switch. Blower access panel must be in place to energize unit.
2 - Transformer provides 24 volt control circuit power.
3 - A heating demand closes the thermostat heating bulb contacts.
4 - The control circuit feeds from "W" leg through the exhaust outlet pressure switch (C.G.A. units only), the air intake vacuum switch (A.G.A. & C.G.A. units) and the limit control to energize the primary control.
5 - Through the primary control the purge blower is energized for approx. 30 sec. prepurge.
6 - At the end of prepurge the purge blower continues to run and the gas valve, fan control heater & spark plug are energized for approx. 8 seconds.
7 - The sensor determines ignition by flame rectification and de-energizes the spark plug and purge blower. Combustion continues.
8 - After approximately 30 to 45 seconds the fan control contacts close & energize the indoor blower motor on low speed.
9 - When heating demand is satisfied the thermostat heating bulb contacts open. The primary control is de-energized removing power from the gas valve & fan control heater. At this time the purge blower is energized for a 30 second post purge. The indoor blower motor remains on.
10 - When the air temperature reaches 90°F the fan control contacts open — shutting off the indoor blower.

**Figure 47–15**
Sequence of operation schematic for Lennox G 14 series unit. (Courtesy of Lennox Industries, Inc.)

They are controlled by an automatic electronic ignition module like any of those previously described. The same requirements and specifications must be followed for satisfactory operation.

### Retrofit, intermittent pilot, gas burner ignition systems.

Retrofit, intermittent pilot, gas burner ignition systems are complete kits consisting of the various parts needed to change a conventional standing pilot system to a cycling pilot control system. See Figure 47–17. These kits are energy-saving controls that can be mounted on furnaces, boilers, and other types of controlled heating systems that use a 24-volt ac control circuit; otherwise, the control system must be changed to 24 volts before the system can be used.

Pilot conversion kits are available for either natural gas or LP gas. Be sure to use the correct one for the job and follow the manufacturer's specifications to ensure satisfactory operation.

### Pilot relight control.

Pilot relight controls generate spark impulses to light the pilot gas. The relight control generates sparks until a pilot flame is sensed between the electrode and the ground. The flame is sensed through flame conduction (ability of a flame to conduct a current). When a flame is sensed and current flows between the electrode and the pilot burner ground, the relight control stops sparking. If the flame goes out during the demand for heat, the relight control begins sparking the instant the flame goes out. See Figure 47–18.

### Flame rectification sensor.

The flame rectification sensor is made from an electrical conducting material supported by a ceramic insulator and mounting bracket. See Figure 47–19. The purpose of the sensor is to detect the presence of a main burner flame. It is mounted on the main burner so that when the main burner gas is lit, the flame

**Figure 47–16**
The 100-percent lockout module. (Courtesy of Johnson Controls, Inc., Control Products Division.)

**Figure 47–17**
Retrofit, intermittent pilot, gas burner ignition system. (Courtesy of Honeywell, Inc.)

**Figure 47–18**
Pilot relight control. (Courtesy of
Adams Manufacturing Company.)

**Figure 47–19**
Flame rectification sensor. (Courtesy of
Honeywell, Inc.)

will surround the sensor and let the electric current flow through the flame to the ground. These sensors do not need temperature to operate. They need only electrical continuity. Therefore, the flame must touch both the sensor and the main burner simultaneously.

**Ignitor sensor.**    Ignitor sensor controls are usually nonprimary-aerated combination pilot burner and ignitor. They are used with the controls recommended by the ignitor sensor manufacturer.

# Summary

- The automatic gas burner ignition system automatically and instantly lights or relights the burner gas as needed for that particular job.
- The low-voltage electricity is supplied by a class 2 transformer and must supply a minimum of 24 volts ac, 60 hertz during the ignition and running cycles.
- Three major types of electronic spark ignitors are (1) those that light the pilot gas with a spark across electrodes, (2) those that light the main burner gas with a spark across electrodes, and (3) those that light the main burner gas with a spark-plug type ignitor.
- The IID system changes the standing pilot by using solid-state electronic circuitry and a flame sensor to replace the thermocouple and pilot safety control normally used on standing pilot systems.
- Heat is not needed with the IID system because flame conduction, or rectification, is used.
- The intermediate cone is most important here because this is where the combustion takes place and where the best flame sensing is possible.
- Ionization causes the surrounding area to become conductive. This conductive property is what is so important to correct flame conduction.
- After ignition of the pilot gas, a dc microamp current flow is conducted through the flame, from the flame to the sensor (the positive probe), and to the pilot tip—which in this case acts as the negative probe—completing the circuit to ground.

- The sensing circuit uses a dc current flow to energize a relay that energizes the main gas valve.
- The gas pressure to the unit should be a minimum of 1-inch WC above the equipment manufacturer's recommended manifold pressure.
- The pilot and sensor are the most important features of the IID application.
- If the pilot flame is hard and blowing, the grounding area of the pilot is reduced so that the necessary current is not flowing. A shutdown of the system follows.
- Positioning of the flame sensor is also critical in pilot installation. Placement of the flame sensor should be such that it will touch the second, or combustion, area of the pilot flame.
- In the HSI system the flame may be sensed by either a remote sensor or an integral sensor.
- When an HSI system is locked out, the circuit to the control module must be broken for at least 30 seconds to reset the control.
- The ignition spark plug and the sensing plug cannot be interchanged because they have different thread diameters.
- The purpose of the 100-percent lockout module is to provide the necessary safety on this type system.
- The purpose of the spark ignitor is to create a spark for direct lighting of the main burner gas.

# Service Call

A customer complains that her central heating unit is not heating. The system reveals that it has an electronic ignition system and neither the pilot nor the main burner is lit. The technician turns the thermostat system switch to the OFF position for 30 seconds to check if the ignition module will reset, and then turns the system on but the ignitor does not produce a spark. The wiring is in good condition. The grounding where the ignitor is fastened to the main burner is rusted. The tech removes the rust to give a good ground, resets the module, and the ignitor sparks and lights the pilot burner. After lighting the main burner gas and ensuring that the system is safe and operating correctly, the technician is satisfied that the system is repaired.

# Student Troubleshooting Problem

A customer complains that his central heating system is not working. The system has an IID electronic ignition system. Neither the pilot nor the main burner is lit. The tech places the thermostat system switch in the OFF position for 30 seconds to see if the module will reset, and then turns the thermostat back on. The ignitor still does not attempt to light the pilot gas. A check of the wiring for the module shows the high tension lead between the module and the spark ignitor is corroded in the socket on the module. The tech cleans the connection and replaces the wire into the socket. The module is also found to be cracked. What could be the problem? How is the system restarted after the repairs have been made?

# Questions

1. What control is used to light a standing pilot?
2. Why are purge time delays used on automatic burner ignition systems?
3. On what change in temperature does a heating thermostat contact close?
4. How hot must the flame sensor be that is used on electronic ignition systems for the system to operate?
5. Where should the flame-sensing probe be located in the flame?
6. How does a flame conduct an electric current?
7. What type of current flows through an electronic ignition system sensing circuit?
8. What is the needed voltage range for 24-volt ignition systems?
9. How is the current flowing through the pilot flame confirmed?
10. On electronic ignition systems, what happens when the pilot flame is blowing?
11. In the HSI system, what is meant by integral sensing?
12. Why is a warm-up time necessary for HSI systems?

# Unit 48: SOLID-STATE OIL BURNER CONTROLS

## Introduction

The main part in an oil burner control circuit is the primary control. It is used to oversee the system and make sure it is safe. When the thermostat demands heat, the primary control ensures that the flame is present in the correct amount of time. It controls the burner motor, ignition transformer, and the oil valve. A sensor in the primary control assembly senses the burner flame on start-up and throughout the burner ON cycle. If the flame fails for any reason, or if a power failure occurs, the primary control shuts down the burner.

**Kwik-sensor cad-cell (protectorelay) burner controls.** Kwik-sensor cad-cell relay burner controls are used on either intermittent or interrupted oil burner operation. They are used to control the ignition of oil burner systems. See Figure 48–1. Cad-cell relays are combined with certain types of controls—be sure to use the proper combination for correct operation. The manufacturer gives the correct supporting controls for the type being used. These controls use a solid-state flame-sensing circuit. They can be manually reset after a system shutdown. Low-voltage terminals are used to simplify the wiring process during installation. Most of them use a transformer for the control circuit low voltage. Each model has its own wiring diagram for its particular use in the system. See Figure 48–2.

**Kwik-sensor combination oil burner–hydronic control.** The kwik-sensor combination oil burner–hydronic controls use an immersion type aquastat controller and an oil burner primary control to give high- and low-limit/circulator control for oil-fired

**Figure 48–1**
Kwik-sensor cad-cell relay. (Courtesy
of White-Rodgers Division, Emerson
Electric Co.)

hydronic heating systems. See Figure 48–3. These controls give intermittent (formerly
called constant) ignition of the fuel oil. This control is used with specific supporting controls. Ask the manufacturer for the correct types as some models mount directly on the
burner whereas others mount externally on the equipment. The controls have an armored
capillary and a remote sensor. Most are capable of multiple-zone control with use of the
correct valves.

**Cadmium sulphide flame detector.**    Cadmium sulphide flame detectors are
photoconductive flame-sensing controls used for staging oil burner systems. These
detectors are mounted so they can sense the oil burner flame and then signal the combination oil burner–hydronic control on flame detection or failure. See Figure 48–4.
Cadmium sulphide flame detectors are glass-to-metal plug-in cells. They are hermetically sealed to prevent deterioration by humidity, soot, or oil fumes. The lead wires are
NEC class 1 type.

*Operation.*    On flame failure, the light-sensitive cadmium sulphide cell, with the
flame-sensing circuitry, causes the combination oil burner–hydronic control to shut down
the main oil burner. These controls have circuits to bypass the cadmium cell for a given
time on reignition of the oil burner or until the flame is sensed by the flame detector. The
oil burner then operates in its normal mode.

**Burner safety controls.**    Burner safety controls are the same safety controls
used on warm-air heating systems and steam or hot water boilers. They will stop the burner
in case of high pressures or temperatures due to the lack of water or airflow. They are wired
into the system in the same way as the fan and limit controls discussed earlier. A typical
wiring diagram is shown in Figure 48–5.

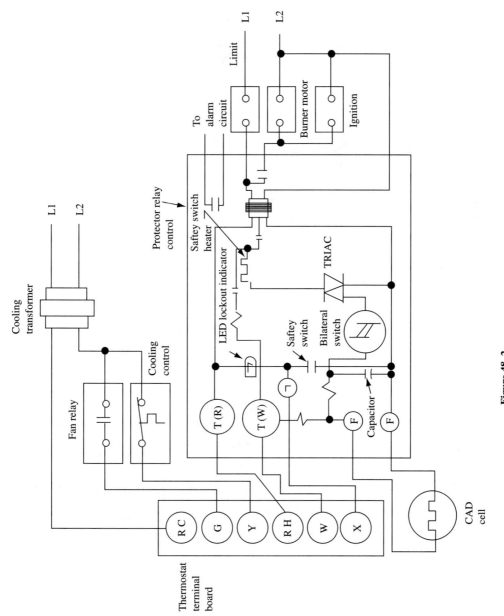

**Figure 48–2**
Typical internal schematic and hookup.

**Figure 48–3**
Kwik-sensor combination oil burner–hydronic control. (Courtesy of White-Rodgers Division, Emerson Electric, Co.)

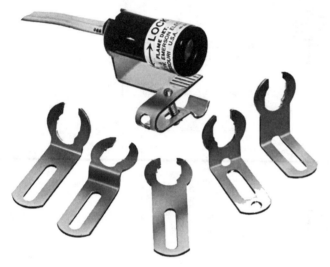

**Figure 48–4**
Flame detector. (Courtesy of White-Rodgers Division, Emerson Electric Co.)

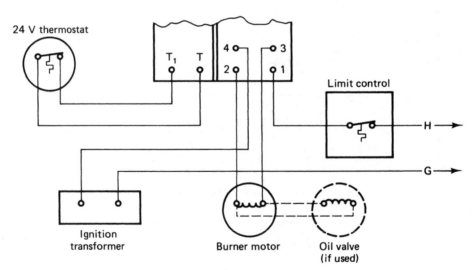

**Figure 48–5**
Wiring diagram with burner safety control.

# Summary

- The fundamental part in an oil burner circuit is the primary control. It is used to sense system operation and to ensure that it is safe.
- A sensor in the primary control assembly monitors the burner flame on start-up and throughout the burner ON cycle.
- Kwik-sensor cad-cell relays are designed to be used in combination on either intermittent or interrupted operation of oil burners.
- Cad-cell relays are used in combination with certain type controls.
- The kwik-sensor combination oil burner–hydronic controls use an immersion type aquastat controller and an oil burner primary control to give high- and low-limit/circulator control for oil-fired hydronic heating systems.
- Cadmium sulphide flame detectors are photoconductive flame-sensing devices used for staging oil burner systems.

# Service Call

A customer complains that his hydronic heating system is not operating. A check of the system shows that the building is heated by an oil burner with a cadmium sulphide flame detector. The system is off. The power to the relay has line voltage. The tech resets the primary control, but the system still does not operate. The location of the cad cell is satisfactory. The glass, however, is covered with soot, so the tech removes the soot from the glass and places the control in the start mode. It starts operating normally. The complete system is now operating satisfactorily. The technician is satisfied that the system is repaired.

# Student Troubleshooting Problem

A customer complains that an oil burner is not heating her residence. A check of the system shows that the protectorelay has the system off. After the relay is reset, the burner tries to start but there is no ignition spark. Voltage is at the transformer primary terminals, but a check of the electronic ignition circuit shows no current flow while the system is trying to start. What could be the problem?

# Questions

1. What does an oil burner primary control do?
2. What is the purpose of the sensor in the primary control?
3. What is the purpose of the kwik-sensor cad-cell relay burner control?
4. When a cad-cell relay shuts down a system, what must be done before it can be restarted?
5. Which control can give multiple-zone control?
6. Which control is a photoconductive flame sensor?

# Unit 49: DISCHARGE GAS TEMPERATURE PROTECTOR

## Introduction

High discharge gas temperature is a primary cause of early compressor failure. The protector helps stop early compressor damage or failure caused by high discharge gas temperatures. They may be used on either rotary or reciprocating compressors.

**Discharge gas temperature protector.** The discharge gas temperature protector is a solid-state control module used for sensing critical compressor discharge gas temperature. See Figure 49–1.

These controls have a built-in four-minute time delay to guard the unit against short cycling. The module has an adjustment dial for changing the cutout temperature from 175°F to 315°F. Discharge gas temperature protectors can be used on either 24-volt or 208/240-volt control circuits. The discharge gas temperature is sensed through an externally mounted thermistor. See Figure 49–2.

The universal mount thermistors are clamped directly to the discharge gas tube and sense the temperature of the refrigerant. An LED warning light indicates a high temperature. See Figure 49–3. During normal operation of both air-conditioning and refrigeration systems, if the discharge gas temperature goes higher than the module temperature set point for 15 seconds, the LED will light. The unit will be stopped until the discharge line temperature drops 65°F below the module set point and a four-minute time delay has passed. When these two conditions are met, the LED will go off and the unit will restart. The module will repeat this cycle until the problem has been corrected.

The major use for the discharge gas temperature protector is to protect residential and commercial air-conditioning and refrigeration compressors against high discharge gas temperatures.

**Applications.** *Residential and Commercial Air-conditioning (24-volt control system).* When the thermostat demands cooling, the unit is started. When the thermostat is

**Figure 49–1**
Discharge gas temperature protector.
(Courtesy of Motors & Armatures, Inc.)

**Figure 49–2**
Externally mounted thermistor.
(Courtesy of Motors & Armatures, Inc.)

satisfied, or another control opens, the compressor is stopped and there will be a lockout of four minutes before the unit can be started again. See Figure 49–4.

*Commercial Refrigeration (208/240 Volt).*    When the thermostat demands cooling, the contactor holding coil is energized. The module will open the control circuit if a switch or another control opens and closes before the four-minute time delay has passed or if the temperature of the discharge line is greater than the module temperature set point. See Figure 49–5. The electrical connections for the discharge gas temperature protector are shown in Figure 49–6.

*Light Commercial Refrigeration.*    For systems that do not use a compressor contactor, when the thermostat demands cooling, the system will operate as usual. The module will open the circuit if a switch opens and closes before the four-minute time delay has passed or if the temperature of the discharge line is higher than the module set point.

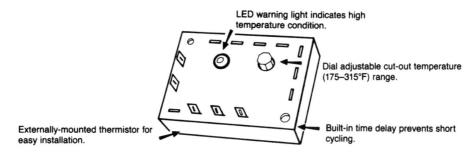

**Figure 49–3**
Discharge gas temperature protector features. (Courtesy of Motors & Armatures, Inc.)

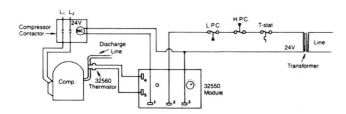

**Figure 49–4**
Twenty-four-volt control circuit using a discharge gas temperature protector. (Courtesy of Motors & Armatures, Inc.)

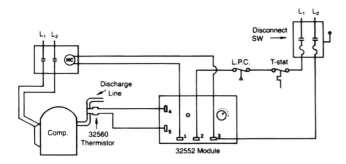

**Figure 49–5**
208/240-volt circuit using a discharge gas temperature protector. (Courtesy of Motors & Armatures, Inc.)

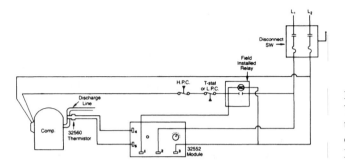

**Figure 49-6**
Wiring diagram for a discharge gas temperature protector without compressor contactor. (Courtesy of Motors & Armatures, Inc.)

# Summary

- High discharge gas temperature is a major cause of early compressor damage or failure.
- The discharge gas temperature protector is a solid-state control module used for sensing critical compressor discharge gas temperature.
- The module has a dial for changing the cutout temperature from 175°F to 315°F.
- A high temperature is shown by an LED warning light.

# Service Call

A customer complains that his commercial refrigeration unit is getting too warm. A check of the system reveals that the compressor with an air-cooled condenser is not running. The unit has a discharge gas temperature protector. The temperature of the compressor is warmer than normal. The LED on the discharge gas temperature protector is lit. The unit is allowed to cool until the four-minute time delay has passed. The unit starts, but the condenser fan does not turn as fast as it should. The technician checks the continuity of the motor and finds it within the recommended range. The fan motor capacitor is weak so the tech replaces it and puts the unit back in operation. The unit is now operating normally. The technician is satisfied that the system is repaired.

# Student Troubleshooting Problem

A customer complains that her residential air-conditioning system is not cooling. The system is an older, three-ton air-cooled unit that has a discharge gas temperature protector. The compressor is not running. The line side of the contactor shows 240 VAC. The contactor contacts are open. When the thermostat is set below room temperature, the contactor still does not close. There is no voltage at the contactor coil control terminals. A further check finds that the discharge gas temperature protector is open. The tech replaces the discharge gas temperature protector and turns the system on, and the contactor closes. The circuit breaker to the unit immediately trips. A continuity test of the compressor windings shows that the start winding has infinite resistance. The tech then replaces the compressor using the correct procedures, turns the unit back on, and it starts running. After the system is recharged with HCFC-22 refrigerant, the suction pressure is 80 psig and the return air temperature is 85°F. The discharge pressure is 390 psig. The outdoor temperature of 95°F. What could have caused the compressor to burn?

# Questions

1. What is the time-delay period of a discharge gas temperature protector?
2. Why is the time delay needed on the discharge gas temperature protector?
3. What is the cutout temperature range of the discharge gas temperature protector?
4. What alerts the technician that the discharge gas protector is open?

# Unit 50: DELAY-ON-MAKE ADJUSTABLE SOLID-STATE TIMERS

## Introduction

Solid-state delay-on-make timers are ideal for staging compressor start-up or delaying motors and other components that would put an overload on the electrical system.

*Operation.*   When the power is turned on, the timing is started for the chosen delay period. After the delay period has passed, the timer makes the circuit. These controls are available in either adjustable or nonadjustable models. See Figure 50–1. Some may be used on a wide range of voltages which allows them to be used on any job. See Table 50–1. Most delay-on-make timers have a simple two-wire connection into the control circuit. See Figure 50–2.

**Figure 50–1**
Delay-on-make adjustable solid-state timers. (Courtesy of Motors & Armatures, Inc.)

**TABLE 50–1**  SELECTION TABLE FOR DELAY-ON-MAKE ADJUSTABLE SOLID-STATE TIMERS

| Mars no. | Time-delay range | Type of adjustment | Input voltage | Hz | Max. amps. | Voltage drop |
|---|---|---|---|---|---|---|
| 32391 | 6 sec . . . 8 min. | dial | 19 . . . 288V AC or DC | 50/60 | 1 | 2.5 at 1 amp. |
| 32394 | 1 sec . . . 1023 sec. | slideswitch | 19 . . . 144V AC or DC | 50/60 | 1 | 2.5 at 1 amp. |
| 32396 | 1 sec . . . 1023 sec. | slideswitch | 80 . . . 277V AC or DC | 50/60 | 1 | 2.5 at 1 amp. |
| 32397 | 10 . . . 1000 sec. | dial | 24 VAC | 50/60 | 6 | 2 at 6 amp. |
| 32398 | 10 . . . 1000 sec. | dial | 120 VAC | 50/60 | 10 | 2.1 at 10 amp. |
| 32399 | 10 . . . 1000 sec. | dial | 230 VAC | 50/60 | 10 | 5 at 10 amp. |
| 32367 | 240 sec. fixed | — | 19 . . . 288V AC or DC | 50/60 | 1 | 2.5 at 1 amp. |

(Courtesy of Motors & Armatures, Inc.)

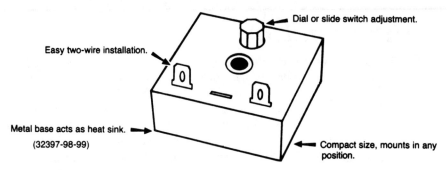

**Figure 50–2**
Delay-on-make adjustable solid-state timer features. (Courtesy of Motors & Armatures, Inc.)

**Uses.**   More popular uses of the delay-on-make timers include: (1) compressor or equipment staging, (2) stopping contactor chatter, (3) evaporator fan delay after defrost, (4) preventing oil burner puffback on steam heating systems, (5) anti-short cycling of compressors, and (6) anti-short cycling for 24-volt air-conditioning control circuits with no interference with thermostat anticipator operation.

**Applications.**   *Commercial and Industrial Compressor and Motor Staging.* In commercial installations where multiple compressors are used for air-conditioning or refrigeration, a start-up of all the equipment at the same time can damage the components because of a low-voltage condition caused by temporarily overloading the electrical system. Starting all the equipment simultaneously can also happen with night shutoff systems or in case of power failure. By installing a delay-on-make timer with a different chosen delay period for each compressor, the starting can be staged or staggered when the circuit is made. See Figure 50–3.

   ***Fan Delay for Commercial Refrigeration System.***   A high-amp timer may be installed in commercial refrigeration systems where the defrost timer does not have a built-in fan delay. See Figure 50–4.

   When the defrost is completed, the timer delays start the evaporator fan for a chosen time to keep from blowing moisture or heat over the product, thus lowering the buildup of frost and food spoilage.

   ***Residential, Commercial, and Industrial Motor Staging without Contactor Coils.*** High-amp delay-on-make timers are used to stage those motors that do not use contactors, thus reducing the locked rotor amperage (LRA). See Figure 50–5.

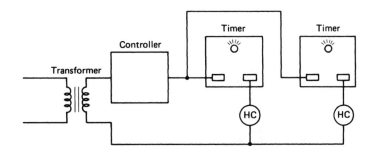

**Figure 50–3**
Delay-on-make adjustable solid-state timer wiring diagram for compressor and motor staging. (Courtesy of Motors & Armatures, Inc.)

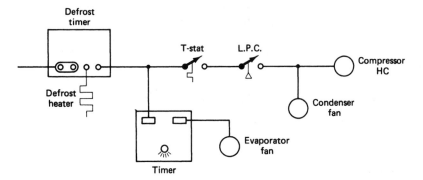

**Figure 50–4**
Delay-on-make adjustable solid-state timer wiring diagram for fan delay. (Courtesy of Motors & Armatures, Inc.)

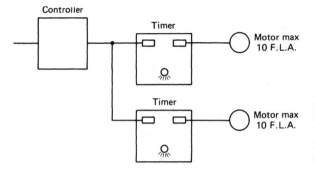

**Figure 50–5**
Delay-on-make adjustable solid-state timer wiring diagram for staging motors without a contactor coil. (Courtesy of Motors & Armatures, Inc.)

# Summary

- Solid-state delay-on-make timers are ideal for staging compressor start-up or delaying the start-up of motors and other components that would temporarily overload the electrical system.
- Most delay-on-make timers have a simple two-wire connection into the control circuit.
- In commercial applications where multiple compressors are used for air-conditioning or refrigeration, simultaneous start-up of all the equipment can damage the components because of low voltage.

# Service Call

A customer complains that the products kept in a commercial refrigeration unit keep spoiling. The refrigeration system has voltage, amperage, and refrigerant pressures and is operating correctly. However, there is no evaporator fan delay after the defrost period which allows warm, moist air to be blown over the stored product. A defrost timer has not been used in this control circuit. After informing the operator of the problem, the technician gets an okay to install an evaporator fan time delay. With the system back in operation, the unit runs for 24 hours before it is checked again. The temperature and humidity inside the cabinet are good. The technician is satisfied that the system is repaired.

# Student Troubleshooting Problem

A customer complains that his commercial refrigeration unit is not working. The system has a delay-on-make timer, high- and low-pressure controls, and the needed overload protectors but the system is not running. Voltage to the contactor line terminals is 240 VAC. The contactor is not pulled in. There is no voltage at the contactor holding coil control circuit terminals. A further check shows that the contacts in the delay-on-make relay are open. The relay heater coil terminals show 240 VAC. The technician replaces the relay and turns the unit on. It starts running and cooling. What could be the problem with the relay?

# Questions

1. How does a time-delay stop contactor chatter?
2. Why is it necessary to delay commercial refrigeration evaporator fan operation after a defrost period?
3. How can the simultaneous start-up of several compressors cause low voltage?

# Unit 51: DELAY-ON-BREAK ADJUSTABLE SOLID-STATE TIMERS

## Introduction

Delay-on-break timers control the time after which a circuit has been broken and before it may be made again. These timers are used to prevent short cycling of air-conditioning, refrigeration, and heat pump compressors.

*Operation.*    When power is applied, the unit is started. When the thermostat is satisfied or a brief loss of power occurs, the unit is stopped. The delay period begins. The compressor will not start until the delay period has passed. These timers are available in both single and adjustable voltage models. See Figure 51–1.

Some time-delay relays have two-wire connection and others have three-wire connection into the control circuit. See Figure 51–2. Models are available that use electronic circuitry to allow continuous current flow through the thermostat anticipator, and are ideal for use on heat pump systems. Several tables are available to help the technician choose the correct model for a specific use, for example, see Table 51–1.

**Uses.**    More popular uses of the delay-on-break solid-state timers include: (1) compressor or equipment staging, (2) stopping contactor chatter, (3) an evaporator fan delay after defrost, (4) preventing oil burner puffback on steam heating systems, (5) anti-short cycling of compressors, and (6) anti-short cycling for 24-volt air-conditioning control circuits having no problems with the thermostat anticipator.

**Applications.**   *Anti-Short Cycling of Residential and Commercial Compressors.* To keep compressors from short cycling, the delay-on-break timer starts a delay period and lockout on the loss of power, or keeps the unit from starting, until the time-delay period has passed. If the thermostat demands during the delay period, the compressor still cannot be started for that period, thereby keeping the compressor from short cycling. See Figure 51–3.

*Easy Two-Wire Connection for Anti-Short Cycling.*    Some delay-on-break timers allow for easy two-wire installation. See Figure 51–4.

**Figure 51–1**
Delay-on-break adjustable solid-state timers. (Courtesy of ICM Corporation.)

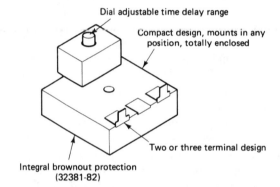

Dial adjustable time delay range

Compact design, mounts in any position, totally enclosed

Two or three terminal design

Integral brownout protection
(32381-82)

**Figure 51–2**
Delay-on-break adjustable solid-state timer features. (Courtesy of Motors & Armatures, Inc.)

**TABLE 51–1** SELECTION TABLE FOR DELAY-ON-BREAK ADJUSTABLE SOLID-STATE TIMERS

| Mars no. | Time-delay range | Type of adjustment | Input voltage | Hz | Max. amps. | Voltage drop |
|---|---|---|---|---|---|---|
| 32392 | 6 sec. . . . 5 min. | dial | 19 . . . 288V AC or DC | 50/60 | 1 | 2.5 at 1 amp. |
| 32387 | 120 . . . 300 sec. | dial | 24 VAC | 50/60 | 1 | 3 at 1 amp. |
| 32388 | 120 . . . 300 sec. | dial | 120 VAC | 50/60 | 1 | 3 at 1 amp. |
| 32389 | 120 . . . 300 sec. | dial | 230 VAC | 50/60 | 1 | 3 at 1 amp. |
| 32381 | 3 min. fixed | — | 24 VAC | 50/60 | 1 | 2.5 at 1 amp. |
| 32382 | 5 min. fixed | — | 24 VAC | 50/60 | 1 | 2.5 at 1 amp. |

(Courtesy of Motors & Armatures, Inc.)

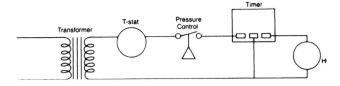

**Figure 51–3**
Delay-on-break wiring diagram for anti-short cycling of compressors. (Courtesy of Motors & Armatures, Inc.)

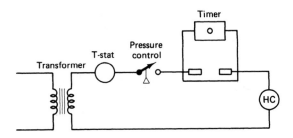

**Figure 51–4**
Delay-on-break wiring diagram for use with two-wire connection for anti-short cycling. (Courtesy of Motors & Armatures, Inc.)

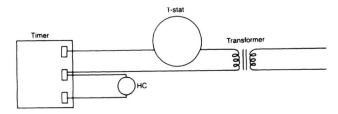

**Figure 51–5**
Delay-on-break wiring diagram for anti-short cycling with brownout protection. (Courtesy of Motors & Armatures, Inc.)

*Anti-Short Cycling with Brownout Protection.* Some delay-on-break timers are used to give brownout protection from heat anticipator-compensated resistor type room thermostats. The delay-on-break lockout interrupts the circuit to keep the compressor from starting when the voltage is below normal values. See Figure 51–5.

# Summary

- Delay-on-break timers control the time during which a circuit has been broken and before it can be made again.
- When the thermostat is satisfied or a short time loss of power happens, the load is stopped and the delay period begins.
- To keep the compressor from short cycling, the delay-on-break timer starts a time delay and lockout on the loss of power, or keeps the unit from starting again until the time-delay period has passed.
- Some delay-on-break timers give brownout protection when heat anticipator-compensated resistor type room thermostats are used.

# Service Call

A customer complains that a commercial refrigeration unit is not keeping the product cold enough. A check of the system shows that the compressor is off, and a further check shows that the delay-on-break timer has the unit off. The voltage to the timer is 240 volts. There is no voltage from the timer to the compressor or no control voltage to the timer motor. When the timer is advanced to close the contacts, there is still no voltage from the timer. The technician deems the timer bad and replaces it with the correct replacement, starts the unit, and checks the voltage, amperage, and refrigerant pressures. After a complete check of the system, the technician is satisfied that it is repaired.

# Student Troubleshooting Problem

A customer complains that her residential air-conditioning unit will not turn off. A check of the system shows that only the condensing unit is running. The system has a delay-on-break timer. The indoor fan is off and the evaporator is iced over. The thermostat temperature lever is set above room temperature. The condensing unit still runs. There is control voltage to the relay control terminals and across the relay contacts. What could be the problem?

# Questions

1. For what are delay-on-break timers used?
2. What happens when the thermostat demands during a delay period?
3. What is the purpose of brownout protection?

# Unit 52: OFF-DELAY/ON-DELAY TIMERS

## Introduction

Off-delay/on-delay timers are used to either keep the unit running for a chosen time after a switch or thermostat opens (off-delay) or start the unit for a chosen time after the power comes on (on-delay), depending on the type of use. Convenient control of a variety of functions is given. Some have a fixed two-minute delay with a 24-volt timer used for off-delay postpurging of gas furnace combustion chambers.

*Operation.* The timer causes a ventilation fan to continue running after the thermostat is satisfied. This delay is to blow all the conditioned air from the distribution ducts. These timers also act as brief on-delays by starting the load for a chosen period when the power is turned on (when wired in series with a switch). See Figure 52–1.

High-amp timers are available that, depending on the use, will do either on-delay or off-delay functions. On-delay can be used for fan delay, pump-down systems, or postpurge

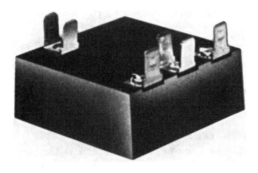

**Figure 52–1**
Off-delay/on-delay timer. (Courtesy of
Motors & Armatures, Inc.)

Fixed 2-minute time delay (32393)

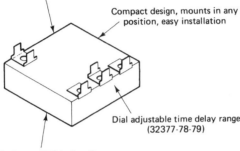

Compact design, mounts in any
position, easy installation

Dial adjustable time delay range
(32377-78-79)

Maximum 10 F.L.A. rating
(32378-79) for direct control of high
amp resistive and inductive loads

**Figure 52–2**
Off-delay/on-delay timer features.
(Courtesy of Motors & Armatures, Inc.)

jobs. See Figure 52–2. Several tables are available to help the technician choose the correct off-delay/on-delay timer for the job, for example, see Table 52–1.

**Uses.**    More popular uses of off-delay/on-delay timers include: (1) residential and commercial air-conditioning, heat pump, and forced-air electric heating systems, (2) hot water heating systems, (3) compressor unloader valves, and (4) residential and commercial ventilation fans.

**Applications.**    *Residential, Light Commercial Off-Delay Postpurging.*    For air-conditioning, heat pump, and forced-air electric heating systems, timers (when wired in parallel with the switch or thermostat) operate as off-delay, causing the fans to run for a chosen time after the thermostat is satisfied. Any conditioned air in the system is blown into the conditioned space. See Figure 52–3.

*Residential, Commercial Hydronic On-Delay.*    For circulator motors, a second motor is often used to help in start-up and to overcome the system inertia. When the timer is installed in series with the switch and the motor, it operates as on-delay, causing the motor to help in start-up for a chosen time and then cycle off. See Figure 52–4.

**TABLE 52-1**  SELECTION TABLE FOR OFF-DELAY/ON-DELAY TIMERS

| Mars no. | Type delay | Time-delay range | Type of adjustment | Input voltage | Hz | Max. amps. | Voltage drop |
|---|---|---|---|---|---|---|---|
| 32393 | off-delay | 2 min. fixed | — | 24 | 50/60 | 1 | 2 at 1 amp. |
| 32377 | off delay/ on delay | 6 sec. . . . 10 min. | dial | 24 VAC | 50/60 | 6 | 2 at 6 amp. |
| 32378 | off delay/ on delay | 6 sec. . . . 10 min. | dial | 120 VAC | 50/60 | 10 | 2.1 at 10 amp. |
| 32379 | off delay/ on delay | 6 sec. . . . 10 min. | dial | 230 VAC | 50/60 | 10 | 5 at 10 amp. |

(Courtesy of Motors & Armatures, Inc.)

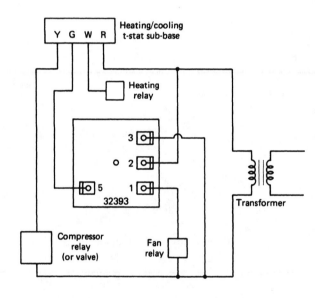

**Figure 52–3**
Wiring diagram for off-delay postpurging using off-delay timer. (Courtesy of Motors & Armatures, Inc.)

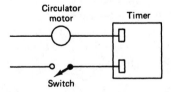

**Figure 52–4**
Wiring diagram for hydronic on-delay using an off-delay timer. (Courtesy of Motors & Armatures, Inc.)

# Summary

- Off-delay/on-delay timers are used to either keep the unit on for a chosen time after which the switch or thermostat opens (off-delay) or energize the unit for a chosen period after the power comes on (on-delay), depending on the type of installation.
- The timer causes a ventilation fan to continue running after the thermostat is satisfied to blow all the conditioned air from the ducts.
- On-delay can be used for fan delay, pump-down systems, or postpurge jobs.

# Service Call

A customer complains that it takes a hot water system an eternity to start blowing warm air. A check of the system shows that the boiler is operating correctly. The technician turns the thermostat below room temperature to stop the circulating pump. After the system becomes static, the tech then turns the room thermostat above room temperature. The circulating pump starts but does not immediately cause the water to circulate through the system. The customer claims that the system has always operated in this way, but permits the tech to install a second circulating pump and an on-delay relay. After completing the work, the tech puts the system back in operation and sets the thermostat above room temperature—almost immediately warm air blows into the space. The technician is satisfied that the system is repaired.

# Student Troubleshooting Problem

A customer complains that a residential heating unit is not heating his home. The system is a forced-combustion air system that uses an off-delay/on-delay timer to control the combustion blower. When the thermostat temperature lever is set above room temperature, the unit does not start. The control circuit has voltage to the relay coil and across the relay contacts. What could be the problem?

# Questions

1. What is the usual amount of time delay with a 24-volt postpurge unit?
2. Can off-delay/on-delay be used for system capacity control?
3. What type of time delay is used to cause the indoor fan to run after the unit is off?
4. Why would delayed fan operation on an air-conditioning unit be helpful?

# Unit 53: BYPASS TIMERS
## Introduction

Bypass timers are used when a switch or control is to be bypassed until the system reaches a certain point in operation (e.g., until a desired pressure level or temperature is reached).

Some bypass timers are adjustable and allow for multivoltage use for the temporary bypass of a control or component during start-up. At the end of the delay period, control is returned to the system control circuit. See Figure 53–1.

**Operation.**    On system start-up, the specific control is bypassed. After the chosen time-delay period has passed, control is returned to the control circuit. See Figure 53–2. Several tables are available to help the technician choose the correct bypass timer for the job, for example, see Table 53–1.

**Uses.**    More popular uses of the bypass timer include: (1) commercial refrigeration systems, (2) commercial air-conditioning systems, (3) chiller systems, (4) hydronic heating systems, and (5) ice makers.

**Applications.**    *Commercial Refrigeration and Air-Conditioning Low-Pressure Bypass.*    These timers bypass the low-pressure control during start-up. When the circuit is made, it starts a delay before going through the control contacts. This keeps the compressor circuit from opening because of false low-pressure readings, as with an outdoor refrigeration unit that is operated on a day with a low outdoor temperature. By installing the bypass timer, the compressor has time to build up pressure before the low-pressure control is placed in the circuit. See Figure 53–3.

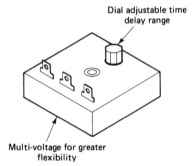

**Figure 53–1**
Bypass timer. (Courtesy of Motors &
Armatures, Inc.

**Figure 53–2**
Bypass timer features. (Courtesy of
Motors & Armatures, Inc.)

**TABLE 53–1**   TABLE FOR SELECTING BYPASS TIMERS

| Mars no. | Time-delay range | Type of adjustment | Input voltage | Hz | Max. amps. | Voltage drop |
|---|---|---|---|---|---|---|
| 32395 | 6 sec . . . 8 min. | dial | 19 . . . 288V AC or DC | 50/60 | 1 | 2 at 1 amp. |

(Courtesy of Motors & Armatures, Inc.)

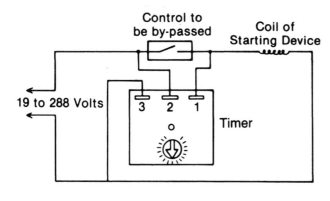

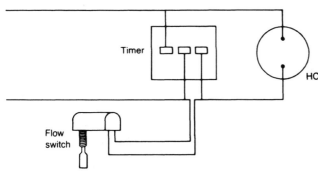

**Figure 53–3**
Wiring diagram for refrigeration and air-conditioning low-pressure bypass. (Courtesy of Motors & Armatures, Inc.)

**Figure 53–4**
Wiring diagram for commercial water flow control. (Courtesy of Motors & Armatures, Inc.)

*Commercial Water Flow Control.*    To avoid nuisance trips of the heating control (HC) because of a brief loss of flow, the timer bypasses the flow switch for a chosen time delay when the switch contacts open. If the flow switch contacts do not reclose before the time delay has passed, the timer will deenergize the HC. To reset the circuit, turn off the power supply and then turn it back on. See Figure 53–4.

# Summary

- Bypass timers are used where a switch or control is to be bypassed until the system reaches a certain level of operation (e.g., until a wanted pressure or temperature is reached).
- On system start-up, the control is bypassed. After the chosen time delay period has passed, the control is placed back in the circuit.

# Service Call

A customer complains that her commercial refrigeration unit will start, then stop after a few minutes of operation. The system has an air-cooled condensing unit on the roof that uses a bypass timer. The outdoor temperature is 45°F. The compressor starts, then stops on the low-pressure control. The bypass timer has control voltage at the clock motor. The inlet of the timer has line voltage, but not the outlet. A check of the timer coil resistance shows that the coil is open. The bypass timer is bad. The tech replaces the timer with the correct replacement and puts the unit back in operation. The compressor runs until the low pressure is built up, then the bypass timer puts the low-pressure control back into the control circuit. After the unit is completely checked and found to be operating satisfactorily, the technician is satisfied that it is repaired.

# Student Troubleshooting Problem

A customer complains that an air-conditioning system in a large office building is not keeping the building cool. The system has an air-cooled unit located on the roof. It has a bypass timer. The outdoor temperature is 45°F. When the thermostat temperature lever is set below room temperature, the unit still does not come on. A check of the control system shows that the bypass timer contacts are open, with no voltage across the relay coil. What could be the problem?

## Questions

1. What is the purpose of a bypass timer?
2. How would a bypass timer be used on a commercial refrigeration system?
3. How would a bypass timer be used on a chilled water system?

# Unit 54: BROWNOUT PROTECTORS AND LOW-VOLTAGE MONITORS

## Introduction

Brownout protectors and low-voltage monitors (sensors) are used in air-conditioning, heat pump, and refrigeration systems to keep from damaging the equipment in case of low voltage or an out-of-phase problem in the power source.

*Operation.*    If a loss of power occurs or if low voltage is sensed, the circuit is opened and a time-delay period is started, allowing the system to equalize before starting the compressor. Brownout protectors are available in both plug-in type and wire-in models. See Figure 54–1. The plug-in model has a signal light that indicates low voltage. It has a five-minute delay-on-break timer. The wire-in type has a five-minute delay-on-make timer. See Figure 54–2. Several tables are available to help the technician choose the correct brownout protector or low-voltage monitor for the job, for example, see Tables 54–1a and 54–1b.

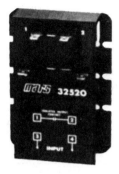

**Figure 54–1**
Brownout protectors and low-voltage monitors. (Courtesy of Motors & Armatures, Inc.)

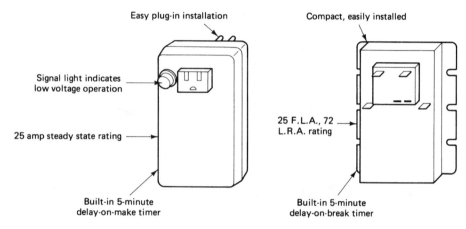

Easy plug-in installation

Compact, easily installed

Signal light indicates
low voltage operation

25 amp steady state rating

25 F.L.A., 72
L.R.A. rating

Built-in 5-minute
delay-on-make timer

Built-in 5-minute
delay-on-break timer

**Figure 54–2**
Brownout protector and low-voltage monitor features. (Courtesy of Motors & Armatures, Inc.)

**TABLE 54–1A** TABLE FOR SELECTING LOW-VOLTAGE MONITORS

| Mars no. | Voltage | Pull-in voltage | Drop-out voltage | Time delay |
|---|---|---|---|---|
| 32520 | 110, 50/60 Hz | 95 | 87 | 5 min. delay-on-make |
| 32522 | 120, 50/60 Hz | 103 | 95 | 5 min. delay-on-make |
| 32524 | 208, 50/60 Hz | 180 | 172 | 5 min. delay-on-make |
| 32526 | 230, 50/60 Hz | 198 | 190 | 5 min. delay-on-make |
| 32528 | 240, 50/50 Hz | 210 | 202 | 5 min. delay-on-make |

(Courtesy of Motors & Armatures, Inc.)

**TABLE 54–1B**   TABLE FOR SELECTING PLUG-IN BROWNOUT PROTECTORS

| Mars no. | Voltage | Pull-in voltage | Drop-out voltage | Time delay |
|---|---|---|---|---|
| 32500 | 110, 50/60 Hz | 95 | 87 | 5 min. delay-on-break |
| 32502 | 120, 50/60 Hz | 103 | 95 | 5 min. delay-on-break |
| 32504 | 230, 50/60 Hz | 198 | 190 | 5 min. delay-on-break |
| 32505 | 240, 50/60 Hz | 210 | 202 | 5 min. delay-on-break |
| 32506 | 230, 50/60 Hz | 198 | 190 | 5 min. delay-on-break |
| 32508 | 240, 50/60 Hz | 210 | 202 | 5 min. delay-on-break |

(Courtesy of Motors & Armatures, Inc.)

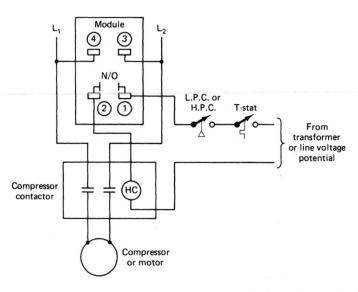

**Figure 54–3**
Wiring diagram using low-voltage protector for commercial refrigeration.
(Courtesy of Motors & Armatures, Inc.)

**Uses.**   More popular uses of the brownout protector and low-voltage monitor include: (1) commercial refrigeration systems, (2) commercial air-conditioning systems, (3) heating system monitors, (4) residential air-conditioning systems, and (5) vending machines.

**Applications.** *Commercial Refrigeration Low-Voltage Protection.*   When power is applied to terminals 3 and 4, a five-minute delay is started before the NO contacts 1 and 2 are closed. See Figure 54–3. If a low-voltage condition occurs during normal operation, contacts 1 and 2 will open and a five-minute delay will be started.

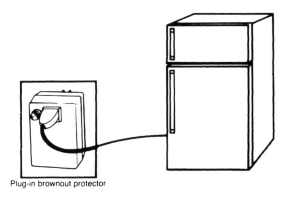

Plug-in brownout protector

**Figure 54–4**
Plug-in appliance, low-voltage protection.
(Courtesy of Motors & Armatures, Inc.)

*Residential Appliance Plug-in Low-Voltage Protection.*   In residential appliances (e.g., air conditioner, refrigerator, freezer), the plug-in protector saves the unit from damage caused by low voltage. Its built-in five-minute delay-on-break timer also keeps the unit from short cycling. See Figure 54–4.

# Summary

- Brownout protectors and low-voltage monitors are used in air-conditioning, heat pump, and refrigeration systems to keep the equipment from being damaged in case of low voltage or out-of-phase problems in the power source.
- The plug-in type has a signal light that indicates a low-voltage condition and has a five-minute delay-on-break timer feature. The wire-in type has a five-minute delay-on-make timer.

# Service Call

A customer complains that his residential air-conditioning system will not come on. A check of the system reveals that it is turned off on the brownout protector. A check of the voltage shows 195 volts on a 240-volt system. The connections are all in good condition. The technician alerts the power company about the problem. After the power company corrects the problem, the voltage is 240 VAC. The voltage and amperage to all motors are now normal. The technician is satisfied that the system is repaired.

# Student Troubleshooting Problem

A customer complains that a residential air-conditioning unit goes off at about 5:30 P.M. almost every day. The system has a low-voltage monitor, high- and low-pressure controls, and overload controls. The time is 11:00 A.M. The system is running. The unit is air cooled. The technician checks the system pressures and for HCFC-22 the suction pressure is 70 psig and the return air temperature is 80°F. The discharge pressure is 325 psig. The outdoor temperature is 95°F. The unit is cooling. What could be the problem?

# Questions

1. What is the purpose of brownout protectors?
2. What happens if either a loss of power or low voltage is sensed?
3. Why should the pressures be equalized before the unit is restarted?
4. What is the most popular time delay on brownout protectors?

# Unit 55: LINE MONITORS
# Introduction

The three-phase line monitor protects motors, compressors, and other equipment from bad electric conditions. They are used to sense the negative cycle element and give a fast response (100 milliseconds) to protect the unit from dangerous electrical phase conditions. Adjustable low-voltage values, a fixed brownout, and an unbalance delay stop a nuisance shutdown of the equipment. See Figure 55–1. Three-phase line monitors are used on both Wye and Delta systems. See Figure 55–2. Several tables are available to help the technician choose the correct line monitors for the job, for example, see Table 55–1.

**Uses.**    More popular uses of the line monitor include: (1) commercial and industrial heating, cooling, and refrigeration systems, and (2) industrial process equipment.

**Applications.**    *Commercial and Industrial Three-Phase Low-Voltage Monitoring.* If a phase loss or reversal occurs during normal operation, the three-phase monitor senses the

**Figure 55–1**
Line monitor. (Courtesy of Motors & Armatures, Inc.)

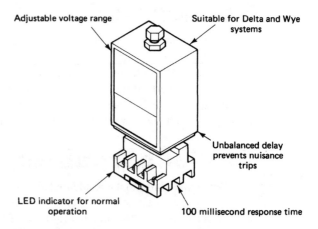

Adjustable voltage range

Suitable for Delta and Wye systems

Unbalanced delay prevents nuisance trips

LED indicator for normal operation

100 millisecond response time

**Figure 55–2**
Line monitor features. (Courtesy of Motors & Armatures, Inc.)

**TABLE 55–1**  TABLE FOR SELECTING LINE MONITORS

| Mars no. | Operating voltage adjustable | Drop-out volts | Pick-up volts | Relay output | Response time |
|---|---|---|---|---|---|
| 32540 | 200 . . . 240 VAC 30, 60 Hz | 180–216 VAC depending on voltage adjustment | 186 . . . 224 depending on voltage adjustment | 8 amp. at 250 VAC | low voltage; 5 sec. loss of phase or phase reversal: 100 milliseconds |

(Courtesy of Motors & Armatures, Inc.)

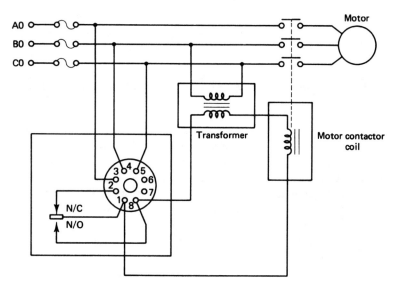

**Figure 55–3**
Wiring diagram for three-phase low-voltage monitoring. (Courtesy of Motors & Armatures, Inc.)

negative part of the power supply and stops the equipment within 100 milliseconds. If a low-voltage condition is sensed, it is monitored for five seconds before the unit is stopped. See Figure 55–3.

# Summary

- The three-phase line monitor protects motors, compressors, and other equipment from bad electric power conditions and is used to sense the negative cycle part and give a fast response (100 milliseconds) to protect the unit from dangerous power phase reversal and phase loss.

- Three-phase line monitors are used on both Wye and Delta systems.
- More popular uses of the line monitor include: (1) commercial and industrial heating, cooling, and refrigeration systems, and (2) industrial process equipment.

# Service Call

A customer complains that her commercial air-conditioning unit will start and run a few minutes and then stop. The system is protected with a three-phase line monitor. When the unit is started, it will run about five minutes and then stop. The voltage at one leg of the three-phase power is 90 volts to ground. The other two legs are 120 volts to ground. The one leg has low power. The tech notifies the power company about the problem. After the power company has corrected the problem, the tech checks the voltage and amperage to all the motors, which are within the normal operating limits. The technician is now satisfied that the system is repaired.

# Student Troubleshooting Problem

A customer complains that the air-conditioning unit in his office building is not cooling. The system is a three-phase unit. Only the outdoor fan is running. The outdoor unit has a line monitor, high- and low-pressure controls, and external overloads. The unit power terminals have 240 VAC three-phase power. Each leg of the three-phase power is checked to ground and two of the legs show 120 VAC to ground. The third leg, the power leg, shows 220 VAC to ground. A check of the compressor motor for continuity shows it to be within the manufacturer's specifications. All the fuses and circuit breakers are good. What could be the problem?

# Questions

1. What is sensed by a three-phase line monitor?
2. How long does it take a line monitor to trip?
3. What type line monitor is used on a Wye system?
4. When a unit is in operation and a low-voltage condition happens, how much time is needed for the monitor to trip?

# Unit 56: PLUG-IN TIME-DELAY RELAYS

## Introduction

Plug-in time-delay relays are used in systems to replace existing relay circuits. They are available in both delay-on-make and delay-on-break models when installed. A load of up to 10 amperes can be switched. The delay-on-make models are DPDT. The delay-on-break has a SPDT switching action. All models are adjustable in 1-second increments from 1 to 1023 seconds. Some models may use an eight-pin octal mounting. An LED light indicates that the control is in the timing mode. See Figure 56–1.

***Operation.***    When power is applied to the delay-on-make models, the time delay begins. At the end of the delay period, the output contacts change positions.

For delay-on-break models, the power must be applied to the input terminals at all times before and during timing. When the start switch closes, the contacts change position. When the start switch opens, the time delay begins. At the end of the time delay period, the output contacts change back to the deenergized position. See Figure 56–2. Several tables are available to help the technician choose the correct plug-in time-delay relay for the particular job, for example, see Table 56–1.

**Uses.**    More popular uses of the plug-in time delay relays include: (1) commercial refrigeration systems and (2) commercial air-conditioning and heating systems.

**Figure 56–1**
Plug-in time-delay relay. (Courtesy
of Motors & Armatures, Inc.)

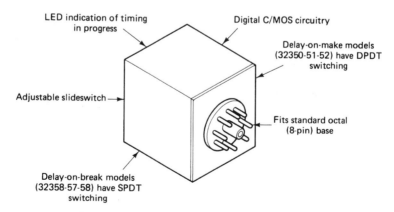

**Figure 56–2**
Plug-in time-delay relay features. (Courtesy of Motors & Armatures, Inc.)

**Applications.**    *Commercial Delay-on-Make Timing.*    The time delay is started when the power is applied to terminals 2 and 7. See Figure 56–3. When the delay period has passed, power on the NC output terminals 4 and 5 switch the NO output terminals 3 and 6. An LED light shows that the timing phase is in progress.

*Commercial Delay-on-Break Timing.*    When the start switch is closed, the NC load output terminal 4 switches to the NO load output terminal 3. See Figure 56–4. The time delay is started when the switch is open. After the delay period has passed, the load output terminals switch back to the normal positions. An LED light shows that the control is in the timing mode.

**TABLE 56–1**    TABLE FOR SELECTING PLUG-IN TIME-DELAY RELAYS

| Mars no. | Type delay | Time-delay range | Type adjustment | Input voltage | Hz | Max. amps | Switch type |
|---|---|---|---|---|---|---|---|
| 32350 | on make | 1 . . . 1023 sec. | slideswitch | 24 VAC | 50/60 | 10 at 240V | DPDT |
| 32351 | on make | 1 . . . 1023 sec. | slideswitch | 120 VAC | 50/60 | 10 at 240V | DPDT |
| 32352 | on make | 1 . . . 1023 sec. | slideswitch | 230 VAC | 50/60 | 10 at 240V | DPDT |
| 32356 | on break | 1 . . . 1023 sec. | slideswitch | 24 VAC | 50/60 | 10 at 240V | SPDT |
| 32357 | on break | 1 . . . 1023 sec. | slideswitch | 120 VAC | 50/60 | 10 at 240V | SPDT |
| 32358 | on break | 1 . . . 1023 sec. | slideswitch | 230 VAC | 50/60 | 10 at 240V | SPDT |
| 93057 | Standard octal (8-pin) base. Can be surface or DIN rail mounted. | | | | | | |

(Courtesy of Motors & Armatures, Inc.)

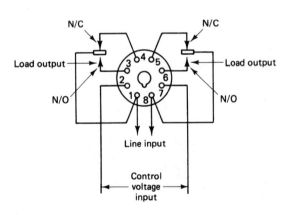

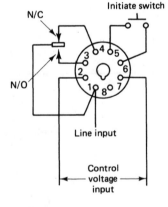

**Figure 56–3**
Wiring diagram for delay-on-make timing using a plug-in time-delay relay. (Courtesy of Motors & Armatures, Inc.)

**Figure 56–4**
Wiring diagram for delay-on-break timing using a plug-in time-delay relay. (Courtesy of Motors & Armatures, Inc.)

# Summary

- Plug-in time-delay relays are used for plug-in applications to replace existing relay circuits. They are available in both delay-on-make and delay-on-break models. Loads up to 10 amperes can be safely switched.
- All models are adjustable in 1-second increments from 1 to 1023 seconds. They may have an eight-pin octal mounting.
- More popular uses of the plug-in time-delay relays include: (1) commercial refrigeration systems and (2) commercial air-conditioning and heating systems.
- Commercial delay-on-make timing delays are started when the power is applied to terminals 2 and 7.
- Commercial delay-on-break timing occurs when the start switch is closed and the NC load output terminal 4 switches to the NO load output terminal 3.

# Service Call

A customer complains that her commercial refrigeration system is not working correctly. The system is controlled by a commercial delay-on-make timer. The LED light on the timer shows that the timer is in the timing mode; however, the LED light never goes off. There is no voltage between terminals 2 and 7 or between terminals 3 and 6. The timer is defective. The technician replaces it with the correct model and puts the unit back in operation. A complete check shows the system is operating correctly, the technician is satisfied that it is repaired.

# Student Troubleshooting Problem

A customer complains that the dairy case in a food store keeps freezing the products. The system has a delay-on-break relay, high- and low-pressure controls, and overload controls. The unit is running and the temperature inside the cabinet is 29°F. The products are almost frozen. The thermostat is set on 38°F. The thermostat contacts are open. The pressure and overload relay contacts are all closed. What could be the problem?

# Questions

1. What is the difference between the various plug-in time-delay relays?
2. What is the heaviest switching load for plug-in time-delay relays?
3. How will the technician know when a plug-in timer is in the timing mode?
4. On a delay-on-make plug-in time-delay relay, when does the time delay begin?

# Unit 57: CURRENT-SENSING RELAYS
## Introduction

The current-sensing relay is an SPDT relay used on electrical systems when it is necessary to electrically separate the operation of the relay from the electrical circuit it is controlling. See Figure 57–1. The relay controls the circuit by passing wires from the circuit being sensed through the sensing loop with the correct polarity. The current flow causes a signal in the sensing loop increased by an internal solid-state circuit to operate the relay. It has two sets of contacts—one is NC, the other is NO.

*Operation.*    When the circuit being sensed is operating normally, the current-sensing relay stays in its normal operating position with its NC contacts closed. If the current draw in the sensed circuit rises above the sensing relay rating, the NC contacts open and the NO contacts close. When the NC contacts open, the circuit through the control loop is broken. When the NO contacts close, an alarm circuit alerts the operator that the system is not working correctly. As the current draw drops to the correct level, the relay returns to its normal operating position. The system is then automatically put back in operation.

**Uses.**    The relay is used to operate parts whose operation must be controlled by the existence or absence of current passing through the relay loop. The minimum amperage necessary to operate the relay is 15 amps. Circuit loads less than 15 amps may be increased by making multiple passes of the sensed wire through the sensing loop to reach the minimum ampere turns for correct operation.

## Summary

- The current-sensing relay is an SPDT relay design used on electrical systems when it is necessary to electrically separate the operation of the relay from the electrical circuit it is controlling.

**Figure 57–1**
Current-sensing relay. (Courtesy of Robertshaw Controls, Uni-Line Division.)

- The relay controls the circuit by passing wires from the circuit being sensed through the sensing loop in the correct polarity. The current flow causes a signal in the sensing loop. The current flow is increased by an internal solid-state circuit to operate the relay.
- The relay is used to operate parts whose operation must be known by the presence or absence of current passing through the loop.

# Service Call

A customer complains that a residential electric heating unit goes on and off. A check of the system reveals that the unit is using a current-sensing relay to start and stop the indoor fan motor. The technician turns the unit on and checks the current through the sensing loop, which shows 13.5 amps flowing through the circuit. A further check shows that the fan motor has just been replaced with a new one, but the new one is not rated for enough current to operate the relay. The tech passes another loop of the sensed wire through the sensing loop to increase the sensed current draw at this point. The unit is completely checked and is operating as it should. The technician is satisfied that the system is repaired.

# Student Troubleshooting Problem

A customer complains that her residential electric heating unit will occasionally fail to come on. The unit has a fan motor current-sensing relay and over-temperature and over-current protectors for the heating elements. The unit is not running. When the thermostat temperature lever is set above room temperature, the fan motor comes on but the heat strips do not. The fan motor is rated for exactly the same amperage as the sensing relay. What could be the problem? How can this problem be solved?

## Questions

1. What type of contact arrangement is used in a current-sensing relay?
2. What is the minimum amperage needed to operate a current-sensing relay?
3. How do current-sensing relays control a circuit?

# Unit 58: SOLID-STATE MOTOR STARTING DEVICES

## Introduction

Solid-state motor starting relays use a self-regulating, conductive ceramic material that increases in electrical resistance as the motor starts, thus quickly reducing the start winding current flow to a milliamp level. When the amperage draw is 10 amps or greater, the relay switches in less than 0.35 second. This connection allows this type relay to be used on refrigerator compressors without being tailored to each particular system. Solid-state relays will start nearly all split-phase, 120-volt hermetic compressors up to 1/3 horsepower. An overload must be used with these relays. See Figure 58–1.

*Operation.*    The solid-state relay is connected in the motor circuit with the ceramic material placed in the electric line to the compressor motor starting terminal. See Figure 58–2. However, the wiring diagram furnished by the compressor manufacturer should always be followed. The relay is connected in series with the motor starting winding. When electricity is applied to the relay, the ceramic material heats up and turns the relay off in about 0.35 second at 10 starting amps or greater. This lowers the current flow through the start winding to a milliamp level until the electric power to the motor is turned off. After the power has been turned off, the ceramic material needs a few minutes to cool before the next starting cycle.

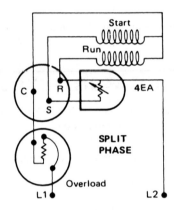

**Figure 58–1**
Solid-state relay connections. (Courtesy of Klixon Controls Division, Texas Instruments, Inc.)

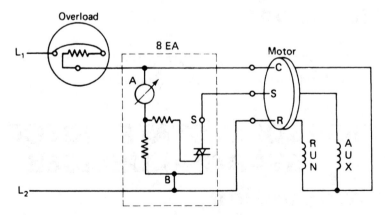

**Figure 58–2**
Solid-state relay connections. (Courtesy of Klixon Controls Division, Texas Instruments, Inc.)

Some of these relays may not be suitable for replacement on capacitor start compressor motors, or for those systems with rapid cycle times.

**Solid-state hard-start kit.**    The solid-state hard-start kit gives the necessary additional starting torque to solve starting problems with permanent split capacitor (PSC) motors. Positive temperature coefficient (PTC) or positive-temperature-coefficient ceramic materials are used to solve most starting problems. The resistance of these materials increases as their temperature rises. At its anomaly temperature, the resistance increase is very sharp. See Figure 58–3.

*Operation.*    The purpose of the PTC start-assist device is to give a surge of current that lasts only while needed to start the compressor motor. This additional current is then lowered so the motor can run as a normal PSC motor. When the electricity is turned on to the motor, current flows through the start winding and through the parallel combination of the run capacitor and the low-resistance PTC. See Figure 58–4.

The low resistance during the starting phase not only increases the start winding current, but also reduces its angular displacement with the current flowing through the main winding. This is an advantage in PSC motors because the phase angle between the starting and running currents is usually greater than 90 percent.

While the surge of current is giving the needed starting torque, the current is also flowing through the PTC and heating it to its high-resistance temperature. Time needed for

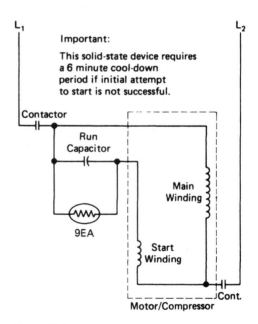

**Figure 58–3**
Solid-state hard-start kit. (Courtesy of Klixon Controls Division, Texas Instruments, Inc.)

**Figure 58–4**
Solid-state hard-start kit connections. (Courtesy of Klixon Controls Division, Texas Instruments, Inc.)

the PTC to heat to its high-resistance state is independent of when the motor starts. It depends on the mass of the PTC, the anomaly temperature, its resistance, and the start winding current. When a 240-volt compressor motor using a start-assist device is first energized at a nominal voltage, the switching time is about sixteen electrical cycles, depending on the current flow. Starting the same compressor motor with 25 percent less voltage increases the switching time to thirty-two electrical cycles, giving additional help under these starting conditions.

The switching times are ideal for the PSC motor because the PTC is in its low-resistance state only long enough to overcome the initial inertia of the motor and compressor. When the PTC switching times are longer than normal, the low-resistance PTC effectively bypasses the run capacitor. The excessive on-time slows down the motor speed while it is trying to overcome the increasing load.

After the PTC has heated to its anomaly temperature, its resistance increases to approximately 80,000 ohms and effectively takes itself out of the electrical circuit without the use of a relay. While the compressor motor is running, only 6 milliamps of current flows through the start device. The low-current draw does not affect the start winding or the running efficiency of the motor. As the motor is stopped, the power to the PTC is also turned off. At this time it starts cooling to the surrounding temperature. If an attempt is made to restart the motor before PTC has cooled to its anomaly temperature, the motor will try to start in the standard PSC mode. It will usually take more than one minute for the PTC to cool below its anomaly temperature so that the start assist will be available for the next start.

It is recommended that a solid-state motor starting relay only be used on compressors up to 48,000 British thermal units (Btus) or motors up to 4 horsepower. However, its use is not limited to this size range.

**Positive-temperature-coefficient starting device.**    PTC resistance starting devices are used on PSC motors to help them start by giving additional starting torque. It is not recommended that these devices be used on systems that use a thermostatic expansion valve or on systems used in short-cycling jobs. They are simple to install, economical to buy, and have a wide range of use. See Figure 58–5.

The material used for the PTC has a steep-slope, positive temperature coefficient that has a cold resistance of about 50 ohms and a hot resistance of about 80,000 ohms.

*Operation.*    The PTC is wired in parallel with the running capacitor. It increases the motor starting torque approximately 200 percent to 300 percent. When the PTC material heats up, it takes the start-assist device out of the starting circuit in approximately 0.2 second. The compressor motor then operates as a standard PSC motor.

*Service.*    When checking this device, allow it to cool to room temperature; then check its resistance with an ohmmeter. When the resistance reading varies from the cold resistance rating, replace the device.

Usually the solid-state relay will have visible signs when it is bad. Sometimes the relay will be severely charred, sometimes it will be severely burned, and sometimes it will only be a light tan color. If any of these conditions are found, replace the relay with the correct replacement.

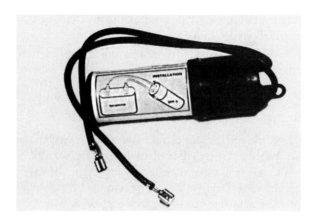

**Figure 58–5**
Positive-temperature-coefficient starting device.

To make sure that the relay is taking the starting winding out of the circuit, check the amperage draw in the starting circuit. After the compressor has started, this amperage draw should drop to the milliamp level. When the relay is pushed directly onto the compressor motor terminals, the total current draw to the compressor must be taken as the compressor is starting. This should be about six times the rated running amps. When the compressor has started, the amperage draw should drop to within the normal running amperage rating.

# Summary

- Solid-state starting relays use a self-regulating, conductive ceramic material that increases in electrical resistance as the motor starts, thus quickly reducing the starting winding current flow to a milliamp level.
- These relays will start nearly all split-phase, 120-volt hermetic compressors up to ⅓ horsepower.
- These relays require an overload.
- This relay is connected in the motor circuit with the ceramic material placed in the electric line to the starting terminal of the compressor motor.
- When the electricity is applied to the relay, the ceramic material heats up and turns the relay off in about 0.35 second at 10 amps or greater.
- The solid-state hard-start kit gives the necessary starting torque to solve starting problems with permanent split capacitor (PSC) motors.
- The purpose of the PTC start-assist device is to give a surge of current that lasts only during the time needed to start the compressor motor.
- The time needed for the PTC to heat to its high-resistance state is independent of when the motor starts. Rather, it depends on the mass of the PTC, the anomaly temperature, its resistance, and the start winding current.

- When the PTC switching times are longer than normal motor starting times, the low-resistance PTC effectively bypasses the run capacitor.
- PTC resistance starting devices are used on PSC motors to help them in starting by applying additional starting torque to the motor.

# Service Call

A customer complains that his domestic refrigerator compressor tries to start but does not. The system has a solid-state start device connected directly to the compressor motor terminals. The technician starts the unit and checks the total starting amperage, which is about six times higher than the running amperage rating of the compressor—but it never drops. It is believed that the solid-state starting device is bad so the tech replaces it with the correct one and puts the unit back in operation. After the compressor starts, the amperage drops to within the normal running amperage rating of the compressor. After checking the complete unit and finding it operating normally, the technician is satisfied that the system is repaired.

# Student Troubleshooting Problem

A customer complains that her air-conditioning unit will not run. The system has a solid-state starting relay, high- and low-pressure controls, and a compressor overload. The unit is not running. When the temperature lever on the thermostat is turned below room temperature, the compressor tries to start but only hums and then goes off on the compressor overload. The voltage is 230 VAC when the compressor is trying to start. When the compressor goes off, the voltage returns to 240 VAC. The current draw when the compressor is trying to start is about six times the rated running amperage. What could be the problem? How can this problem be solved?

# Questions

1. How do solid-state starting devices work?
2. Is there a need for an overload to be used with solid-state start devices?
3. How much current flows through the starting circuit after a solid-state device has opened?
4. What is the time span between when a unit using a solid-state device is stopped and when it can be restarted?
5. What is meant by positive temperature coefficient?
6. What is the solid-state device switching time when there is a 25-percent voltage drop?
7. How much resistance should a cold solid-state start device have?
8. On what type systems should solid-state start devices not be used?

# Unit 59: SOLID-STATE HEAT PUMP DEFROST CONTROLS

## Introduction

The solid-state defrost control method uses a solid-state control board and two thermistors. See Figure 59–1. A thermistor is a device in which its electrical resistance changes with a change in temperature. One thermistor is near the outlet of the outdoor coil to sense the coil temperature at this point. The other thermistor senses the outdoor air temperature as it enters the coil.

*Operation.*   When the outdoor air temperature drops below 45°F, frost will probably start forming on the outdoor coil. Frost causes an increase in the temperature difference between the two thermistors. When this temperature difference reaches 15°F to 25°F, the defrost cycle is initiated.

In this type control system, the coil of the outdoor fan and defrost relay are operated by a 24-volt dc coil. It is energized through terminals R and R of the solid-state defrost control board. See Figure 59–2. This type system uses a high-pressure control to terminate the defrost cycle. When the refrigerant pressure inside the outdoor coil reaches 275 psig, the defrost cycle is terminated. See Figure 59–3. This amount of pressure shows that the temperature (about 124°F) of the outdoor coil is high enough to melt all the frost from its surface. This control is in the refrigerant line in the outdoor unit.

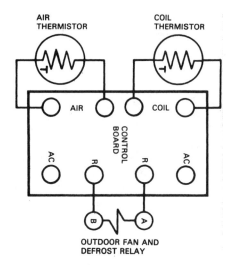

**Figure 59–1**
Solid-state defrost control. (Courtesy of Lennox Industries, Inc.)

**Figure 59–2**
Solid-state defrost control location. (Courtesy of Lennox Industries, Inc.)

**Figure 59–3**
Defrost termination switch. (Courtesy
of Robertshaw Controls Co., Uni-Line
Division.)

# Summary

- The solid-state defrost control method uses a solid-state control board and two thermistors.
- One thermistor is located near the outlet of the outdoor coil to sense the coil temperature at this point. The other thermistor senses the outdoor air temperature as it enters the coil.
- This type system uses a high-pressure control to terminate the defrost cycle.

# Service Call

A customer complains that a heat pump system is not heating his home. A check of the system reveals that the outdoor coil is iced over. The system has not had a defrost cycle in some time. The defrost cycle is controlled by a solid-state defrost control. The board is operating correctly. A check of the thermistors shows that the one at the outlet of the coil has corrosion between it and the refrigerant line. The tech then cleans the tube and thermistor and puts an anticorrosion gel on both the line and the thermistor to stop future corrosion. The tech then reconnects the thermistor to the line, defrosts the coil, and puts the unit back in operation. The complete unit is now operating normally. The tech covers the outdoor coil to cause it to ice over and puts the unit in the defrost mode. The unit goes into defrost and when all the frost has melted from the coil the high-pressure termination switch causes the unit to go back into the heating mode. The technician is satisfied that the system is repaired.

# Student Troubleshooting Problem

A customer complains that a heat pump unit is not heating her residence. A check of the system reveals that the outdoor coil is completely clear of ice, and the indoor fan is running but the outdoor fan is not. The heat strips are energized. A time and temperature defrost control is used. The termination sensing element is correctly fastened to the outdoor coil. There is voltage to the motor terminals of the defrost time clock. The defrost timer contacts are closed. What could be the problem? (Refer to Units 41 and 43.)

# Questions

1. What is the purpose of the two thermistors on a solid-state defrost control?
2. At what temperature does frost usually start forming on the outdoor coil of a heat pump?
3. At what temperature does the solid-state defrost control switch the unit into defrost?
4. At about what temperature is the defrost cycle initiated?

# Unit 60: SOLID-STATE MOTOR PROTECTORS

## Introduction

Solid-state motor protectors are quite popular with equipment manufacturers because of their rapid response time and accuracy in operation. A thermistor is used to measure the operating temperature of compressor motor windings and to signal the control circuit when an overheated condition is present.

Heat is the major cause of electric motor failure. The heat of a hermetic compressor motor is due to the load being cooled and the heat caused by motor operation. The life of a motor is largely decided upon by the conditions under which it operates; therefore, it is necessary that hermetic motors be correctly protected to prevent a costly replacement. Solid-state motor protectors can give this protection.

*Operation.*    A single-module, three-sensor system is one of the most popular solid-state motor protectors in use. See Figure 60–1. It shows a single module using a single thermistor. The thermistors are placed in the hot spot of the motor winding to sense the highest heat inside the motor. On single-phase motors, usually only one thermistor is used—it is in the motor run winding. These are fast acting and are usually quite accurate. The oil and refrigerant inside the system do not affect the thermistor insulation or its operation.

A three-phase motor uses three thermistors—one in each winding because each has a power source of its own. See Figure 60–2. These thermistors have a positive temperature coefficient. As the temperature of the thermistor increases, its resistance also

**Figure 60–1**
Solid-state model, three-sensor system.
(Courtesy of Robertshaw Controls Co., Uni-
Line Division.)

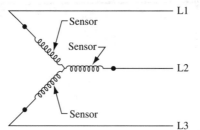

**Figure 60–2**
Three-phase motor sensor location.

increases. When the resistance of the thermistor reaches a given value, the module will open a set of NC contacts to open the control circuit to the compressor contactor holding coil. The module connections into the circuit are shown in Figure 60–3.

Connections made to the module include the thermistor connections, the control circuit connections, the manual reset connections, and the power supply connections. The module should be wired into the circuit as recommended by the equipment manufacturer.

**Troubleshooting.**   The steps used to troubleshoot the module and the thermistors are simple. Only a few checks are required to learn if the unit is good or if it is bad and must be replaced. The technician first checks to see if the control circuit contacts are open or closed. When the control contacts are closed, the problem is in another control. When the contacts are open, the technician checks three other areas: the module, the thermistors, or another type of overload. If the motor has been operating in an overloaded condition, it must be checked to make sure it is not damaged. The motor condition should be checked before replacing the module or the thermistors.

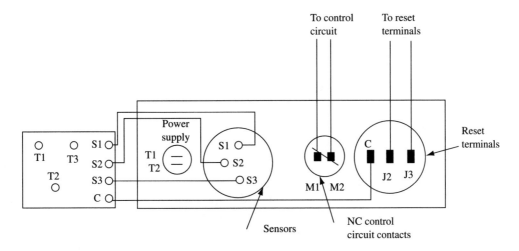

**Figure 60–3**
Module wiring connections.

To check the thermistors, use an ohmmeter and take the following steps.

1. Disconnect the electric power from the unit.
2. Check the maximum test voltage at 6 volts.
3. Never short the thermistors because they could be damaged.
4. When fuses are installed in the circuit, do not bypass them or replace them with larger sizes.
5. Use the ohmmeter to check the resistance of the thermistors. Be sure to zero the ohmmeter for the range to be used.
   a. When the thermistor resistance is shown to be above 95 $\Omega$, the thermistor must be allowed to cool and reset, then be rechecked.
   b. If the thermistor is found to have infinite resistance (open), install a 75-$\Omega$, 2-watt resistor.
   c. When the thermistor is good, the only thing left to do is replace the module.

Most compressor manufacturers use their own method of protecting their compressors in case a thermistor goes bad. They use a discharge line thermostat to open the control circuit when the discharge gas temperature reaches the maximum limit. It will sense the discharge gas temperature and stop the compressor when it reaches an unsafe level.

# Summary

- Solid-state motor protectors are quite popular with equipment manufacturers because of their rapid response time and accuracy in operation.

- Heat is the major cause of electric motor failure. The heat of a hermetic compressor motor is due to the load being cooled and the heat caused by motor operation. The life of a motor is largely decided upon by the conditions under which it operates.
- When the temperature of the thermistor increases, its resistance also increases.

# Service Call

A customer complains that a commercial air-conditioning system is not cooling. The system's compressor is not running, which is controlled by a solid-state module and three thermistors. The technician checks the thermistors and all three are good. The resistance of the motor windings is also within the manufacturer's specifications. When the module is replaced and the system turned on, it starts running and cooling. The complete unit is now in good operating condition. The voltage and amperage to the compressor are within the normal operating range. The technician is satisfied that the system is repaired.

# Student Troubleshooting Problem

A customer complains that a commercial refrigeration frozen food unit is off and all the products are defrosting. The system has solid-state motor protectors and high- and low-pressure controls. The compressor is not running. The line side of the contactor has 240 VAC. The contactor is not pulled in. There is no voltage at the control circuit terminals of the contactor holding coil. The cabinet thermostat is demanding cooling. The compressor is very hot to the touch. What could be the problem? What could cause this problem?

# Questions

1. In a solid-state motor protector, what is used to measure the motor temperature?
2. What causes most electric motor problems?
3. Why must thermistors be protected from refrigerant and oil?
4. Why do thermistors have a positive temperature coefficient?

# Unit 61: SOLID-STATE HEAT PUMP CONTROLS

## Introduction

Most manufacturers of residential heat pump systems use solid-state boards to control operation of the complete system. Solid-state controls are used on the high-efficiency models to help in energy conservation and comfort conditions inside the home.

*Operation.*    During normal operation, the solid-state control board controls the equipment operation. It determines when a defrost cycle is needed and causes the unit to go into the defrost mode. Sensors are used to decide when the refrigerant pressures are either too high or too low and will control the unit to correct the condition or stop unit operation. The speed of the condenser fan is sometimes controlled from this board. Each equipment manufacturer usually has a solid-state control board of their own design, thus when making any checks on the board, the specific equipment manufacturer's recommendations must be followed to keep from damaging the board or its components.

Usually equipment manufacturers will also have a system tester for their particular equipment. It should be used for best results.

# Summary

- Most manufacturers of residential heat pump systems use solid-state boards to control operation of the complete system.
- During normal operation, the solid-state board controls equipment operation.
- Equipment manufacturers usually have a solid-state control board of their own design. Thus, when making any checks on the board, the equipment manufacturers' recommendations must be followed to keep from damaging the board or its parts.

# Service Call

A customer complains that a residential heat pump system is not heating his home. The system is controlled by a solid-state control board. The compressor is not running and none of the strip heaters are on. The control board, checked according to the equipment manufacturer's recommendations, is bad. The technician replaces the board with the correct one and starts the unit. It starts heating the home. A complete check of the unit shows it is operating satisfactorily, the technician is satisfied that the system is repaired.

# Student Troubleshooting Problem

A customer complains that a heat pump system is not cooling her home. The system is controlled by an electronic board. Nothing is running. After the test terminals on the board are jumpered, the unit still does not start. After the high- and low-pressure controls are reset, the unit will not start. When the thermostat temperature lever is set below room temperature, nothing happens. The line side of the contactor shows 240 VAC, but the contactor holding coil terminals show no control voltage. What could be the problem?

# Questions

1. Why should the specific equipment manufacturer's recommendations be followed when checking a solid-state control board?
2. Why do equipment manufacturers design their own solid-state control boards?

# Unit 62: SOLID-STATE HUMIDISTATS

## Introduction

Humidistats are used to control humidification equipment on heating systems and on other systems that need humidity control at a certain level. Adding humidity to some buildings is necessary because the air gets drier when heated. Today, materials used in solid-state humidity controllers change in electrical resistance when the amount of moisture present in the air changes. Only a few of these materials are used in solid-state humidity controllers.

**Solid-state type controllers.**   With a change in humidity, the sensors used in electronic humidity controllers change in electrical resistance. These sensors are made of two electrically conductive materials separated by a humidity-sensitive hygroscopic insulating material, such as polyvinyl acetate or polyvinyl alcohol. Also, a certain salt solution is often used.

Electronic humidity controllers are most popular where close humidity control is needed. In some jobs it may be used for controlling the fan speed to provide more accurate humidity control.

*Operation.*   When air passes over the sensing element, the element detects any small changes in humidity. The electrical resistance between the two conductors changes with the amount of moisture present in the air. In most jobs the sensor is placed in one leg of a Wheatstone bridge, where the output signal can be used to control the equipment or give a readout. Generally, some type of signal amplification is needed.

There are several ways to wire these controls into a circuit. However, they should be wired so that the humidifier operates only when the indoor fan is running, in order to keep moisture from collecting on the furnace heat exchanger during the OFF cycle. One suggested wiring diagram is shown in Figure 62–1. This diagram provides a safety—the water solenoid cannot be energized unless both sources of electricity are available.

## Summary

- Humidistats are used to control humidification equipment on heating systems and other jobs that need close humidity conditions.
- Today, materials used in solid-state humidity controllers change in electrical resistance when the amount of moisture in the air changes.
- With a change in humidity, the sensors used in electronic humidity controllers change in electrical resistance.
- Electronic humidity controllers are most popular in applications where close control of the humidity is needed.

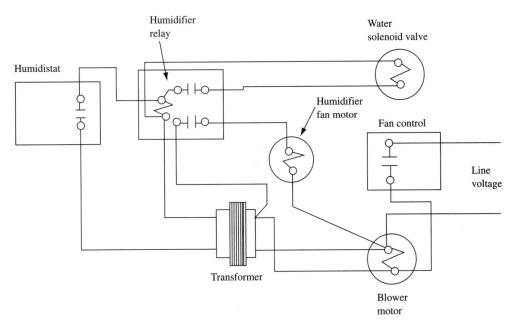

**Figure 62–1**
Typical humidifier wiring diagram.

# Service Call

A customer complains that the paper in a print shop will not feed through the machine correctly. A check of the system shows that the relative humidity inside the print shop is 30 percent, which is too low for printing needs. The humidistat's contacts are stuck open, keeping the humidification equipment from operating. The tech replaces the humidistat with a correct replacement, puts the unit back in operation, and the humidification equipment starts operating. The technician returns to the job after about 24 hours and the humidity level is about 51 percent, which is a satisfactory humidity level for the print shop. The unit is now operating correctly. The technician is satisfied that the system is repaired.

# Student Troubleshooting Problem

A customer complains that the windows in her residence are sweating and moisture is running down the walls. A check of the system shows that the humidifier is working correctly. The humidistat is set on 55 percent RH. The outdoor temperature is 29°F. What could be causing the problem? What will solve the problem?

# Questions

1. Why are humidification systems needed?
2. How do electronic humidity controllers operate?
3. In what situations are electronic humidity controllers most popular?
4. When should a humidifier operate?

# Unit 63: SOLID-STATE CRANKCASE HEATERS

## Introduction

Crankcase heaters are used to keep the correct temperature inside the compressor crankcase. This helps keep liquid refrigerant from collecting in the oil during cool or cold weather conditions or during the compressor OFF cycle. See Figure 63–1. Liquid refrigerant mixing with the oil could possibly cause broken valves and galled bearings, pistons, and piston rings—damage that would require compressor replacement.

*Operation.* The solid-state crankcase heater is energized when the compressor motor is deenergized due to a change in the resistance through the solid-state material. The purpose of this material is to keep a constant temperature inside the crankcase. When the compressor is not running, the oil and the crankcase cool down. The heat is conducted away from the solid-state material and thus lowers its resistance. Because the material tries to keep a constant temperature, more current flows to produce heat which replaces that being taken away. In this way, when the crankcase is cooler, more current will flow through the solid-state material to warm the crankcase.

Usually enough heat is created when the compressor is running to stop refrigerant migration to the crankcase. As the crankcase gets warmer during operation, less heat is

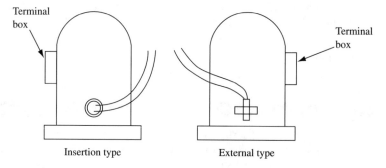

Insertion type                    External type

**Figure 63–1**
Solid-state crankcase heaters.

needed from the crankcase heater. The solid-state material lowers the amount of current flowing to the heater. When the crankcase temperature reaches about 100° F the current flow through the crankcase heater is reduced significantly. This actually turns off the heater.

**Checking the heater.**    The solid-state crankcase heater is almost a maintenance-free device. When the heater fails to work, check for line voltage to the terminals. If voltage is found and the heater still does not operate, replace it with a new one. A good way to check the heater is to touch the heater sheath on the protrusion made when the heater was assembled. When the compressor has been off long enough for the crankcase to cool down, the heater should be warm to the touch.

# Summary

- Crankcase heaters are used to keep the correct temperature inside the compressor crankcase in order to keep liquid refrigerant from collecting in the oil during cool or cold weather conditions.
- The solid-state crankcase heater is energized when the compressor motor is deenergized due to a change in the resistance through the solid-state material. The purpose of this material is to keep a constant temperature inside the crankcase.
- A good way to check the heater is to touch the heater sheath on the protrusion made when the heater was assembled. When the compressor has been off long enough for the crankcase to cool down, the heater should be warm to the touch.

# Service Call

A customer complains that his residential air-conditioning system is not cooling. A check of the system reveals that everything is running; however, the suction line is only slightly cool to the touch. The technician checks voltage, amperage, and pressures while the unit is running. The voltage is 240 volts at the compressor terminals. The amperage is about half of what is anticipated for this size compressor. The suction pressure is 100 psig for HCFC-22. The discharge pressure is 150 psig. The unit is equipped with a solid-state crankcase heater. The technician feels the heater and finds that it and the crankcase are cool to the touch. The crankcase heater terminals show 240 VAC, but with no current flow. The crankcase heater is deemed bad and the compressor has slugged liquid refrigerant and oil that damaged the compressor valves. The tech recovers the refrigerant from the unit, changes the compressor and the needed driers and filters, and replaces the crankcase heater. After the system is evacuated and recharged with refrigerant, the tech starts the unit and it begins cooling the home. The operation of the complete unit is now satisfactory. The technician is satisfied that the system is repaired.

# Student Troubleshooting Problem

A customer complains that a residential cooling system is not cooling. The system has a wraparound solid-state crankcase heater. The line side of the contactor has 240 VAC. The contactor holding coil terminals fail to show 24 VAC. The contactor is not pulled in. The compressor is very warm to the touch. When touching the compressor, the tech feels a slight electrical shock. The compressor motor shows no grounded windings. Current is flowing through the crankcase heater. What could be causing the problem?

# Questions

1. What is the purpose of a crankcase heater?
2. When is the crankcase heater normally energized?
3. When will the greatest amount of current flow through the crankcase heater?
4. When does the solid-state material stop the current flow to the crankcase heater?

# Unit 64: ECONOMIZER CONTROL PACKAGE

## Introduction

Larger air-conditioning systems incorporate every effort to conserve energy in order to lower the cost of operation. Even 10 percent energy savings lowers monthly electric bills and provides a financial savings. This of course allows more profits for the company using the equipment.

**Economizer use.** The economizer package can be used to retrofit an existing system or it may be installed with a new system. Sensors are put at critical points throughout the building. These sensors send a signal to the solid-state economizer module sensing the temperature at these crucial points. The module interprets this signal and sends the needed signal to the air-conditioning, heating, and ventilating system to make the necessary changes in the temperature at that point. Several types of economizer packages are available. The one chosen will generally depend on system needs and preference of the design engineer.

The economizer module controls the amount and type of air delivered to the area needing some change in temperature. When the economizer receives a signal showing that the space is too cool, it will reduce the amount of cooled air and increase the amount of warmed air, or outside air to the space. It will follow this process for each sensor used and will keep the wanted temperature needed by that sensor regardless of the conditions of the other spaces in the building. Usually the most effective way of lowering the power usage is to reduce the air being treated by the equipment and to bring in outside air when the outdoor conditions will give the wanted comfort.

## Summary

- Larger air-conditioning systems incorporate every effort to conserve energy in order to lower the cost of operation.

- Sensors are put at critical points throughout the building. They send a signal to the solid-state economizer module which measures the temperature at these crucial points.
- The module interprets this signal and sends the needed signal to the air-conditioning, heating, and ventilating system to make the necessary changes in the temperature at that point.

# Service Call

A customer complains that a large air-conditioning system is too expensive to operate. A check of the system shows that all the equipment is operating correctly. The system has an economizer package. The economizer is not working correctly. After the module is replaced and the unit starts operating normally, the technician is satisfied that the system is repaired.

# Student Troubleshooting Problem

A customer complains that the air-conditioning unit in a large office building is not cooling the space. The system has an economizer package and everything is running. The outdoor temperature is 98°F and the outdoor relative humidity is 85 percent. The economizer dampers are fully open to the outside air. The economizer control is operating correctly—it is set to close the outdoor dampers at 75°F with 45 percent relative humidity. What could be the problem?

# Questions

1. What is the purpose of an economizer package?
2. How is the temperature controlled throughout the building using an economizer?
3. How does the economizer control the temperature inside the building?

# Unit 65: INTEGRATED FURNACE CONTROL MODULES

## Introduction

Some control manufacturers include furnace control modules that may be interchanged with another manufacturer's control. This greatly simplifies the problem of trying to stock a module for each manufacturer's equipment.

*Operation.*   The integrated furnace control module is equipped with a microprocessor that continually senses, analyzes, and controls the gas burner operation, the indoor fan motor, and the induced draft blower motor.

They are equipped with an LED indicator light that will glow continuously or flash on and off to alert the user about a problem with the heating system. For these modules to operate correctly, all wiring made in the field must be according to the equipment manufacturer's recommendations.

**Incorrectly wired modules.**   Following are some examples of the problems that can occur when incorrect wiring is installed.

1. When the electrical polarity of the power supply is reversed, a heating cycle may begin, but the system will be locked out after five or six seconds of operation.
2. The indoor fan motor will operate in the heating speed when the system is in the cooling mode if the Y wire in the control circuit is not connected to the Y terminal on the control module.
3. When the heating unit is not correctly grounded electrically, there will be nuisance furnace lockouts.

# Summary

- The integrated furnace control module has a microprocessor that continually monitors, analyzes, and controls the gas burner operation, the indoor fan motor, and the induced draft blower motor.
- For these modules to operate correctly, all wiring made in the field must be done according to the equipment manufacturer's recommendations.

# Service Call

A customer complains that a central residential heating unit will only operate occasionally. A check of the system shows that the heating unit is operating unpredictably and that the LED light on the control glows continuously. The burners will light and then, when the indoor fan motor starts, they will go out. The control module has a bad ground. The tech correctly grounds the module to the ground wire inside the unit wiring cable, and starts and completely checks the unit. The unit is operating normally. The technician is satisfied that the system is repaired.

# Student Troubleshooting Problem

A customer complains that a furnace is not heating his residence. The system has an integrated furnace control module. The customer says that this is the second time in a week that the furnace has quit working properly. The furnace is now operating correctly. What could be the problem?

# Questions

1. What is the purpose of the integrated furnace control module?
2. How can the user know when a problem occurs with the heating unit?
3. What will happen if the cooling control wire is not connected to the monitor?

# 5 Six-Step Troubleshooting Procedure

## Introduction

Because of the huge growth in the air-conditioning and refrigeration industry to meet the needs and demands of the equipment user for more accurate and efficient equipment control, many different types of electronic control systems and control devices are now available. These need preventive and corrective upkeep for continuous operation.

## Unit 66: SIX-STEP TROUBLESHOOTING PROCEDURE

### Troubleshooting

*Troubleshooting* is a term that we hear often, but what does it mean? Troubleshooting is sometimes misunderstood to mean simply fixing a unit when it does not work correctly. This repair is only a small part of the complete procedure. In addition, the troubleshooter can check equipment performance by comparing theoretical knowledge with the current operating efficiency. This evaluation must be made both before and after the repairs are made. Performance data and other general information for the different units are available to help in making a wise comparison of the operation of specific units.

The function of almost any type of installation depends almost entirely on the correct and continuous operation of many different types of controls and circuits. This need clearly supports the fact that the main job of a technician is the upkeep of these controls and equipment.

# Maintenance (Upkeep)

The term *maintenance* refers to all actions that a person does on an actual unit to keep the equipment in good working condition or restore it to good working condition. This upkeep includes inspection, testing, servicing, repair, and rebuilding. The correct upkeep of equipment cannot be done by an untrained person. The upkeep must be done by a person who is thoroughly familiar with the equipment. Proper upkeep requires the knowledge of how the equipment does its job, or rather, "the theory of operation."

A logical or systematic approach to troubleshooting is necessary in a technician's overall knowledge of the basics of the system on which the work is performed. Many hours have been lost because the technician took time-consuming, unplanned methods of trouble analysis. The troubleshooting analysis procedure presented in this section has been developed to make a path for the tech to follow toward a goal of effective upkeep and optimum operational use. Understanding the idea and basic importance of the suggested troubleshooting steps explained in the following pages makes the technician's ability to troubleshoot any unit using electronics for system control much easier.

# Logical Approach

Before entering into a discussion of troubleshooting details, let us establish the basic element on which satisfactory trouble analysis is based. This basic element—so often overlooked in almost every attempt at troubleshooting—is a logical approach.

**Logic.**    *Logic* is the system of principles of reasoning used in any branch of knowledge or study. Examining the definition of logic and its relation to this subject, it would be well to remember the words "principle of reasoning." In a broad sense, reasoning is logic.

When the complicated nature of most of today's electronic systems is examined, it should seem that the personnel who are given the job of keeping the equipment running must have specific training. These technicians are far from superhuman beings in understanding and maintaining such equipment. So what is the secret to their capability? It is simply and basically the fact that they have learned to *think* logically.

After learning the fundamental theories related to basic electronic circuitry, the technician will be ready to learn how these theories may be used to form a complete system designed to do a certain job. Armed with this knowledge and the logical approach, the technician can effectively divide any piece of equipment and test it in an orderly and professional manner. This procedure will save many wasted hours, compared with when a haphazard method is used.

# Six-Step Procedure

At this point a standardized approach toward electronic equipment troubleshooting and maintenance procedures will likely save many hours of needless equipment downtime and costly repairs by improper upkeep techniques. Another equally important point is to always keep electronic equipment ready to operate as it was designed and specified by the equipment manufacturer.

The six-step procedure is as follows:

1. Symptom recognition
2. Symptom clarification
3. Listing of probable faulty functions
4. Localizing the faulty function
5. Localizing trouble to the circuit
6. Failure analysis

**Step 1: Symptom recognition.**    Symptom recognition is the first step in the logical approach to trouble evaluation. To repair a unit, the technician must first learn whether it is working correctly. All units are designed for a specific job or group of jobs according to the needs established by the equipment manufacturer. This requires that a certain type of performance be done at all times. If it were possible to know when the equipment is performing poorly, it would be equally possible to keep the unit in the best condition. Therefore, recognizing trouble symptoms is the first step in unit troubleshooting.

A *trouble symptom* is a sign or an indicator of some problem in the unit. *Symptom recognition,* therefore, is the act of recognizing such a sign when it appears.

When you have a fever or a headache, you know your body has a problem somewhere. When your car's engine makes a knocking sound, you know that some part of the engine is not working correctly. Similarly, when a technician learns that a unit is not operating as it should, that operator knows that a problem exists somewhere in that unit.

*Normal and Abnormal Performance.*    Because a trouble symptom is an indication of an unwanted change in the unit operation, there must be some standard of normal performance to serve as a guide. By comparing the present operation with the normal, the technician knows that some trouble is present and makes a decision as to just what the symptom is.

Your normal body temperature is 98.6°F. A change above or below this temperature is an abnormal condition—a trouble symptom. If your body temperature is 102°F, by comparing this with the normal, it can be said that the symptom is an excess temperature of 3.4°F. Thus, the symptom has been exactly defined.

The normal television picture is a clear, correctly contrasted picture of an actual scene. It should be centered within the vertical and horizontal limits of the screen. If the picture suddenly begins to roll vertically, this should be recognized as a trouble symptom because it does not act in the expected normal manner.

The normal sound from a superheterodyne receiver is a clear duplicate of the sender's voice. If it sounds like the sender is talking from the bottom of a barrel filled with water, the receiver operator knows that this distortion is a trouble symptom.

*Performance Evaluation.*    While performing their assigned jobs, most electronic devices give information that an operator or technician can either see or hear. These senses of hearing and sight, therefore, allow the symptoms to be identified as either normal or abnormal unit operation. The display information may be the only job of the unit or it may be a secondary job to permit an efficiency evaluation.

Electrical information, to be presented as sound, must be applied to a loudspeaker or a headset. A visual display appears when the information is sent to a cathode-ray tube or to a meter built into the unit control circuit, which can be seen by the operator. Pilot lights also give a visual indication of unit operation.

Having knowledge of the normal equipment displays will help the technician to recognize an abnormal display. The ability to recognize these trouble symptoms is the first of the six-step troubleshooting procedure.

*Equipment Failure.*     Electronic equipment failure is the simplest of trouble symptoms to recognize. Equipment failure means that either the entire unit or some part of the unit is not operating, and will therefore show no performance display on the test instrument. The lack of all sound from a superheterodyne receiver when all the controls are in their correct positions shows a complete or partial failure. Similarly, the lack of a visible trace or picture on the screen of a cathode-ray tube, when all controls are correctly set, shows some form of unit failure. If a technician has seen the plate current reading on the tuning meter of a broadcast transmitter and the reading suddenly drops to zero, the technician has observed a failure.

*Degraded Performance.*     Even if the audible and visual signs are present, the unit may not be operating normally. When the unit is doing its job but giving the operator information that does not correspond with the design specifications, the performance is said to be *degraded.* Such performance must be corrected as quickly as a unit failure. This performance may range from nearly perfect to barely operating.

If a person was sick but went to work anyway, chances are that person's performance would be degraded during the illness. The person would still be doing the job, but not at full capacity.

*Know Your Equipment.*     Before troubleshooting an electronic unit, the technician must have a complete and thorough knowledge of its normal operating traits. Remember that any electronic unit, no matter how complex, is built by using the basic electronic circuits and devices. These components are combined so the wanted performance is given; therefore, a knowledge of electronic fundamentals will allow a person to analyze the performance of any unit. This basic knowledge can be supplemented by consulting equipment handbooks, instruction books, and upkeep orders related to each unit.

The information needed to evaluate unit performance is usually given by audible signals or visual displays on test instruments. However, unless the technician puts this knowledge of the unit to use in understanding these displays, their purpose will have little meaning. Unless personal knowledge is used to identify or question reported trouble symptoms, much time will be wasted getting involved in doing unnecessary troubleshooting work.

**Step 2: Symptom clarification.**     As a second step, the obvious or not so obvious symptom should be further defined. Most electronic circuits or systems have operational controls, additional indicating instruments other than the main indicating device, or other built-in aids for evaluating performance. These aids should be used at this point to see whether they will affect the unit being worked on or give additional data that further define the symptom.

Breaking out the test equipment and unit diagrams, and going headlong into testing procedures on only the original recognition of a trouble symptom, is an unrealistic approach. Unless the trouble symptom is first defined, the technician can quickly and easily be led away from the problem. The result, as before, would be loss of time, unnecessary waste of energy, and perhaps even a total dead-end approach.

Symptom clarification involves gaining a more detailed knowledge of the trouble symptom. Recognizing that the fluorescent screen of a cathode-ray tube is not lighted is not enough information to decide exactly what would be causing the trouble. This symptom could indicate that the cathode-ray tube is burned out, that a problem exists in the internal circuitry related to the tube, that the intensity control is turned too low, or even that the equipment is not turned on. Think of the time that might be wasted if the technician tears into the unit and begins testing when all that is needed is to turn the ON-OFF switch on, adjust the intensity control, or simply plug in the main power cord.

As discussed, the primary reason for placing symptom clarification as the second step in the six-step procedure is that many similar trouble symptoms can be caused by various unit faults. To continue efficiently, the technician must make a good decision as to which fault is probably causing the specific symptom in question.

Symptom clarification requires adequate knowledge of unit operation, which leads to an awareness and the ability to detect trouble symptoms. This knowledge is assumed throughout the remaining steps of the logical troubleshooting procedure. It cannot be overemphasized that both knowledge of how a unit works and a systematic approach to troubleshooting have equal importance. Having one without the other leaves the technician unprepared.

The purpose of symptom clarification is to help the technician to fully understand what the symptoms are and what they really mean. This elaboration is required to gain further insight into the problem.

**Step 3: Listing of probable faulty functions.**    The performance of the third step depends on the information gathered in the two previous steps and applies to units that have more than one functional area or unit. It allows the troubleshooter to mentally select the functional unit (or units) that is probably causing the problem, as is shown by the information obtained in steps 1 and 2. The technician makes the selection by logically thinking where the trouble could be located to produce the information just gathered.

The term *function* refers here to an electronic operation done by a specific area (or unit) of a piece of equipment. Frequently the terms *function* (an operational subdivision of a piece of equipment) and *unit* (its physical subdivision) are the same. A functional unit consists of all the parts needed for the unit to do its job, regardless of how these parts are packaged. Within this text the terms *function, unit,* and *functional unit* are used interchangeably, although in some equipment one or more circuits of a given function may have been built into a unit other than that having the title of the function.

The technician cannot confer with a piece of equipment the way a doctor confers with a patient. The technician must learn directly from troubleshooting by checking the information gathered and by using knowledge of how the equipment works electronically. Faulty unit or function selection requires the technician to use logic similar to that used by a medical doctor, auto mechanic, and other "technical doctor" when they search for the cause of an illness or problem.

Assume that a person who is continually bothered with a headache finally goes to the doctor. If the doctor elaborates on the symptom by checking the eyes, nose, and throat, taking temperature, and listening to the heartbeat, but then promptly sends the person to the operating room to have a foot amputated, the diagnosis would be questioned. Instead of taking such an illogical step, the doctor will decide from examination whether the most probable cause is poor eyesight. Only after making such a decision will the doctor prescribe a possible remedy.

Similarly, the technician who completes the first two steps of the six-step procedure and then picks just any test or repair procedure in an attempt to correct the trouble is a poor troubleshooter. The technician must first check the information gathered, and then, using knowledge of the equipment operation and the aids given in the correct technical manuals, make a technically sound decision about what is the probable cause of the recorded symptoms.

The millions of cells and thousands of parts in the human body could cause a lot of trouble to the medical doctor when making the diagnosis if each part or cell had to be checked separately to find the exact cause of the illness. Instead, the body is divided into functional groups, each containing many related parts. The doctor then relates the symptoms of the illness with the normal performance of the functional groups. Any evidence of abnormal performance gives the doctor a clue to the exact cause of the illness.

The abnormal performance indications noted in steps 1 and 2 should also give the technician some clues to the probable location of an electronic problem. Electronic equipment can have as many as 10,000 circuits, or 70,000 individual parts. Finding the faulty part by checking each of the 70,000 parts in turn is highly remote. The size of the task can be reduced by a factor of seven by checking the outputs of each circuit rather than checking each part separately.

However, performing 10,000 tests is still a big job. By dividing the 10,000 circuits into their normal groupings of electronic functional units—seven, a dozen, or two dozen—the technician can reduce the job to a practical number of tests. Whether the equipment contains thousands, hundreds, or only a few circuits, logical reasoning shows that the troubleshooting problem can be repaired more quickly and accurately by reducing the total number of circuits into a smaller number of groups.

**Step 4: Localizing the faulty function.**    The first three steps in the systematic approach to troubleshooting have dealt with the inspection of apparent and not so apparent equipment performance problems, plus a logical selection of the probable faulty functional units. Up to this point, no test equipment other than the controls and indicating devices physically built into the equipment have been used. No dust covers or equipment doors have been removed to allow access to any of the parts or internal adjustments. After evaluating the symptom information, the technician makes mental decisions as to the most probable areas in which the malfunction could be located.

Localizing the faulty function means that the technician must learn which of the functional units of the multiunit equipment is really at fault. This search is done by the technician methodically checking each faulty functional unit selection until finding the faulty unit. If none of the functional units on the list of selections shows poor performance, the tech must backtrack to step 3 and reevaluate the symptom information. Occasionally returning to step 2 (symptom clarification) may be necessary to retrieve more symptom-detailed data.

For step 4 the technician will use any factual equipment knowledge and skill in testing procedures. Using standard or specialized test instruments and understanding the test data will be important throughout this step and the remaining troubleshooting steps.

***Testing a Faulty Functional Unit Selection.***    The reason for step 4 is to learn which functional unit of the electronic equipment is responsible for the problem. The selection of any unit as the probable cause should always be based on equipment knowledge and basic electronic principles. In step 3 it was pointed out that there may be only a few or there may be many possibilities for problem functional unit selections. The number of selections depends entirely on the type of equipment and on the information gathered in steps 1 and 2 of this troubleshooting procedure. It is important to use a logical approach when deciding which faulty functional unit to test first.

***Factors to Consider.***    The simultaneous elimination of several functional units as the cause of the trouble should be an important part in learning which faulty unit to test first. This requires an inspection of the operational characteristics of the equipment in question to see whether correct test results from any one of the selections could also eliminate other units listed as possible causes.

Test point accessibility is also an important factor affecting the technician's logic in choosing a faulty functional unit for further testing. A *test point,* as described in some instruction books, is a special jack located at some accessible spot on the equipment, such as a front panel or chasis. The jack is electrically connected (directly or by means of a switch) to some important operating voltage or signal voltage. Actually, any point where wires join or where parts are connected can serve as a test point.

Another factor to consider is experience and history of repeated failures. Experience with similar equipment and related trouble symptoms, and the possibility of unit failure based on records of repeated failures, should have some bearing on the choice of a first test point. However, the selection should be based mainly on logical decisions formed from data gathered in the previous steps, without undue emphasis on personal experience and history of the equipment.

In summary, then, the first factors to be considered in selecting the first test point, listed in their usual order of importance are as follows:

1. The functional unit that will give the best information for eliminating other units, based on the data gained in steps 1, 2, and 3 of the six-step procedure, if a certain unit is obviously not the cause.
2. Accessibility of test points. For example, a test point might be avoided as a first choice because the equipment must be disassembled to get to this test point.
3. Experience and history of repeated failures. This factor is carefully considered in the light of data gathered in steps 1, 2, and 3.

***Analysis of the Tests.***    Once the units have been isolated, what happens if the first check does not pinpoint the faulty unit? Here the technician has either made an error in making one check or the results of the check were misunderstood, leading the technician in the wrong direction. This shows the importance of writing down the results. Having the information written down makes re-checking the location of the error much easier.

If a final check shows the suspected units to be satisfactory, it will be necessary to reevaluate the information learned from the previous checks. The question now is how far back to go.

All the information could be ignored and the procedure started again with step 1; however, this should not be necessary because the problem was basically confirmed when the trouble was first reported. Returning to step 2 would allow a reevaluation of the meter readings or other evidence that was present when the operating controls were used. A return to step 3 would permit a review of the list of faulty units previously prepared to ensure that a possible faulty unit was not overlooked.

*Trouble Verification.*    Having isolated the trouble to the actual faulty unit, it is now necessary to consider whether a fault in this unit could logically produce the trouble symptom indicated.

To find the faulty functional unit, proceed from symptom information to actual location. To verify the located functional unit, go in the reverse direction.

Step 4 involved the testing of equipment on a limited basis; that is, it has used only those tests that are necessary to isolate a faulty functional unit. A logical use of equipment knowledge and symptom analysis, together with the four factors—simultaneous elimination of several functional units, test point accessibility, personal experience, and history of repeated failures—help the technician to list the faulty functional unit selections made in step 3 and pick the most logical one for the first test. This logic was then used in the systematic selection of all the following test points. At each point, a new bit of information enabled the technician to narrow the trouble area until the faulty unit was located.

The completion of this step as presented in this text should leave no doubt about which functional unit is at fault. However, as a final check of the search, as discussed, verify the isolated faulty functional unit by backtracking and matching the theoretical trouble symptoms with those actually present.

**Step 5: Localizing trouble to the circuit.**    Steps 1 and 2 of the six-step troubleshooting procedure gave the troubleshooter initial diagnostic information. This information begotten from operating the controls gives visual evidence, such as meter readings or scope displays, so that the effects of the trouble can be further evaluated. Step 3 applies this information and the technician's equipment knowledge to select the functional units in a multiunit set which are most likely causing the trouble. Actual tests were done in step 4 so the faulty part of the equipment could be isolated.

In step 5 extensive testing will be done to localize the trouble to a specific circuit within the faulty functional unit. To do this, first isolate a group of circuits within a functional unit and arrange them according to a common electronic subfunction. Once the faulty circuit group has been located, the tests can be done to isolate the faulty circuit(s).

This procedure follows the same reasoning used throughout the six-step troubleshooting procedure, which is the continuous narrowing of the trouble area by making logical decisions and making logical tests. Such a process reduces the number of necessary tests, which not only saves time but also lessens the possibility of error.

**Step 6: Failure analysis.**   The recognition, verification, and descriptive information given in steps 1 and 2 help the technician to make a logical and good estimate of the selection of the faulty functional unit in step 3. In step 4 simple input–output tests were done. Step 5 continued deeper into the circuits, including the equipment being tested, and required more extensive tests to be done.

The final step in the six-step troubleshooting procedure requires that certain branches of the faulty circuit be tested to learn which is the faulty part. These branches are the interconnected networks related with each element of the transistor, electron tube, or other active device in the faulty circuit.

Step 6 calls for the faulty circuit parts to be repaired or replaced so that the equipment can be returned to maximum efficiency. However, locating the faulty part does not complete step 6. The cause of the failure must also be determined. It is quite possible that another failure will occur, and unless all faults are corrected, the trouble will repeat itself. The final step in failure analysis requires that certain records be kept. These may also point out consistent failures that could be caused by a design error. When this step has been finished satisfactorily, make the necessary repairs.

*Schematic Diagrams.*   Schematic diagrams show the detailed circuit arrangements of electronic parts (represented by symbols) that comprise the complete circuits within the equipment or unit. These diagrams show what is inside the unit and give the final picture of the equipment.

*Voltage and Resistance Charts.*   Once the faulty circuit has been isolated, voltages and resistances of the various circuit branches must be measured to learn which parts within the circuit are at fault. The measurement results must be compared with voltage and resistance charts or tables for evaluation. This information may be on the apron of its associated foldout schematic diagram, or it may be on a separate page in the service manual. The normal voltage and resistance reading to ground (or other test point) for each tube socket pin is given. Also listed are the conditions needed to watch the gain reading, such as control settings and equipment conditions.

*Types of Circuit Trouble.*   Step 1 explained that a piece of equipment was not running. Equipment failure results in a complete lack of working information that would normally be seen or heard by the equipment operator. The second type of abnormal performance discussed was degraded performance.

Regardless of the type of trouble symptom, the actual fault can usually be traced to one or more of the circuit parts—resistors, capacitors, etc.—within the unit. The actual fault may also be classified by the degree of failure. The complete failure or abnormal performance of a part, of course, falls in line with the previous use of these terms. These types of faults are easily covered.

A third degree of malfunction that is not always as obvious is the intermittent part malfunction. Intermittent, by definition, refers to something that alternately ceases and begins again. This definition applies to electronic part malfunctions. The part operates normally for a time, then fails completely or operates on a degraded level for a while, and then returns to normal operation. The cycling nature of this problem aids in learning that it exists. However, locating the faulty part is often difficult, because while the testing is being done in the circuit in which this part is located, it may be operating normally. Thus the technician will pass it as satisfactory, only to be faced with the problem again when the cycle completes itself.

*Isolation of Faulty Parts.*    The first step in isolating a faulty part within a circuit is to use the same input-conversion-output (ICO) method used in the previous steps. The output signal should be analyzed to help in making a correct selection of the parts or branch of the circuit that may cause the bad output. The voltage, duration, and shape of the output waveform may show the possible open or shorted parts or out-of-tolerance values. This step has two functions—it reduces to a minimum the number of test readings needed, and it helps indicate whether the faulty part, when located, is the sole cause of the problem.

The second step in isolating a faulty part is a visual inspection of the parts and leads in the circuit. Often this inspection will show burned or broken parts or bad connections. Open filaments in electron tubes may also be found in this check.

Voltage measurements at transistor leads or at the pins of electron tubes can be compared with normal voltages listed in available voltage charts, which provides valuable assistance when locating the trouble. This check will often help isolate the trouble to a single branch of the circuit. A separate branch is generally associated with each pin connection of the transistor or electron tube. Resistance checks at the same points are also useful in helping to find the trouble. Suspected parts can often be checked by a resistance measurement.

When a part is suspected of being defective, a good part may be substituted for it in the circuit. Keep in mind, however, that a fault not found in the circuit may also damage the new part. Another factor to consider before taking this step is that some circuits are critical and therefore substituting parts (especially transistors or electron tubes) may alter the circuit limits.

In some types of equipment the circuits are specifically designed for part substitution. For example, the plug-in circuit module is being used in many electronic circuits. It contains all the necessary parts (transistors, capacitors, and inductors) for a circuit branch or even the entire circuit. Once the trouble is traced to a module, replacement of the module is the only way of correcting the fault.

*Systematic Checks.*    Probable findings should always be checked first. Next, because of the safety practice of setting a voltmeter to its highest scale before making measurements, the points having the highest voltages should be checked (transistor collector and electron tube plate and screen grid). Then the parts having smaller voltages should be checked in a falling order of their applied voltage, that is, the transistor emitter and base or the electron tube cathode and control grid.

Voltage, resistance, and waveform readings are seldom identical to those listed in the service manual. The most important question is How close is close enough? In answering this, many factors are to be considered. Tolerances of the resistors, which greatly affect the voltage readings in the circuit, may be 20 percent, 10 percent, or 5 percent; in some critical circuits, precision parts are used. The tolerances marked or color coded on the parts are therefore one important factor. Transistors and electron tubes have a wide range of characteristics and will cause variations in voltage readings. The accuracy of the test instruments must also be considered. Most voltmeters have accuracies of 5 to 10 percent, whereas precision meters are more accurate.

For proper operation, critical circuits require voltage readings within the values specified in the manufacturer's technical manual; however, most circuits will operate satisfactorily if the voltages are off a small percentage. Important factors to consider are the symptoms and the output signal. If no output signal is produced at all, the technician should suspect that a large variation of voltages will be found in the trouble area. Trouble that results in circuit performance just out of tolerance, however, may cause only a slight change in circuit voltages.

***Finding the Faulty Part.***    The voltage and resistance checks discussed earlier show which branch within a circuit is at fault. The next step is to isolate the trouble to a particular part (or parts) within the branch. One way of doing this is to move the test probe to different points, where two or more parts are joined electrically, and measure the voltage or resistance between the test point and the ground. Generally, however, the correct values (particularly voltage) will be difficult to learn from these points on a schematic diagram and may not be available at all. Thus, this procedure should be used for making resistance checks to find shorts and openings in a branch. A better check to use when voltage readings are not normal is a thorough check of the value of each resistor, capacitor, and inductor in the branch. The instruments needed for these measurements include impedance bridges and Q-meters.

***Review of Previous Data.***    A review of all the symptom and test information gathered thus far will help isolate other faulty parts, whether the malfunction of these parts is due to the isolated problem or to some entirely unrelated cause (multiple malfunction).

To learn if a multiple malfunction exists, start with the isolated part and find out what effect the breakdown of this part has on equipment operation. If the isolated problem can cause all the normal and abnormal symptoms and indications gathered thus far, we can logically assume that it is the only part at fault. If not, the technician's knowledge of electronics and of the equipment itself is invaluable to be able to learn what other problems must happen to give all the symptoms and test data.

This final step brings an end to the six-step troubleshooting procedure. Each step narrows the trouble to the trouble area until the problem part is discovered. Review the steps to ensure that multiple problems do not exist and to verify the cause of the problem. Record the necessary actions taken.

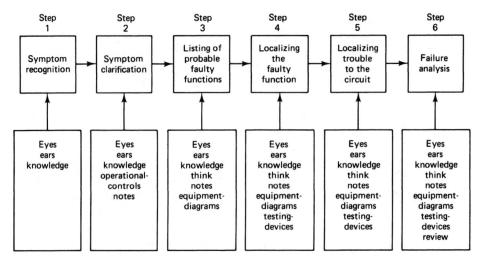

**Figure 66–1**
The six-step troubleshooting procedure.

By replacing the faulty part and rechecking equipment operation, the equipment can be returned to operation with the knowledge that the troubleshooting duties have been completed.

Although not directly connected with the troubleshooting steps outlined here, the technician should reorder any parts used in making the repair(s). Proper record keeping helps return the equipment to a satisfactory operating status once the trouble has been located.

The six-step troubleshooting procedure is summarized in Figure 66-1.

# 1 Lab Workbook

# CONTROL SYSTEM COMPONENTS
## Introduction

It is necessary that the air-conditioning and refrigeration technician be familiar with the parts that comprise a control system. Without this knowledge it will be more difficult to do the job. With a complete knowledge of controls and how they fit into the overall operation of the system, the technician will make better judgments on their use and replacement.

## Reference

Unit 3

## Materials

1. Transformer
2. 24 VAC thermostat
3. 24 VAC solenoid valve
4. 24 VAC contactor
5. Low-voltage wiring
6. Line-voltage wiring
7. Any necessary terminals

# Tools

1. Tool box or pouch
2. Ammeter

# Procedure

1. Use the line-voltage wiring to connect the transformer primary side to the line-voltage source. The voltage must match that which is needed by the transformer.
2. Use the low-voltage wiring to connect the R terminal on the transformer secondary to the R or V terminal on the thermostat.
3. Use the low-voltage wiring to connect the Y or C terminal on the thermostat and to connect the control circuit parts in electrical series, except for the contactor.
4. Use the low-voltage wiring to connect the contactor holding coil in electrical parallel to the solenoid valve coil.
5. Set the thermostat lower than room temperature.
6. Measure the current flowing through the control circuit. Record: _____
7. Measure the current flowing through the solenoid valve circuit. Record: _____
8. Measure the current flowing through the contactor coil circuit. Record: _____
9. Is this the same amount of current flowing through the entire circuit? _____
10. Can the heat anticipator be set to match the current flow? _____
11. Have the instructor check your work. Instructor's initials: _____
12. Disconnect all the parts and return them to their correct place.
13. Clean the work area.

# 2 Laboratory

# FUNDAMENTALS OF ELECTRIC CONTROL CIRCUITS

## Introduction

It is common practice to use electricity in carrying the thermostat's measurement of some change in the space temperature to the controlled parts of the system. The technician must be familiar with this process, or troubleshooting systems will be quite difficult and confusing.

## Reference

Unit 7

## Materials

1. Modulating type 24 VAC thermostat
2. Line-voltage to 24 VAC transformer
3. Modulating motor
4. Line-voltage wiring
5. Low-voltage wiring
6. Any necessary connectors

# Tools

1. Voltmeter
2. Tool box or pouch

# Procedure

1. Connect the primary side of the transformer to the line voltage. The line voltage must match that needed by the transformer.
2. Connect the R terminal on the secondary side of the transformer to the thermostat R or V terminal.
3. Connect the other thermostat terminals to the modulating motor according to the manufacturer's wiring diagram.
4. Turn on the line voltage to the transformer.
5. Check the voltage to ensure that it is the correct type.
6. Set the thermostat to the room temperature. What does the motor do? _____
7. Set the thermostat above room temperature. What does the motor do? _____
8. Set the thermostat below room temperature. What does the motor do? _____
9. Have the instructor check your work. Instructor's initials: _____
10. Disconnect all system parts and return them to their correct place.
11. Clean the work area.

# 3 Laboratory

# STARTERS AND CONTACTORS
## Introduction

In the air-conditioning and refrigeration industry, the largest switching load for the control system is the compressor motor. The various fans, water pumps, and other needed machinery wired in electrical parallel with the compressor motor also add to the load requirements of the starter or contactor. It is the job of the technician to understand how these controls and system work so that proper upkeep and repairs can be made.

## Reference

Unit 16

## Materials

1. Line voltage to 24 VAC transformer
2. Starter or contactor with 24 VAC holding coil
3. 24 VAC heat-cool thermostat
4. Small electric motor
5. Line-voltage wiring
6. Low-voltage wiring
7. Any necessary connectors

# Tools

1. Voltmeter
2. Ammeter
3. Tool box or pouch

# Procedure

1. Connect the line voltage to the transformer primary terminals.
2. Use low-voltage wiring to connect the R terminal on the transformer secondary to the R or V terminal on the thermostat.
3. Use low-voltage wiring to connect the C or Y terminal on the thermostat to one terminal on the contactor or starter holding coil.
4. Use line-voltage wiring to connect the line voltage to the line-voltage terminals on the starter or contactor.
5. Use line-voltage wiring to connect the motor according to the motor manufacturer's recommendations.
6. Turn on the line voltage to the transformer and to the contactor or starter.
7. Set the thermostat below room temperature. What did the starter or contactor do? _____
8. Check the motor amperage and record. _____
9. Set the thermostat above room temperature. What did the starter or contactor do? _____
10. What did the motor do? _____
11. Have the instructor check your work. Instructor's initials: _____
12. Disconnect all the parts and replace them in their correct place.
13. Clean the work area.

# 4 Laboratory

# ELECTROMAGNETIC RELAYS
## Introduction

The electromagnetic relay is used for many purposes throughout the air-conditioning and refrigeration industry. Each new and imaginative relay need is specified by the equipment manufacturer. When a technician understands the operation of these relays, it is then much easier to learn their specific use in a control system. Keeping up with the different relays and their uses takes much time and study, but the technician must do it to stay competitive.

## Reference

Unit 17

## Materials

1. 24 VAC heat-cool thermostat
2. Line voltage to 24 VAC transformer
3. 24 VAC relay
4. Line-voltage fan motor
5. Line-voltage wiring
6. Low-voltage wiring
7. Any necessary connections

# Tools

1. Voltmeter
2. Ammeter
3. Tool box or pouch

# Procedure

1. Use the line-voltage wire to connect the line voltage to the primary terminals.
2. Use low-voltage wire to connect the transformer R terminal to the thermostat R or V terminal.
3. Use low-voltage wire to connect the G thermostat terminal to one of the coil terminals on the fan relay.
4. Use low-voltage wire to connect the other fan relay coil terminal to the C transformer secondary terminal.
5. Use line-voltage wire to connect one line-voltage terminal on the relay to a fan motor terminal. Connect the other fan motor lead to the common line-voltage wire.
6. Turn on the line voltage.
7. Set the thermostat fan switch to the ON position. What does the fan motor do? _____
8. Check the amperage draw of the fan motor. Record: _____
9. Turn the thermostat fan to the AUTO position.
10. Set the thermostat system switch to cool.
11. Set the thermostat below room temperature. What happens to the fan motor? _____ Why? _____
12. Have the instructor check your work. Instructor's initials: _____
13. Disconnect all wiring from the parts. Replace the parts and wiring in their correct place.
14. Clean the work area.

# 5 Laboratory

# THERMAL RELAYS
## Introduction

Thermal relay contacts and terminals are the same as the electromagnetic relay. The method of pulling the movable contact against the stationary contact is the major difference between them. The thermal relay uses a bimetal blade with a heater coil wound around it. The bimetal heater leads are connected to the control circuit power supply through the thermostat. They are used on electric heaters and other jobs that need some type of delay period during their operation.

## Reference

Unit 18

## Materials

1. 24 VAC heat-cool thermostat
2. Line voltage to 24 VAC transformer
3. Double-pole single-throw thermal relay
4. Small, line-voltage heating element
5. Line-voltage wiring
6. Low-voltage wiring
7. Any needed connections

# Tools

1. Tool box or pouch
2. Voltmeter
3. Ammeter

# Procedure

1. Use line-voltage wire to connect the line-voltage terminals of the transformer to the line-voltage power supply.
2. Use low-voltage wire to connect the transformer secondary terminal R to the thermostat R or V terminal.
3. Use low-voltage wire to connect the H or W thermostat terminal to one low-voltage lead on the thermal relay.
4. Use low-voltage wire to connect the other low-voltage lead on the relay to the ground terminal on the transformer secondary.
5. Use line-voltage wire to connect one lead of the fan motor and the heat strip to one line-voltage terminal on the thermal relay.
6. Use line-voltage wire to connect one wire of the line voltage to the other line-voltage terminal on the thermal relay.
7. Use line-voltage wire to connect the remaining terminals on the fan motor and the heat strip to the other side of the line voltage.
8. Turn on the line voltage.
9. Set the thermostat system switch on the heat position.
10. Set the thermostat above room temperature. Quickly check the voltage and amperage to the fan motor and the heat strip after they come on. Leave the heat strip on for only a few seconds or it may burn out or blow the overheat protector. Record the amperage and voltage. A. _____ . V. _____ .
11. Have the instructor check your work. Instructor's initials: _____
12. Disconnect all the parts.
13. Replace all the parts in their proper place.
14. Clean the work area.

# 6 Laboratory

# POTENTIAL STARTING RELAYS
## Introduction

Potential relays are operated by voltage (potential) that is generated in the start winding of the compressor motor. These type relays are operated only with voltage that is sensitive to electromagnetic coils. Potential relays are the cause of many unnecessary compressor replacements, because the technician does not understand how to troubleshoot them. The technician should take the time to learn how they operate and apply this knowledge when working on these systems.

## Reference

Unit 19

## Materials

1. Compressor motor
2. Potential relay to fit the compressor motor
3. 24 VAC contactor
4. 24 VAC heating-cooling thermostat
5. Line voltage to 24 VAC transformer
6. Starting capacitor
7. Line-voltage wiring
8. Low-voltage wiring
9. Correct connectors and terminals

# Tools

1. Voltmeter
2. Ammeter
3. Tool box or pouch

# Procedure

1. Use line-voltage wire to connect the line voltage to the primary side of the transformer.
2. Use line-voltage wire to connect the line voltage to the terminals on the line side of the contactor.
3. Use line-voltage wire to connect one side of the line voltage to relay terminal 5 and to the common compressor motor terminal.
4. Use line-voltage wire to connect the other side of the line voltage to the compressor motor run terminal.
5. Use line-voltage wire to connect the compressor motor run terminal to terminal 1 on the potential relay. Place the start capacitor in this line.
6. Use line-voltage wire to connect the potential relay terminal 2 to the compressor motor start terminal.
7. Use low-voltage wire to connect the transformer terminal R or V on the thermostat.
8. Use low-voltage wire to connect the Y or C terminal on the thermostat to one side of the contactor holding coil terminal.
9. Use low-voltage wire to connect the other terminal on the contactor holding coil to the common side of the transformer.
10. Turn on the line voltage.
11. Place the ammeter on the wire to the compressor motor common terminal.
12. Connect the voltmeter leads on the common and run terminals of the compressor motor.
13. Set the thermostat below room temperature.
14. Watch and record the starting voltage and amperage. A. _____ . V. _____
15. Did the voltage drop during the start period? _____
16. What was the starting amperage? _____
17. What was the running amperage? _____
18. Set the thermostat above room temperature.
19. Turn off the line voltage to the unit.
20. Remove the start capacitor from the start circuit. Use line-voltage wire to connect the wiring between the compressor motor run terminal and terminal 2 on the relay.

21. Place the ammeter on the compressor motor common wire.

22. Connect the voltmeter leads between the compressor motor common and run terminals.

23. Turn on the line voltage.

24. Set the thermostat below room temperature.

25. Quickly observe the starting amperage of the compressor motor. Record. _____

26. Quickly observe the voltage during the starting period. Record. _____

27. Was there any difference in the voltage and current readings when the start capacitor was in the circuit? _____ Why? _____

28. Set the room thermostat above room temperature.

29. Place the ammeter on the wire between relay terminal 2 and the compressor motor start terminal.

30. Set the thermostat below room temperature.

31. Quickly read to starting and running current. Record. S. _____. R. _____.

32. Set the thermostat above room temperature.

33. Have the instructor check your work. Instructor's initials: _____

34. Turn off the line voltage.

35. Disconnect the line voltage from the source.

36. Remove the wiring from all parts.

37. Place all parts in their correct place.

38. Clean the work area.

# 7 Laboratory

# SIZING POTENTIAL STARTING RELAYS
## Introduction

Many times it is necessary that the technician learn the correct potential relay for the compressor motor. When the relay is missing or is so badly damaged that the numbers cannot be read, some technicians try to use the shotgun approach in selecting a potential relay. This is the time that a technician must use the knowledge gained in this laboratory exercise.

When the shotgun approach is used, the relay may work for a while and then either go bad or burn out the compressor motor. If the wrong relay is used, it may not work at all.

## Reference

Unit 19

## Materials

1. Line to 24 VAC transformer
2. 24 VAC thermostat
3. 24 VAC contactor
4. Compressor motor
5. Jumper wire with a spring-loaded switch and clips
6. Line-voltage wire
7. Low-voltage wire
8. Any needed connections

# Tools

1. Voltmeter
2. Tool box or pouch

# Procedure

1. Use line-voltage wire to connect the line-voltage side of the transformer to the voltage source. Ensure that the power is turned off.
2. Use low-voltage wire to connect the R terminal on the secondary side of the transformer to the R or V terminal on the thermostat.
3. Use low-voltage wire to connect the Y or C terminal on the thermostat to one terminal of the contactor holding coil.
4. Use low-voltage wire to connect the other contactor holding coil terminal to the ground terminal on the transformer secondary.
5. Use line-voltage wire to connect the power source to the line-side terminals on the contactor.
6. Use line-voltage wire to connect the load-side terminals of the contactor to the common and run terminals of the compressor motor.
7. Place the jumper wire with the clips and switch between the compressor motor run and start terminals.
8. Place the voltmeter between the compressor motor start and run terminals. Ensure that the meter is set on a scale high enough to keep from damaging the meter.
9. Turn on the line voltage.
10. Push the switch to the closed position.
11. Set the thermostat below room temperature.
12. Just as the motor starts turning, release the switch so it will open the starting circuit.
13. When the compressor motor reaches its full running speed in approximately two or three seconds, read and record the voltage. _____
14. Set the thermostat above room temperature.
15. Have the instructor check your work. Instructor's initials: _____
16. Turn off the line voltage.
17. Disconnect all the parts.
18. Place all the parts in their correct place.
19. Clean the work area.

# 8 Laboratory

# PRESSURE CONTROLS
## Introduction

A compressor motor can be protected from over-current damage or possible damage to the bearings from the lack of refrigerant and oil by using pressure controls. These controls are used to sense high or low refrigerant pressures. The equipment manufacturer will usually recommend settings for both controls to best protect the equipment. The technician must be able to replace, install, or adjust these controls to maintain correct compressor motor protection.

## Reference

Unit 23

## Materials

1. Working refrigeration system or a stand made so that pressure can be put on the controls
2. High-pressure control
3. Low-pressure control, or combination high/low-pressure control

## Tools

1. Refrigerant gauge manifold
2. Tool box or pouch

# Procedure

1. Install both the high- and low-pressure controls on the system.

2. Start the system.

3. Assuming that HCFC-22 is the refrigerant type, set the low-pressure control to open its contacts at 45 psig and close its contacts at 75 psig.

4. Set the high-pressure control to open its contacts at 375 psig and close its contacts at 350 psig.

5. Install the refrigerant manifold gauges.

6. Slowly close off the suction service valves and read the pressure when the low-pressure control opens the circuit. Record. _____

7. Slowly open the suction service valve and read the pressure when the low-pressure control closes the circuit. Record. _____

8. Place a piece of cardboard over the condenser and read the pressure when the high-pressure control opens the circuit. Record. _____

9. Remove the cardboard and read the pressure when the high-pressure control closes the circuit. Record. _____

10. Are these the pressures on which the pressure controls were set? _____

11. Make any needed adjustments to bring the control to the needed settings, and recheck.

12. Have the instructor check your work. Instructor's initials: _____

13. Remove all gauges and controls as needed.

14. Return all parts to the correct place.

15. Clean the work area.

# 9 Laboratory

# TEMPERATURE CONTROLS
## Introduction

Any building, regardless of location, age, or architecture, can now be equipped with year-round comfort and safety with a well-designed, properly installed, and carefully controlled air-conditioning system. Keeping perishable foods at the necessary temperature would also be a problem if automatic temperature controls had not been designed and manufactured. The technician should be familiar with the different type temperature controls available to provide the customer with correct service and care.

## Reference

Unit 25

## Materials

1. 24 VAC heat-cool thermostat
2. Line voltage to 24 VAC transformer
3. 24 VAC gas valve
4. Low-voltage wiring
5. Line-voltage wiring
6. Necessary connections

# Tools

1. Plumb bob or small level
2. Tool box or pouch
3. Ammeter
4. Thermometer

# Procedure

1. Mount the thermostat on a solid board and level it according to the manufacturer's installation instructions.
2. Use line-voltage wire to connect the primary side of the transformer to the line voltage.
3. Use low-voltage wire to connect the R terminal on the transformer to the R or V terminal on the thermostat.
4. Use low-voltage wire to connect the H or W terminal on the thermostat to one terminal on the gas valve.
5. Use low-voltage wire to connect the other side gas valve terminal to the other terminal on the transformer secondary.
6. Turn on the line voltage.
7. Place the ammeter on the low-voltage wire between the thermostat R or W terminal to the gas valve.
8. Set the thermostat above room temperature.
9. When the gas valve is energized, read the amperage draw through the wire. Record. A. _____
10. If the ammeter is on the lowest scale and the current cannot be read, what could be done to get a current reading? _____
11. Now do what you wrote for question 10.
12. What is the current draw now? _____
13. Using the correct formula, does this match the heat anticipator setting on the thermostat? _____
14. Set the heat anticipator to match the current draw.
15. Place the thermometer as close as possible to the thermostat sensing element.
16. Turn off the line voltage.
17. After the thermometer reading is steady, move the thermostat setting until the contacts just make. Check the thermostat temperature and match it to the thermometer reading. If they do not match, the thermostat must be calibrated. If it cannot be recalibrated and the temperature difference is more than 3°F, the thermostat should be replaced. Be sure to move away from the thermostat quickly to keep from affecting the reading by the heat from your hands.

18. If it is an adjustable thermostat, adjust it to read what the thermometer is reading.
19. Have the instructor check your work. Instructor's initials: _____
20. Disconnect all the parts.
21. Return all the parts to their proper places.
22. Clean the work area.

# 10 Laboratory

# FAN CONTROLS
## Introduction

Automatic controls are on the job 24 hours a day to operate heating and cooling systems safely and economically and to provide comfortable living conditions. One such control is the fan control. The fan control is an operating control that starts and stops the indoor fan when the temperature inside the heat exchanger reaches a certain temperature. The technician should fully understand the fan control so that it can be adjusted correctly, and certain problems can be solved by making other adjustments.

## Reference

Unit 30

## Materials

1. Adjustable fan control
2. Operating furnace or station setup with means of heating the fan control sensing element

## Tools

1. High-range thermometer
2. Tool box or pouch

# Procedure

1. Install the fan control.

2. Wire the furnace according to the furnace manufacturer's recommendations.

3. Set the fan control to close its contacts at 135°F.

4. Set the control to open its contacts at 100°F.

5. Insert the thermometer into the discharge air plenum as close as possible to the fan control sensing element.

6. Set the thermostat above room temperature.

7. The main gas flame should come on.

8. When the thermometer scale reaches 135°F the control should close the fan circuit. If it does not, adjust the control so that it will.

9. Set the thermostat below room temperature. The main burner flame should go out.

10. When the thermometer scale reads 100°F the control should open the fan circuit. If it does not, adjust it so that it will.

11. Did any of the control adjustments need more than 5°F to match the thermometer reading? _____

12. If an adjustment of more than 5°F was needed it is good practice to replace the control.

13. Have the instructor check your work. Instructor's initials: _____

14. Disconnect all the wiring and place the fan control in the correct place.

15. Clean the work area.

# 11 Laboratory

# LIMIT CONTROLS

## Introduction

Limit controls are used on all types of warm-air furnaces to keep the discharge air plenum temperatures from getting too high and possibly causing a fire or ruining the equipment. They have either NC SPST or NC SPDT switch contacts, depending on their specific use. The contacts on the SPST type open on a rise in temperature and close on a fall. SPDT contact types open the NC contacts on a rise in temperature and at the same time close the NO contacts. The NO contacts are used to make the fan motor circuit on horizontal and counterflow furnaces when the temperature inside the furnace gets too high. These are safety controls that only operate when an overheated condition occurs. The technician should be familiar with what causes the controls to operate so that repairs can be made quickly and economically.

## Reference

Unit 31

## Materials

1. SPDT limit control
2. Operating furnace or workstation setup to provide heat to the control

## Tools

1. Tool box or pouch
2. High-range thermometer

# Procedure

1. Install the limit control.
2. Insert the thermometer into the discharge air plenum as close as possible to the limit control sensing element.
3. Adjust the limit control to open the circuit at 190°F.
4. Set the thermostat above room temperature.
5. The main gas should come on.
6. Block a part of the return air through the furnace.
7. Read the temperature on the thermostat.
8. When the thermometer scale reaches 190°F, the limit control should stop the main burner flame. If it does not, adjust it so that it will.
9. What happened when the temperature reached 190°F? _____ . Is this normal for this type limit control? _____
10. Remove the airflow blockage and allow the furnace to cool down. At what temperature did the limit control start the main burner? _____. Is this normal for a limit control? _____
11. Have the instructor check your work. Instructor's initials: _____
12. Remove the limit control from the furnace.
13. Replace all parts in their proper place.
14. Clean the work area.

# 12 Laboratory

# DISCHARGE GAS TEMPERATURE PROTECTOR

## Introduction

High discharge gas temperature is a major reason for early compressor failure. The protector helps in stopping compressor damage or failure caused by high discharge gas temperatures. They may be used on either rotary or reciprocating compressors. The technician should be familiar with the reasons for problems that cause this control to operate. It is a safety control that operates only when a problem occurs with the unit.

## Reference

Unit 49

## Materials

Air-cooled condensing unit equipped with a discharge gas temperature protector

## Tools

1. Refrigerant gauge manifold set
2. Tool box or pouch
3. Ammeter
4. Voltmeter
5. Piece of cardboard as large as the condenser

# Procedure

1. Correctly install the refrigeration gauge manifold set.
2. Place the ammeter on the wire to the compressor motor common terminal.
3. Place the voltmeter leads across the load side of the contactor contacts.
4. Start the unit.
5. What are the starting voltage and amperage? Record. V. _____.
   A. _____
6. What are the running voltage and amperage? Record. V. _____.
   A. _____
7. What are the pressure readings? H. _____. L. _____
8. Place the cardboard over the condenser and read the refrigerant pressures.
9. When the compressor motor stops, record the voltage. _____.
   Amperage. _____. Suction pressure. _____. Discharge pressure.
   _____
10. When the compressor motor stopped, was there a change in any of the readings taken on the start-up of the unit? _____ . What were they?
   _____
11. Remove the cardboard and allow the unit to run and cool down.
12. Have the instructor check your work. Instructor's initials: _____
13. Remove the gauges, voltmeter, and ammeter from the unit.
14. Clean the work area.

# 13 Laboratory

# DELAY-ON-MAKE ADJUSTABLE SOLID-STATE TIMERS

## Introduction

Solid-state delay-on-make timers are used for staging compressor start-up or delaying motors and other components that would place an overload on the electrical system. They may also be used to delay the start of a compressor that has been stopped so that the refrigerant pressures would have time to equalize, thus reducing the starting load on the compressor motor. Almost all equipment manufacturers are using this type relay to protect their equipment from an overloaded start. The technician should be familiar with these controls and their operation to allow more efficient and effective service.

## Reference

Unit 50

## Materials

1. Delay-on-make adjustable electronic timer
2. Line-voltage wire
3. Low-voltage wire
4. Line voltage to 24 VAC transformer
5. 24 VAC heat-cool thermostat

6. Electric motor or compressor
7. 24 VAC contactor
8. Needed terminals

# Tools

1. Tool box or pouch
2. Voltmeter
3. Wristwatch with a second hand
4. Small level or plumb bob

# Procedure

1. Mount the thermostat on a stable board.
2. Use line-voltage wire to connect the transformer primary to the line voltage.
3. Use line-voltage wire to connect the line voltage to the line-side terminals on the contactor.
4. Use line-voltage wire to connect the compressor to the load side of the contactor to the common and run terminals on the compressor motor.
5. Use low-voltage wire to connect the R terminal on the transformer to the R or V terminal on the thermostat.
6. Use low-voltage wire to connect the thermostat terminal Y or C to terminal 1 on the solid-state time-delay relay (use the control manufacturer's wiring recommendations).
7. Use low-voltage wire to connect the time-delay relay terminal 3 to one of the holding coil terminals on the contactor.
8. Use low-voltage wire to connect the other terminal on the contactor holding coil to the other terminal on the transformer secondary.
9. Turn on the line voltage.
10. Set the thermostat below room temperature.
11. Check the amount of time before the compressor or motor starts. Record.

_____

12. Set the thermostat above room temperature.
13. Turn the adjustment knob on the control toward the longer direction.
14. Set the thermostat below room temperature.
15. Check the amount of time before the compressor motor starts. Record.

_____

16. Set the thermostat above room temperature.
17. Turn the adjustment knob on the control toward the shorter time direction.

18. Set the thermostat below room temperature.
19. Check the amount of time before the compressor or motor starts. Record.
   _____

20. Have the instructor check your work. Instructor's initials: _____
21. Disconnect all the parts.
22. Replace all the parts to their proper place.
23. Clean the work area.

# 14 Laboratory

# CURRENT-SENSING RELAY
## Introduction

The current-sensing relay is an SPDT relay used on electrical systems when there is a need to electrically separate the relay-controlled voltage from the circuit it is controlling. The relay controls the circuit by wires from the circuit with the correct polarity passing through the sensing loop and being sensed. The current flow generates a signal in the sensing loop that is increased by an internal solid-state circuit to operate the relay. It uses two sets of contacts—one is NC, the other is NO. The technician should be familiar with the operation of these relays, because when not sized correctly they can cause a lot of problems.

## Reference

Unit 57

## Materials

1. Current-sensing relay
2. 24 VAC heat-cool thermostat
3. Line voltage to 24 VAC transformer
4. Electric motor rated for 12 amps or less
5. 24 VAC relay
6. 24 VAC fan relay
7. Line-voltage wiring
8. Low-voltage wiring
9. Needed connectors

# Tools

1. Tool box or pouch
2. Ammeter

# Procedure

1. Mount the thermostat on a solid board and level.
2. Use line-voltage wire to connect the primary side of the transformer to the line-voltage source.
3. Use low-voltage wire to connect the R terminal on the transformer to the R or V terminal on the thermostat.
4. Use low-voltage wire to connect the F or G terminal on the thermostat to one terminal of the fan relay coil.
5. Use low-voltage wire to connect the other fan relay coil terminal to one side of the NC contact terminals on the current-sensing relay.
6. Use low-voltage wire to connect the other NC terminal on the current-sensing relay to the other transformer terminal.
7. Use low-voltage wire to connect the NO contact terminal to one terminal on the 24 VAC fan relay coil.
8. Use low-voltage wire to connect the other 24 VAC fan relay coil terminal to the ground terminal on the transformer.
9. Use line-voltage wire to connect one motor terminal to the line voltage source.
10. Use line-voltage wire to connect the other motor terminal through the sensing loop on the current-sensing relay and to the NO terminal on the fan relay.
11. Use line-voltage wire to connect the other NO contact terminal on the fan relay to the line-voltage source.
12. Turn on the line voltage.
13. Move the thermostat fan switch to the ON position. What happened? _____
14. Check the current draw through the sensing loop on the sensing relay. Record. _____
15. Turn off the line voltage and make another pass of wire through the sensing loop.
16. Turn on the line voltage.
17. What happened? Record. _____
18. Have the instructor check your work. Instructor's initials: _____
19. Disconnect all the parts.
20. Return all the parts to their proper place.
21. Clean the work area.

# 15 Laboratory

# SOLID-STATE CRANKCASE HEATERS

## Introduction

Crankcase heaters are used to keep the temperature inside the compressor crankcase high enough to stop liquid refrigerant from collecting in the oil during cool or cold weather conditions or during the compressor OFF cycle. Liquid refrigerant mixing with the oil could possibly cause damage requiring compressor replacement. The technician should be able to determine if the crankcase heater is what caused the problem.

## Reference

Unit 63

## Materials

1. Solid-state crankcase heater or unit equipped with one
2. 24 VAC room thermostat
3. Line to 24 VAC transformer
4. Contactor with 24 VAC holding coil and a set of NC contacts and two sets of NO contacts
5. Line-voltage wiring
6. Low-voltage wiring
7. Needed connectors

# Tools

1. Tool box or pouch
2. Ammeter
3. Voltmeter
4. Ohmmeter

# Procedure

1. If a unit is not available with an electronic crankcase heater already on it, use steps 2 through 8.
2. Mount the thermostat on a solid board and level it.
3. Use line-voltage wire to connect the primary winding of the transformer to the line-voltage source.
4. Use low-voltage wire to connect the R terminal on the transformer secondary to the R or V terminal on the thermostat.
5. Use low-voltage wiring to connect the Y or C thermostat terminal to one of the contactor holding coil terminals.
6. Use low-voltage wiring to connect the other holding coil terminal to the G terminal on the transformer secondary.
7. Use line-voltage wiring to connect one of the crankcase heater leads to the line-voltage source.
8. Use line-voltage wiring to connect the other crankcase heater lead through the contactor NC contact.
9. Use line-voltage wiring to connect the power source to the line terminals on the contactor.
10. Use line-voltage wiring to connect the contactor load-side terminals to the common and run terminals of the compressor motor.
11. Turn on the line-voltage power source.
12. Use the voltmeter to check the voltage to the heater leads.
13. Use the ammeter to check the current draw through the heater. If the current is too low for the meter to read accurately, wrap the lead around the meter tong and measure again. Record. _____
14. Slightly heat the crankcase heater while measuring the current flow. Record. _____
15. Set the thermostat below room temperature.
16. Measure the voltage and amperage of the crankcase heater. Record.
    A. _____.V. _____.
17. Was there any change in either the voltage or the ammeter? Record.
    V. _____. A. _____.

18. Set the room thermostat above room temperature to stop the compressor.
19. Continue to measure the voltage and amperage.
20. What happened? Record. _____
21. Is this normal? Record. _____
22. Have the instructor check your work. Instructor's initials: _____
23. Turn off the line-voltage power source.
24. If you put the system together, disassemble it. If not, do not make any wiring changes unless told to do so by the instructor.
25. Replace all the parts in their proper place.
26. Clean the work area.

# Glossary

## A

**Accumulator**    A shell placed in the suction line near the compressor suction connection to keep liquid refrigerant from entering the compressor by vaporizing the liquid. For example, an accumulator is used on some heat pump systems and commercial refrigeration systems to catch any liquid refrigerant that passes through the evaporator coil. It will hold the liquid until it is all vaporized. There is usually an oil pickup opening that will let the oil return to the compressor so that lubrication of the parts can be kept at a maximum.

**Actuator**    That part of a regulating valve that changes fluid, thermal, or electrical energy into mechanical motion to open or close valves, dampers, etc. For example, an actuator is the modulating motor that changes the position of the controlled component to satisfy the demand of the thermostat or other controller.

**Adjustable differential**    A means of changing the difference between the control cut-in and cut-out settings, for example, the number of pounds between when the low-pressure control will open and close its contacts. The control is set to open its contacts at 15 psig and close them at 30 psig. The control can be adjusted to close the contacts at 35 psig or open them at 10 psig.

**Air-sensing thermostat**    A thermostat that operates because of air temperature changes. The sensing bulb is located in the air stream. For example, a room thermostat used to control a residential air-conditioning system is an air-sensing thermostat. The bimetal senses the air temperature and causes the contacts to either open or close as the air temperature needs.

**Ambient compensated**    Such a control is used so that varying the air temperature at the control does not have much effect on the control setting. For example, this type control operates between the difference in temperature of the indoor and outdoor air. When in the cooling mode and the outdoor air temperature drops, the amount of cooling for the building is lowered. Likewise, when the outdoor air temperature rises, the amount of cooling to the building increases. The opposite is true for the heating operation.

**Ambient temperature**    The temperature of the surrounding air in the area around the control body and power element. For example, the ambient air temperature is the temperature of the air surrounding the outdoor unit. Ambient air could also be the air that surrounds the indoor unit, for example, the return air temperature.

**Ampacity**    The current capacity of an electrical conductor, depending on what type of material it is made from, the conductor size, and the type of insulation on the conductor. For example, a number 10 AWG conductor is rated at 30 amps for 100 feet of length. When more than 100 feet is used, the current-carrying capacity is reduced because of the increased resistance. The conductor manufacturer's specifications must be used to keep from overloading the circuit.

**Amplifier**    An electronic device that increases the electron flow in a circuit. For example, some of the signals sent to a controller are so small they are not usable. An amplifier is used to increase the number of electrons that flow to the controller so it can use them to operate a control.

**Amps**    The measure of electric current flowing through an electrical circuit. For example, when an electric motor is running, or trying to run, the amperage or current flowing through the conductor can be measured with an ammeter. If the current flowing to an electric motor is measured as 10, it is usually expressed as 10 amps.

**Aquastat**    A control used on hydronic heating and cooling systems to measure the temperature of the fluid in a pipe or boiler and make a change in the operation of the controlled equipment. For example, an aquastat is usually placed at the outlet of a boiler or chiller barrel to measure the temperature of the water as it leaves the equipment. When in the cooling mode, it will cause the equipment to operate to keep a temperature of about 45°F. When in the heating mode, the aquastat tries to keep a boiler outlet water temperature as needed by the building. This temperature usually varies with the outdoor temperature.

**Armature**    That part of an electric motor, relay, or generator that is moved by magnetism. For example, the armature of an electric motor is the part that turns inside the motor housing and is connected to some external load such as a fan, pump, or any other component that needs moved in order to cause the wanted action. The armature of a relay is that part that carries the movable contact into or away from the stationary contact to make or break an electrical circuit.

**Automatic changeover**    The automatic change from one operation to another; otherwise, any changes must be done manually. For example, an automatic changeover control senses the water temperature inside the pipe. When the temperature rises to a certain degree, the control will change the operation of the thermostat to the heating mode. Likewise, when the temperature of the water inside the pipe drops to a certain degree, the automatic changeover control will change the thermostat action to the cooling mode. These temperatures will usually vary from system to system, depending on the design engineers' calculations and system operating traits.

**Automatic control**    The control of the various operations of a unit without manual adjustment. For example, an automatic control is one that causes the unit to operate as wanted without someone manually making the adjustments.

**Automatic flue damper**    A damper placed at the outlet of the draft diverter on gas-burning equipment, which opens when operation is needed and closes when the need has been met. These dampers are used to conserve gas by keeping the venting of flue gas from the equipment during the OFF cycle. For example, automatic flue dampers are used to operate on the temperature of flue gases. They will open when the temperature rises to a certain degree and will close when the temperature has dropped to a certain degree to keep the furnace from venting all the hot gases to the atmosphere. This process would require that the furnace heat exhaust air every time heating is wanted.

**Automatic recycle**    The contacts in this control automatically return to their original position after actuation when the pressure or temperature returns to normal. For example, when the gas fails to light within a definite period, an electronic ignition system will reset its contacts to start another trial for ignition. This will happen for a certain number of tries then it will lock out the system. Each control usually has a different number of tries.

**Auxiliary contacts**    A set of contacts used to perform a secondary function, usually in relation to the operation of the main contacts. For example, if the main contacts in a starter or contactor fail to function correctly, the set of auxiliary contacts will close to alert the user that a problem exists.

# B

**Bellows**    A corrugated metallic diaphragm with a metal cup. Often a complete power unit is referred to as a bellows. For example, the bellows in a pressure control is what allows the pressures to change in the control to operate the contacts. Bellows are sometimes used in thermostats in place of bimetal strips.

**Bimetal**    Two pieces of metal welded together so that they act as one piece. Bimetals are used in thermostats, thermometers, fan and limit controls, and other heat-sensing devices. A bimetal will bend in one direction when heated and return to the original position when cooled to the original temperature. For example, the bimetal used in thermostats will tend to straighten out when heated, taking the movable contact with it so that it will touch the stationary contact when the needed drop in temperature has been reached. When the bimetal is cooled, it will tend to wind into a tighter spiral and either open the heating contact or move toward the position that will change the movable contact. This will then make contact with the cooling stationary contact to start the cooling equipment.

**Boiler**    A closed vessel in which steam is generated or water is heated. For example, the steam or hot water used in a large air-conditioning system or industrial process is heated by a boiler. The boiler is a heating device connected to pipes that carry the steam or hot water to the place where it is wanted. A boiler may be gas fired, oil fired, or electrically heated. Some boilers are either gas or oil fired.

**Boiler pressure control**    A control used to operate a boiler burner in response to the pressure inside the system. It is used to cycle the burner on and off as the pressure rises and falls. For example, the boiler pressure control is used on a steam boiler. It will open the circuit to the gas valve when the pressure reaches that which is wanted for operation. When the pressure falls to the control set point, the contacts will close the circuit to the main gas valve and start the boiler making steam. The settings of these controls will vary from system to system. The wanted control setting for each system must be kept or the system may not operate correctly.

**Blower control**    The blower control is a relay used for automatic operation of an indoor blower in response to the needs of another circuit. The contacts are closed on demand from the temperature control circuit. For example, on an electric heating unit when the first stage heat strips are energized, the control will start the blower motor to start air blowing through the heating element and keep it from getting too hot and possibly burning out.

**Break point**    The temperature at which the refrigerant charge of the element has completely vaporized. At temperatures above the break point of the fill, there is very little internal pressure change in the element, about 1 psig for each 10°F temperature change. For example, when a bellows type thermostat has reached its maximum operating temperature, the refrigerant charge in the power unit is completely vaporized. The thermostat will not respond to any further change in temperature, which is to prevent damage to the thermostat.

**British thermal unit (Btu)**    The amount of heat required to raise the temperature of one pound of water 1°F. For example, a ton of refrigeration means that the temperature removal of the unit is equal to the melting of one ton of ice in 24 hours, which is equal to 12,000 Btu/h.

**Burner**    A device in which the mixing of gas or oil and air is done. For example, in a residential heating unit, the burner is the part where the fuel and air are mixed and passed through the burner ports so it can be lit and produce heat. Burner ports are where the flame is setting.

**Bypass**    A pipe or duct usually controlled by a valve or damper for bypassing the flow of a fluid. For example, a bypass valve is used on water-cooled condensers so that part of the water can be sent back to the cooling tower and not pass through the condenser. The amount of bypass is regulated by a modulating motor, a pressure control, and a three-way valve.

# C

**Capacitor**    A device used to boost the voltage to a motor. Running capacitors are used in the start winding to increase the running torque of the motor. Starting capacitors are used in the start winding to increase the motor starting torque. For example, a starting capacitor is used on systems in which the refrigerant in the system is not equalized to unload the system before its next start-up. Starting capacitors are taken out of the circuit when the motor has reached about 75 percent of its normal running speed. Running capacitors are used to boost both the starting and running torque and, when the motor is running, to lower the current draw and reduce the operating costs. Capacitors must be sized correctly or they may cause damage to the motor winding.

**Check valve**    A valve that allows flow in only one direction. For example, check valves are used in heat pump systems to bypass the flow control device during one cycle and make it flow through it during the other cycle.

**Circuit**    The tubing, piping, or electrical wiring that allows flow from the source of energy, through the circuit, and back to the energy source. For example, in air-conditioning control systems, the circuit receives the electron flow usually from one terminal of a 24 VAC transformer through the wiring and controls and back to the other terminal of the transformer. This process will continue until the circuit is broken by a control or by damage to the circuit wiring.

**Circuit breaker**    A safety device used to protect an electric circuit from overload conditions. For example, a circuit breaker is usually located in the power supply panel for the building. The circuit breaker is sized to match the load capacity of the circuit. Should an overload condition occur, the circuit breaker will trip and stop the flow of current through the circuit. It must be reset before the equipment will operate.

**Closed circuit**    A completed (made) electrical circuit through which the electrons are flowing. For example, when a thermostat closes its contacts the contactor holding coil is energized because of the flow of electrons through the control circuit. When the contactor coil pulls in the contactor armature, the contacts to the compressor motor, fan motor, or any other equipment are energized to start running.

**Coil**    A coil of wire placed within close relation to a movable armature. When the coil is energized electrically, it causes the armature to move in the needed direction. Coils are used in every relay, contactor, and starter in electrical systems. For example, the coil in a starter is generally referred to as the *holding coil* or *contactor coil*. They produce the magnetic field needed to pull in the armature to open or close contacts as the equipment operation needs.

**Combination fan and limit control**    A control that cycles both the fan and main burner off and on in response to the temperature of a designated place in the heating equipment. For example, the fan part of the control is used to start the fan when the temperature inside the heat exchanger reaches about 135°F to 150°F. It will stop the fan when the heat exchanger drops to about 100°F.

The limit part of the control is a safety control. It will shut down the main burner when the temperature reaches about 180°F to 200°F. It will restart the main burner when the temperature drops to about 160°F. Both manual and automatic models are used. It is generally preferred that the limit control never operate.

**Commercial refrigeration**    A reach-in or service refrigerator of commercial size with some means of cooling. For example, in a grocery store all the perishable products are kept in cases and cabinets to keep them from spoiling. The case or cabinet is cooled by a refrigeration unit that has the capacity to bring the load and the case or cabinet down to a suitable temperature.

**Compressor unloaders**    Controls placed on a compressor to reduce or increase its pumping capacity in response to the system needs. Unloaders open or close in response to the suction pressure. For example, unloaders are generally used on larger air-conditioning and refrigeration systems. When the needed capacity drops to a certain point, each unloader takes away the pumping capacity of a certain number of cylinders until the maximum capacity reduction is reached. When the load needs more capacity, the unloaders will start loading the cylinders until the needed capacity has been reached. Thus, the system is always producing some refrigeration but not continuous full capacity. This will lower the electric bill and give some continuous cooling.

**Condenser fan**    The fan that forces air over an air-cooled condenser coil. For example, the condenser fan is placed in an air-cooled condensing unit to remove heat from the hot refrigerant vapor after it leaves the compressor discharge. The fan can be cycled to keep the head pressure within the needed range for correct refrigerant pressures.

**Conductor**    A substance or body capable of conducting electric or heat energy. For example, a conductor in an electric circuit is the wire that carries the electricity to the point of use. The wire from the thermostat to the compressor contactor holding coil is a conductor.

**Contact Rating**    The capacity of electrical contacts to handle the current and voltage to the circuit. For example, a motor operates on 240 VAC and draws 10 amps. The motor control contacts must be able to handle this load or they will shortly burn out and the control must be replaced. The contact amperage and voltage rating must meet or exceed the ratings of the circuit.

**Contactor**    The control used to energize a heavy electrical load such as a compressor or heating element in an electric heating unit. Contactors are usually energized and deenergized from a different circuit than the one being controlled. For example, a contactor is used in a residential cooling unit to start and stop the compressor and condenser fan motor. This is done by the 24 VAC control circuit, but it controls the 240 VAC circuit to the necessary components. The same is true for an electric heating unit. The 24 VAC control circuit is used to open or close the 240 VAC circuit to the heat strips.

**Control**    Any device used for regulation of a machine in normal operation—manual or automatic. A control is usually responsive to temperature or pressure, but not to both at the same time. For example, the thermostat in a refrigeration or air-conditioning system is a control that is responsive to temperature changes. When the temperature in the conditioned space rises to the thermostat cut-in temperature, the equipment will start running. When the temperature drops to the thermostat cut-out temperature, it will open its contacts to stop the equipment.

**Control module**    Contains a PC card, motor, wiring harness, and switches to receive signals from inputs, including several thermistors, processing them and responding correctly to cause the equipment to operate. For example, the control module on an air-conditioning system is generally located in the condensing unit. It has control over the complete operation of the system. The manufacturer's information should be used in testing these cards to keep from damaging them.

**Control system**    The components used for the automatic control of a given process. For example, the control system on a residential air-conditioning system consists of the thermostat, transformer, contactor holding coil, fan relay, safety controls, fan and limit controls, and the needed pressure controls. These all comprise the control system.

**Cooling anticipator**    A resistor placed inside the thermostat in electrical parallel to the cooling bimetal. It causes the cooling unit to cycle on before the room temperature actually rises to the cut-in setting of the thermostat. For example, when the thermostat is in the cooling mode and the room temperature reaches the OFF setting, its contacts will open. The control circuit current then starts flowing through the cooling anticipator, which is a carbon resistor placed in the thermostat close to the cooling bimetal. When the current is flowing through it, heat is produced inside the thermostat. This heat will cause the thermostat to demand cooling before the temperature actually reaches the thermostat ON set point. This reduces the wide swings in temperature inside the space. Some technicians say this also helps to control the humidity inside the space.

**Cooling tower**    A device that cools water to the outdoor wet-bulb temperature by evaporating some of the water. For example, the cooling tower is located outside where it can get plenty of air movement through it, or it is equipped with a fan to pull the air through the tower. The air passing through the tower evaporates some of the recirculating water to cool the remaining water to the wet-bulb temperature of the outdoor air. The cooled water is then pumped back through the water-cooled condenser where it changes the hot refrigerant vapor into cooler liquid refrigerant.

**Cross ambient fill**    A vapor-pressure element large enough to ensure that liquid is in the bulb regardless of whether the bulb is warmer or colder than the control. This happens when the temperature at the control and capillary are alternately above and below the needed temperature of the sensing bulb. For example, a cross ambient fill is used in thermostats that have mechanical means to keep them from expanding or contracting so much that they will be ruined during the excessive temperatures. The cross ambient fill ensures that pressure remains inside the power unit at all times so that correct control is possible.

**Current**    The flow of electrical energy in a conductor that happens when the electrons change positions. For example, current is the amount of electrons flowing by a given point in a given period of time. This flow can be measured with an ammeter.

**Current relay**    A switching relay that operates on a predetermined amount of electrical flow (or lack of current flow). For example, current relays are used on smaller refrigeration compressor motors such as domestic refrigerators and freezers. The relay has NO contacts. All current to the main winding must pass through the relay coil. When the compressor motor starts maximum current is flowing through the relay coil. When the current flow increases to the pick-up current of the relay it will close its contacts. The electricity is now also sent to the start, or auxiliary, winding of the motor. As the compressor motor increases in speed, the current flow drops to the point that gravity will open the contacts to take the start winding out of the circuit. The motor now operates without the start winding.

**Cut-in setting**    The point at which the control contacts close to make the electrical circuit. For example, a low-pressure control is used on a refrigeration system. When the compressor is off, the refrigerant pressure inside the system starts rising. When it reaches the cut-in setting of the low-pressure control, the contacts will close to start the compressor motor running. When a system using CFC-12 is cooling a dairy case, the control cut-in setting will be about 35 psig. When 35 psig is reached, the control contacts close to start the compressor motor.

**Cut-out setting**    The point at which the control contacts open to break an electrical circuit. For example, when a low-pressure control is used as a temperature control on a dairy case and the refrigerant pressure drops to a certain point, the control will open its contacts to stop the compressor motor. On a dairy case this pressure will be about 30 psig for CFC-12. The compressor will stay off until the refrigerant pressure rises to the cut-in pressure setting of the control.

# D

**Damper**    A device used to control the flow of air. For example, dampers are used in multizone units to direct the air to the place where it is needed and direct it away from the point where it is not needed. When both the air-conditioning unit and the boiler are operating simultaneously, dampers may be used to more closely control the air temperature within a given space by both heating and cooling to the air. This is done in a quantity to satisfy the needs of the particular space.

**Defrost**    To remove accumulated ice from the evaporating coil. For example, during the operation of a heat pump system in the heating mode or a commercial refrigeration system, ice will form on the evaporating coil. This ice must be removed to keep the unit operating as economically as possible.

**Defrost control**    The control that determines when a system needs defrosting. It initiates and in some cases terminates the defrost cycle on heat pump and commercial refrigeration systems. For example, the defrost control senses the amount of frost on the evaporating coil. When the predetermined amount has accumulated, the defrost control will start the system to defrosting. When the coil is clear of frost the defrost termination control will put the system back into the original operating mode.

**Defrost cycle**    The refrigeration cycle in which the ice accumulation is melted from the evaporating coil. For example, when the defrost control senses that a defrost is needed it will actuate the correct controls to place the unit in the defrost cycle.

**Defrost timer**    A control connected into the electric circuit to start the defrost cycle and keep it on until the ice has melted. For example, when the defrost timer reaches the time a defrost cycle is needed, it will place the unit in the defrost cycle and keep it in this mode until the allotted time has passed. Then it will cause the unit to return to its normal operating mode.

**Differential**    The difference between the cut-in and cut-out settings of a control. For example, the differential is the number of degrees or pounds of pressure between the cut-in and cut-out points of a control. In a thermostat the differential is the number of degrees the temperature must change before the thermostat contacts will either open or close depending on whether the system is heating or cooling. The differential of a pressure control is the number of pounds per square inch the pressure must change before the control contacts will either open or close, depending on whether the pressure is rising or falling.

**Differential screw**    An adjusting screw used to change the difference between the cut-in and cut-out settings of a control. For example, when changing the differential of a pressure control, the differential screw is turned to the needed cut-in or cut-out of the control. This screw only changes the differential, not the range of the control.

**Direct spark ignitor**    A method by which the main burner gas is lit by a spark across the ports of the main burner rather than lighting by a pilot burner. The spark is generated by a high-voltage transformer in the ignition circuit. For example, the spark of an electronic ignitor is directed across the ports of the burner where the gas-air mixture comes out. The spark has a high temperature so the gas will be lit. When the flame is established the sensing circuit stops the spark. The system then operates normally.

**Double pole**    Two single-pole contacts operating at the same time. For example, double-pole contacts may have one pole that is NO and one pole that is NC; or they may both be NC or NO, depending on the need for the relay. A pole consists of one movable contact and one stationary contact.

**Double throw**    A set of contacts that make in one direction and break in the other direction. For example, a double-throw relay has two stationary contacts and one movable contact. The movable contact will touch either one or the other stationary contact to make a circuit. When one circuit is made the other is broken. This is a common pole configuration in fan relays used on heating and cooling systems.

**Drop-out voltage**    The voltage or point at which the pull of the electromagnet is not strong enough to keep the armature pulled in. For example, a potential motor starting relay is pulled in when the compressor motor is running and producing the correct amount of voltage. When the electric power is turned off to the motor it begins to slow down. When it slows enough, the voltage in the start winding will drop to the drop-out voltage of the relay coil. The coil will then allow the spring force on the armature to pull the armature away from the coil.

**Dry-bulb temperature**    The temperature shown on a regular thermometer that indicates the heat of the air and water vapor. For example, a dry-bulb thermometer is the one normally used to measure supply and return air temperatures of a conditioned space. It does not show how much moisture is in the air, only the temperature.

**Dummy terminals**    The extra terminals that do not connect to the electrical components of the control, but rather are used for convenience in wiring. For example, on a potential relay terminals 3, 4, and 6 are dummy terminals. They have no internal connections or any effect on the operation of the relay.

# E

**Electric heating**    A heating system in which the source of heat is electricity. For example, an electric heating unit uses the flow of electrons through a high-resistance wire to give the needed heating. The air flowing through the heat strips is warmed by them. The unit is controlled by a thermostat as is any other type of heating system.

**Electromagnet**    A coil of wire wound around a soft-iron core. When electric current flows through the coil an electromagnet is formed. For example, the core of a contactor holding coil is an electromagnet. When the control circuit is made to the coil it creates an electromagnetic field around itself. The electromagnetic field attracts the armature to the coil face in order to open or close the contacts.

**Enthalpy**    The total amount of heat contained in a quantity of a substance, including both sensible and latent heat. For example, in air-conditioning work the enthalpy of the outdoor air is used to open or close the outside dampers. When the enthalpy is low the controller opens the dampers. Having the outdoor air cool the building is one way to save on the electric bill.

**Evaporator**    The part of a refrigeration system in which the refrigerant boils and picks up heat. For example, the evaporator is the refrigerant coil placed in the case of a commercial refrigeration or air-conditioning unit. As the air flows through the evaporator, the refrigerant absorbs heat from the air. The heat is then taken to the condenser where it is removed to the outside air. The evaporator coil is also known as the *evaporating coil*.

# F

**Fail-safe control**    A control used such that a component failure will cause the control to take the safest action, thus protecting the system on which it is installed. The situation is usually a contact-open condition. For example, a fail-safe control is installed in the electric circuit to sense any abnormal operation of the equipment. When an unsafe condition is sensed it will open a set of contacts to shut down the system and prevent any damage.

**Fan**    An enclosed propeller that produces motion in the air; commonly means anything that moves air. For example, the fan causes the air to move inside the building. This fan is what moves the air through the equipment where it is either cooled or heated as needed.

**Fan control**    A switch used to cycle the indoor fan on and off in response to the temperature inside the furnace heat exchanger. The control is usually set to start the fan at about 135°F and stop it at about 100°F. For example, a fan control is an operating control that operates the fan in response to the temperature inside the heat exchanger. The fan control provides a delay before the fan is started so that cold air will not be blown into the conditioned space.

**Fixed differential**    The factory-set difference between the control cut-in and cut-out setting. It cannot be changed in the field. For example, a furnace limit control has a fixed differential. When the cut-out setting is changed, the cut-in setting is also changed. The differential for this control is usually about 20°F. The range can be changed but not the differential.

**Fixed setting**    Exhibits no convenient means for changing the control settings after the control leaves the factory. For example, many high-and low-pressure controls are factory set to meet the equipment manufacturer's wants. Some thermostatic expansion valves are factory set to what the equipment manufacturer wants. They have the differentials built into them and cannot be changed in the field.

**Flame rectification**    A way of operating flame safety equipment by ionizing an area by a pilot flame. This flame completes a circuit in the safety circuit and lets the equipment operate. For example, in a flame rectification system, the flame must touch both the pilot burner and the sensing probe or the ignition system will shut down the entire system even though a pilot flame is burning. When the flame does not touch both at the same time and continuously, an open circuit is sensed and the system will shut down. The temperature of the sensing probe has nothing to do with the operation of these systems.

**Flux (electrical)**    The electric or magnetic lines of force in a region. For example, the number of lines of flux around an energized magnetic coil determines the strength of the field of flux that surrounds the coil. These lines of flux are what pulls the armature into the relay or contactor coil.

**Full-load amperage**    The amount of current flowing in an electric circuit when the unit is operating at full capacity. For example, all electric motors have a full-load amperage rating. When the motor is operating at maximum capacity this amount of current will be used by the motor. In an air-conditioning or refrigeration system, the compressor will operate at full-load amps when it is first started. After the load has cooled down, the current draw of the compressor motor will drop.

**Furnace**    That part of the heating unit where combustion occurs. For example, the furnace is actually the heat exchanger. In the sections of the heat exchanger is where the gas or oil is burned. These sections are connected to the flue vent so the products of combustion can be taken outside.

**Fuse**    An electrical safety device consisting of a strip of fusible metal used to keep from overloading an electric circuit. For example, the fuses are placed in a disconnect box and the circuit passes through them. When an overload occurs, the fusible metal strap burns apart and breaks the circuit, thus stopping the flow of electricity. The fuse must be replaced before the system will operate correctly. Do not jumper fuses. This could cause further damage to the equipment or property.

# G

**Gas valve**    A valve placed in the pipeline to control the flow of gas to a burner. For example, a gas valve is controlled by the thermostat. When heat is needed the thermostat will cause the gas valve to open and let gas flow to the burner so the space can be heated. When the space temperature is satisfied, the thermostat will cause the gas valve to close and stop the flow of gas, thus stopping the heating process.

**Glow coil**    Lights the gas flame in a gas furnace when heated to a red-hot temperature. It is usually used to light the pilot burner gas. For example, the glow coil is placed so that it is in the flow of gas and air from the pilot burner. It is heated by electric current flowing through it. When the sensor detects a flame the glow coil is turned off by a relay so it will last longer.

**Ground**    An intentional or accidental connection from a power source to the earth, or a connecting body that serves in place of the earth to complete an electrical circuit. For example, in an electric motor, when the winding insulation breaks down and allows the bare wire to touch the metal frame of the motor housing, a ground is created. This will usually cause the fuses to blow or the circuit breaker to trip. This grounded condition can be found with an ohmmeter by measuring the resistance from the motor terminals to the motor housing or some other grounded part of the system. This condition can happen in any part of the circuit. Sometimes it is hard to find when it happens inside a conduit or some other inaccessible place.

# H

**Hard-start kit**    A group of parts used to increase the starting torque of an electric motor. It consists of a starting capacitor, a starting relay, and the necessary wiring to install the kit. For example, a hard-start kit is used on refrigeration compressor motors in areas where the electric power is not always up to standard. The starting capacitor and starting relay are added to a PSC compressor motor to change it to a CSCR motor. A kit may be used or the separate parts may be bought and installed on the unit. This will help in starting and help prevent compressor burnout and compressor downtime.

**Head pressure**    The pressure against which the compressor must pump the refrigerant vapor. For example, when a compressor is running and the head pressure is taken, this is the pressure the compressor must pump before any refrigerant is circulated through the system.

**Head pressure control**    A pressure-operated control that opens the electric circuit when the head pressure gets higher than the control cut-out setting. For example, the head pressure control is connected to the discharge of the compressor. It constantly measures the head pressure. When the pressure rises to the cut-out setting of the control, it will stop the compressor to prevent damage to the compressor or the motor. When the pressure drops it will either restart automatically or must be manually reset before operation can continue.

**Heat pump**    An electrically operated unit used to remove heat from one location and transfer it to another. A heat pump is used for both heating and cooling. It reverses the refrigerant cycle when heating is needed. For example, in the cooling cycle the heat pump operates like any other cooling system. When heating is needed the direction of the refrigerant flow is reversed to direct the hot, compressed refrigerant to the indoor coil rather than the outdoor coil. The refrigerant is condensed and the heat removed from it is used to heat the space.

**Heating anticipator**    A resistor placed inside the thermostat and placed in series with the temperature control circuit during the heating cycle to cause a false heat inside the thermostat. This causes the thermostat to stop the heating equipment before the temperature inside the space actually reaches the thermostat set point. For example, the heating anticipator is an adjustable wire-wound resistor placed in series with the heating control circuit. It must be adjusted to match the current draw through the circuit during operation. When this resistor is not adjusted to match the circuit current draw, the heating equipment will not operate correctly.

**Heating control**    Any control that controls the transfer of heat. For example, a heating control is a gas valve that controls the gas flow into the heat exchanger. When it opens heat can be transferred, when it closes heat cannot be transferred.

**Hermetically sealed**    A compressor or other device enclosed in a gas-tight housing. For example, a hermetically sealed compressor, sometimes referred to as a *tin can,* is most commonly used in refrigeration systems today. Usually they are built in sizes up to and including 7.5 tons. The internal parts of a hermetically sealed compressor motor cannot be serviced without special cutting equipment. It is usually much faster and less expensive to buy a replacement compressor.

**Hertz (Hz)**    The frequency in cycles per second of an AC power source. In the United States this power is usually 60 Hz. For example, 60 Hz power means that the cycles of the alternating power changes direction 60 times per second. Thus, it is positive 30 times per second and negative 30 times per second.

**High-pressure control**    A control used to sense the discharge pressure of a compressor and to stop the compressor motor when this pressure reaches a maximum pressure to prevent damage. For example, when the discharge pressure of a compressor reaches, for example, 375 psig for HCFC-22, the control contacts will open and break the control circuit and cause the compressor motor to stop. The contacts will remain open until the pressure has dropped to the cut-in setting of the control. If it is automatic reset, the compressor will automatically start again. If it is a manual reset, it will remain off until the control is manually reset to prevent overloading the motor and perhaps causing damage.

**Holding coil**    That part of a magnetic starter, contactor, or relay that causes the control to operate when energized. For example, the coil in a starter or contactor that causes the contacts to close is called a holding coil.

**Horsepower rating**    A rating in terms of a motor. Underwriter's Laboratories consider 1 horsepower equivalent to 746 watts. Most ratings now refer to electrical ratings in amps. For example, a motor that uses 746 watts of power when running at full load is considered to be a 1-horsepower motor. In electrical terms a 1-horsepower motor on 120 VAC will draw 12 amps. When operating on 240 VAC the amperage rating will be 6 amps.

**Hot-gas defrost**    A method of evaporator defrosting that uses the hot-gas discharge gas to remove the frost from the coil. For example, when the outdoor coil of a heat pump needs defrosting, the reversing valve changes the direction of refrigerant flow so that the hot gas will enter the evaporating (outdoor) coil to melt the frost or ice from it. When the frost or ice has melted, the reversing valve will shift positions and direct the hot gas to the indoor coil to heat the space.

**Humidity controller**    A control used to start and stop humidifying equipment in response to the amount of relative humidity inside the conditioned space. For example, the humidistat placed on the wall of a building will start the humidifier when the relative humidity (RH) drops to the set point and stop the humidifier when the RH rises to the setting of the humidistat. The humidistat usually has a scale indicating the maximum RH for each degree of outdoor air temperature.

# I

**Ignitor sensor**    Used on flame-safe equipment to light the pilot-burner gas and then sense if a flame is present. If no flame is present or the flame is not good enough to light the main burner gas, the control system will not open the main gas valve. For example, almost all new heating equipment has ignitor sensors installed at the factory. Depending on the type and make, the sensor also acts as the ignitor. It is placed in the area where the pilot flame, or the main burner flame, will surround it when lighted. A major type in residential and commercial heating equipment is the hot surface ignitor (HSI) system. In this type, the surface is electrically heated to a high enough temperature to light the pilot or the main burner gas. When the flame is sensed and is satisfactory for correct gas ignition, the electricity to the HSI is turned off and the surface then becomes a sensor for the control system.

**Inherent motor protection**   A safety limit device built inside a motor or other equipment; it protects against over temperature, over current, or both. For example, the internal thermostat in a hermetic compressor motor senses both current flow and the temperature of the motor winding. When too much current flows through the winding, it causes more heat than normal. This higher temperature will cause the control to open either the control circuit or the common terminal to the motor and stop it before damage can occur. They are also used for over-temperature protection in cases when the system loses part of its refrigerant charge and is not being cooled enough to prevent damage to the motor windings from heat.

**Intermittent pilot**   A pilot that is lit each time the thermostat demands heat. When the thermostat is satisfied, the pilot and main burners are turned off simultaneously. For example, this system is commonly called an intermittent ignition device (IID) system. When the thermostat demands heat the IID is energized to light the pilot flame. When the flame is established, the ignition goes off but the pilot stays burning. The ignition system is deenergized until the next time the thermostat demands heat, then the process starts again.

**Internal overload**   A protective control placed inside the motor winding at a place to protect the motor from overheating, over current, or both. For example, the internal overload is placed inside the motor winding at the hottest spot in the winding. It is used to protect the motor from all means of too high a temperature. Usually when the internal overload goes bad the compressor must be replaced. Make sure the overload is bad and not just tripped before replacing the compressor.

# L

**Limit control**   A control with its sensing element located inside the furnace heat exchanger to sense an over-temperature condition. When an over-temperature condition is sensed, the main burner gas will be turned off to prevent a possible fire or damage to the equipment or the building. This is not an operating control, but rather a safety control that operates only when a hazardous condition exists. For example, the sensing element of a limit control is placed in the hottest position in the heat exchanger. When the temperature inside the heat exchanger rises to 180°F to 200°F, the limit control will break the control circuit to stop the flow of gas to the burners. Some limit controls have a DPST switch that will break the control circuit and make the indoor fan circuit to blow the hot air from the furnace.

**Locked-rotor amps (LRA)**   The current that is needed the instant electric power is applied to start the motor; also, the current draw when the motor cannot start turning. For example, when the electricity is first applied to a motor the inertia must first be overcome before the motor will start turning. This first amperage will be equal to about six times the running amperage. As the motor starts turning, the needed amperage will drop to the normal running amperage rating of the motor.

**Low-ambient kit**   A kit consisting of the needed parts to cycle the condenser fan motor to keep the head pressure high enough to feed the correct amount of refrigerant to the flow control device. For example, the low-ambient kit consists of a high-pressure control that opens its contacts on a fall in pressure. This control stops the fan motor so the head pressure will rise to the normal operating limit to keep the needed pressure drop across the flow control device. The pressure control is set to cycle the condenser fan off before the head pressure drops low enough to interfere with operation of the flow control device. It will bring the fan back on when the needed control difference is reached.

**Low-pressure control**    A control used to stop the compressor motor when the suction pressure drops low enough to cause damage to the compressor or motor. The control contacts are usually located in the control circuit, but in some cases they may be in the line-voltage circuit to the compressor motor. For example, the low-pressure control senses the pressure in the low side of the system. When the pressure drops to a point indicating the pressure has dropped enough to keep the unit from cooling correctly and keep up lubrication of the compressor and motor, the control contacts open to stop the motor. This drop in pressure is usually because of a refrigerant loss or a reduced airflow over the evaporator.

**Low-water cutoff**    A control used to stop the main burners of a boiler if the water inside the system drops to a dangerous level. For example, this is usually a float-operated control with NO contacts. When the water level is high enough for correct operation the contacts are closed by the float that is raised and lowered by the water level. If the water level drops enough to let the contacts open, the circuit to the main gas valve will be broken so the main gas valve will close and stop heating the boiler. These controls are mainly used on steam boilers, but are sometimes used on hot water boilers, as backup safety controls.

# M

**Manual reset**    A safety control-set mechanism that must be manually reset if the control locks out. For example, a manual reset control may be used as a high-pressure, low-pressure, or any other control the manufacturer thinks is necessary. The unit must be checked out before operation is continued. These controls have a reset button so that resetting them is not difficult. When the control is reset, the unit should be completely checked out and any repairs made before the unit is allowed to continue operating.

**Modulating control**    A control that makes corrections in small increments, rather than the complete ON or OFF operation. For example, a modulating control is primarily used on systems that need some conditioning all the time. The control is usually done by a modulating motor, step controllers, and a modulating type thermostat. As the thermostat changes its needs, a signal is sent to the modulating motor to move a certain number of degrees in rotation. This change *may* bring on or stop a stage in the equipment.

# N

**Normally closed (NC)**    A switch or valve that stays closed when there is no power to it, or it is deenergized. For example, a control placed on a workbench with no external force on the operating mechanism and its contacts are closed is called an NC control.

**Normally open (NO)**    A contact pair that is open when the control is deenergized. For example, a relay is equipped with NO contacts when placed on a workbench without being connected to any electricity and the contacts are open.

# O

**Oil-failure control**    A control used to stop the compressor if the oil pressure drops to a dangerously low level to keep from damaging the compressor. For example, a pressure differential control is placed between the oil pump outlet and the suction pressure connection. It senses the oil pressure and the suction pressure. It will stop the compressor if the net oil pressure drops below about 40 psig difference between the oil pressure and the suction pressure. When this differential is reached, a heater in the control is energized. When a certain number of seconds pass without a rise in the net oil pressure the control will stop the compressor.

**ON-OFF control**    Used on jobs to control the equipment operations. The system is either on or off (two position) as opposed to proportional or modulating. For example, this type control is generally used in residential air-conditioning systems. When the thermostat demands operation, the complete unit starts running at full capacity. When the thermostat is satisfied, all the equipment stops running, with the exception of the indoor fan. It may be either in the AUTO position where it will cycle on and off with the equipment or it may be in the ON position so it will run all the time.

**Open circuit**    An electrical circuit without a continuous path for the current to flow through. This open circuit may be caused by an open switch or a broken circuit such as a blown fuse or tripped circuit breaker. For example, any time the circuit is opened by either a control contact or by damage to the wiring so that no current can flow through it, the circuit is said to be broken or open.

**Outdoor reset control**    A control used to sense both the indoor and outdoor air temperature and change the operating cycle of the equipment. The cycle is in response to the demands of the space being heated by a hot water boiler and is controlled by an outdoor reset control. For example, when the outdoor temperature drops, the control acts to make the boiler heat the water a little bit hotter for each degree of temperature change. Also, if the outdoor temperature rises, the control acts to cause the boiler to heat the water less. The two reasons for using this control are economy and comfort. When the water does not need to be heated to the maximum temperature, the boiler will use less fuel. Also, the cooler water will allow cooler air to be blown into the space when the hotter water is not needed. It also allows for better air circulation.

**Overload protector**    A control used to protect ungrounded conductors from motor over-current by opening its contacts. It stops unsafe running conditions and protects the motor from burnout. For example, on most smaller compressor motors the overload protectors are placed inside the motor windings to sense the hottest point in the winding. When the conditions cause the compressor motor to draw more current than it was designed, the motor winding will overheat and cause the overload protector to open the circuit and stop the motor. When the motor winding cools to the temperature that will let the contacts close, the overload protector will restart the compressor motor. This condition should not be allowed to continue because the motor could be damaged or the overload protector may be ruined. In either case the compressor motor will need to be replaced before the system will operate again.

**Overload relay**    A thermal control that opens its contacts when the current through the heater coil is more than the specified value for the specified time. For example, an overload relay is usually placed on the side of a contactor to make it a starter. The overload may be mounted somewhere other than on the contactor. It has a resistance heater inside the relay through which the current to the motor passes. When the current flow is more than the wire is designed to handle, the overload will get hot and let the contacts open the control circuit to the holding coil. When the overload has cooled enough, the relay can be manually reset to start the equipment.

# P

**Pilot burner**    A small burner used to light the main burner on heating equipment. It is sometimes used to provide the heat to operate the pilot safety devices used on gas-burning equipment. For example, on most modern heating equipment, the pilot burner is lit by an electronic ignition control. When the flame is established, the electronic ignition system will automatically shut down. The pilot will then light the main burner when the thermostat demands heating. On older standing pilot systems, the pilot may be manually lit or lit once by an electric ignition system. The pilot remains lit until the system is shut down for the cooling season.

**Pilot-duty rating**    An electrical rating for controls used to energize and deenergize pilot circuits such as the holding coil of a motor contactor. For example, the extra set of contacts in a control will only carry a small amount of current without burning out. They are primarily used to energize another contactor coil such as a water pump contactor, a combustion fan on large heating equipment, a warning signal circuit to alert the operator that a problem is present, or any other part that uses a low current flow.

**Pilot safety**    A control used to stop unsafe operation of the main burner if the pilot is not large enough to light it safely. For example, a pilot safety control is one that is usually operated by a thermocouple. If the pilot flame is too small or if something else is not operating correctly to allow safe system operation, the pilot safety circuit will shut down the heating equipment. It will usually stay off until the problem is repaired.

**Positive-temperature-coefficient starting device**    A compressor motor starting control that increases in resistance with an increase in temperature. When the resistance is increased to a given level, the current flow through the device is reduced to a small trickle, effectively taking the starting components out of the circuit. For example, a positive temperature coefficient (PTC) is a compressor motor starting control. When cool, it practically has no resistance, which allows more current to flow through the start winding of the motor. When the compressor motor starts, current flows through the PTC and causes it to heat up. As the temperature rises, the resistance of the PTC increases and the current flow is reduced. These controls may be used on several different size compressor motors without problems. When this type control fails, the compressor will not usually start until the control is replaced.

**Potential relay**    A relay used to start a motor needing a high starting torque. The relay coil is energized by voltage generated in the motor starting winding. For example, almost all CSCR motors are equipped with a potential motor starting relay. Its purpose is to let extra current flow to the starting winding to give the needed torque to start the motor and its load. After the motor has reached about 75 percent of its normal running speed, the potential relay contacts open and the motor operates as a PSC motor until it stops and tries to start again. Then the relay contacts are closed to allow the electric current to the start winding to help in starting the motor again.

**Potentiometer**    A wire-wound coil used for measuring or controlling by sensing small changes in electrical resistances. For example, a modulating thermostat has a potentiometer that exactly matches the potentiometer in the modulating motor. When the thermostat moves in one direction on the potentiometer, the motor moves in the other direction to balance the resistance on both sides of the control circuit. These controls can result in exact temperature inside the conditioned space.

**Power**    The voltage used for operating a device. Use of the word *power* commonly refers to electrical power or voltage. For example, when the power is turned on to the refrigeration unit, it starts running. Power is the source of energy that causes the equipment to run. It also powers some of the controls used in its operation.

**Proportional controls**    *See* Modulating control.

**Protectorelay**    A control used on oil burners to allow safe ignition of the oil on start-up. It is a heat-sensing control mounted on the unit flue stack to sense the heat coming from the flame. If no heat is detected, the protectorelay will shut the unit down. For example, when the thermostat demands heat, the oil burner is energized. When the oil is lit, heat begins to flow up the stack. The protectorelay senses this heat and when enough is detected, the control contacts will shift to the operating position so the unit will continue heating. If enough heat is not sensed the unit will be shut down until the control is reset and the problem repaired.

**Pull-in voltage**   The amount of voltage that causes a relay armature to seat on the pole face. For example, suppose the pickup voltage of a potential relay is 270 VAC. When the voltage generated in the start winding reaches 270 VAC the armature will be pulled into the pole face and open the starting relay contacts, taking the starting components out of the circuit. The pull-in may be used to either close a set of contacts or open them, depending on the use and design of the relay.

# Q

**Quick connects**   The terminals of a switch that are usually connected to a push-on connector rather than the normal screw terminals. For example, terminals called *spade connectors* are quick connect terminals. They are usually easier to work with than the screw type, especially in close working places.

# R

**Range**   A change within the limits of the settings of a pressure or temperature control. For example, when both the cut-in and cut-out points of a control are changed equally, the range of the control has been changed. Also, the range of a pressure, or any other control, may be considered the highest and lowest settings on which the control can be set.

**Range-adjusting screw**   A screw used to change the operating set points of a control; changes are limited to those within the control range. For example, the range-adjusting screw is used to change both the cut-in and cut-out settings of a control equally. If the cut-in and cut-out pressure of a low-pressure control are set at 30 psig cut-in and 20 psig cut-out, and the range adjusting screw is turned to change the cut-in setting to 15 psig, the cut-out setting would be changed to 5 psig cut-out pressure.

**Redundant gas valve**   A valve that has two seats, both of which must be open before gas can flow through; however, it takes only one seat closed to stop the flow of gas. This valve is a safety control to keep from overheating the equipment because a gas valve stuck open. For example, all new gas furnaces are required to have redundant gas valves installed at the factory. It takes both seats to be open before the furnace will operate. If something should happen, however, such as an open limit control contact, all gas flow to the furnace will stop and the unit will not heat. It is possible for one seat to stick in the open position, but it is unlikely that both seats will stick open at the same time.

**Relative humidity**   The ratio of the amount of water vapor actually present in the air to the greatest amount possible at the same temperature. For example, if a conditioned space has a relative humidity of 45 percent, it would be holding only 45 percent of the moisture vapor it could hold at that temperature.

**Relay**   An electromagnetically operated control used for switching an electrical circuit with one circuit controlling another, usually of higher voltage and current usage. Consider an indoor fan relay as an example of a relay using a circuit of one voltage (24 VAC) to control line voltage (240 VAC or 120 VAC). These two circuits and voltages never come into contact with each other.

**Remote bulb thermostat**   A thermostat that is sensitive to changes in temperature at a place away from the thermostat body. For example, a remote bulb thermostat is mainly used on walk-in refrigerated coolers. The thermostat may be mounted outside the cooler—out of the way so it will not be damaged or so that it can be serviced easier. The bulb is mounted inside the cooler to sense the temperature there. The thermostat is the same use as any other type, but it is located outside the conditioned space.

**Reversing (four-way) valve**    A valve used on heat pump systems and commercial refrigeration systems to reverse the flow of refrigerant through the system, directing the hot gas to the evaporating coil to remove any ice accumulation. For example, when a refrigeration system is operating, moisture is taken from the air and collects on the cooling coil surface in the form of ice or frost. When 0.125 inch of ice has accumulated the efficiency of the unit is decreased 25 percent. Therefore, to keep the unit operating at peak efficiency the coil must be periodically defrosted. At this point the defrost control signals the reversing valve to reverse the refrigerant flow so the hot discharge gas will go to the evaporating coil and melt the frost accumulation. When all the frost is melted, the reversing valve will reverse the direction of refrigerant flow and the normal cooling or heating cycle will continue.

**Run winding**    The winding in an electric motor that gives the power for turning the rotor during operation. It is made of larger wire than the start windings. For example, the run winding is made of the largest wire inside the motor. It provides the torque to keep the motor running after it has started. It will have the least resistance of the two windings (run and start).

# S

**Sensor**    An electronic device that undergoes a physical change or a characteristic change as the surrounding conditions change. For example, most large air-conditioning and refrigeration control systems include a temperature or humidity sensor or both. As the temperature or humidity changes, the resistance of the sensor also changes to cause the system to do a given job when the measured conditions reach a given value.

**Sequencer**    A control used to start or stop equipment in a certain sequence every time. It is normally used on multistage systems for capacity control. For example, a large commercial air-conditioning system used to give the most economical comfort possible is equipped with step controllers to increase or decrease the equipment capacity. The step controllers are operated by a modulating type motor and a matching thermostat. When the thermostat changes, even slightly, the modulating motor will also move in the direction needed by the thermostat to either increase or decrease the equipment capacity to fit the load. This will keep a part of the equipment running continuously, but it will not be operating at full capacity which will reduce the operating costs.

**Set point**    The setting at which the wanted control action occurs. For example, the set point is the position on the controller where the indicator is set. If a thermostat is set on 75°F, this is the set point of the control.

**Short circuit**    An electrical condition that happens when one part of the circuit is in contact with another part and causes all or part of the current to take the wrong path. For example, when a motor has a shorted winding, the electric current is bypassing a part of the winding and the motor will not usually run. It will also draw high current. A short circuit in the electrical wiring will usually cause a fuse to blow or a circuit breaker to trip.

**Single pole**    One set of two electrical contacts. These two contacts make or break on switch action. For example, a single pole consists of one movable and one stationary contact. When a relay or some other control causes the movable contact to touch the stationary contact, a circuit is made. When the control causes the movable contact to move away from the stationary contact, the circuit is broken.

**Single throw**    A set of contacts that make or break in only one direction of operation. For example, a set of single-throw contacts, when caused to move by another control, will either make or break the contacts. A contactor has single-throw contacts.

**Solenoid valve**    A valve used to open or close a line carrying a fluid. These valves are generally used on pump-down refrigeration systems and on gas- or oil-fired units. For example, when a furnace pilot is lit and the flame has been detected, the main gas valve—a solenoid valve—will open to let gas to the main burner on demand from the thermostat. Solenoid valves can be used to control the flow of water or any other type of fluid on demand from another controller.

**Solid-state start relay**    A compressor starting control that makes use of solid-state parts to start a compressor motor. For example, a solid-state control used to replace the usual starting relay on a compressor motor. It is commonly called a PTC relay. When the compressor first starts, the PTC material is cool and will allow full current to flow to the motor start winding. As the current flows through the PTC, it gets warmer. This increase in PTC temperature increases its resistance to the flow of current. When the motor reaches about 75 percent of its full running speed, the PTC material has enough resistance to effectively take its effect from the starting circuit and the motor runs like a PSC motor.

**Standing pilot**    A pilot burner that stays burning after it has been lit regardless of the system demands. For example, standing pilots are used mostly on older type furnaces. They are usually lit by either a person with a pilot lighter or by a pilot ignition system. When these pilots were lit they stayed lit until turned off or until something happened in the pilot circuit to cause it to be unsafe for lighting the main burner. In such case the problem must be repaired before the furnace would operate again.

**Start winding**    The auxiliary winding in an electric motor used to give the extra torque for starting and help cause some torque when the motor is running. For example, the start winding is the smaller wire of the two windings. The size of the wire causes electric current flowing through both it and the main winding to be slightly out of phase. This out-of-phase condition causes the motor to have more starting torque; however, the smaller wire cannot carry the same current load as the main winding. Therefore, it is taken out of the starting circuit. When a running capacitor is used, a small current flows through both the run capacitor and the start winding to help give the needed running torque. If the full current were allowed to flow through the start winding continuously, it would overheat and burn out.

**Starter**    A control used for switching heavy current, voltage, or both from another circuit. They are commonly used to start compressor motors and other heavy current—using components. For example, starters are usually controlled by another circuit such as the control circuit. The control circuit is usually 24 VAC; however, it may also be line voltage (120 or 240 VAC). The most common use of a starter is to start compressor motors and other motors that draw more current than normal relay contacts can withstand. Sometimes a starter will have auxiliary parts such as overloads, auxiliary contacts, or pilot-duty contacts, depending on the use and design of the equipment.

**Starting relay**    A relay used to direct the electric current to the auxiliary (start) winding in an electric motor during the starting period. For example, a potential relay is a starting relay. It has NC contacts. When the motor starts turning, the NC relay contacts let full current flow to the start winding. When the motor reaches about 75 percent of its normal running speed, the voltage generated in the start winding will energize the electromagnetic coil in the potential relay and pull the armature to the coil, thus opening the contacts and stopping the flow of current to the start winding. The relay contacts will stay open until the motor stops, then the magnetic field in the relay coil drops below the drop-out voltage and the relay contacts close, ready for the next start cycle.

**Step controller**    *See* Sequencer.

# T

**Temperature**     The measure of heat intensity. For example, the numerals on a thermometer indicate the heat intensity. If the temperature is 80°F there is more heat intensity in the air than if it was 60°F.

**Terminal**     An electrical connection such as a screw terminal or a quick connect terminal. For example, a terminal is the place where the electric wiring is connected to a part to cause it to give some function within the system. The terminal is where the wiring connects to the compressor motor.

**Terminate**     To complete an event or stop an operation. For example, *terminate* is most commonly used when referring to the defrost cycle of a refrigeration or heat pump system. When the defrost is completed, the termination control puts the system back in normal operation.

**Thermal overload element**     The alloy piece used for holding an overload relay closed; it melts when the current draw is more than the element is rated. For example, the current flowing to a motor is rated at a given amount of current. The thermal element in the overload relay is sized to allow this much current to flow through. When the current is greater than what the overload element is rated, the element will heat the relay causing it to open its contacts and break the control circuit.

**Thermal relay**     A relay actuated by the heating effect of an electrical current; it is sometimes referred to as a *warp switch*. For example, a thermal relay is commonly used on electric heating units to allow the delayed or sequencing before the heating elements can come on. This is to keep from having a large load on the electric line and cause blown fuses or tripped circuit breakers. They are usually heated by the control circuit voltage (24 VAC); however they can be heated by line voltage (120 or 240 VAC), depending on the equipment design.

**Thermostat**     A temperature-sensing control used to control the operation of heating and cooling equipment in response to the temperature of the conditioned space. For example, the room thermostat in a residential air-conditioning system controls the operation of both the furnace and the air-conditioning unit. It controls in response to the room temperature where it is located.

**Time clock**     A timing control used to control the ON-OFF cycles of equipment in response to the time of day. Time clocks have contacts that make and break circuits at a given time each day automatically. For example, the defrost control used on a commercial refrigeration system is a time clock. It has trippers that trip switches at the same time every day to either start or stop the defrost cycle. The timing is usually set to start the defrost period during a time when there is less case usage. This allows the unit time to bring the case back to normal operating temperature before the heavy use begins.

**Time-delay relay**     A control used to give a time interval between operations of components. For example, a time-delay relay is used in a refrigeration system when the compressor is stopped to allow time for the refrigerant pressures to equalize before attempting a compressor restart. This relay protects the compressor motor from overload, low voltage, or any other malfunction that may occur.

**Timer**     A clock-operated control used to open and close an electric circuit on a regular schedule. For example, a commercial air-conditioning unit is to be turned off at 7:00 P.M. and back on at 7:00 A.M. The clock has trippers screwed on the timer face at the exact times needed. When the time clock reaches these points, the electric circuit is either opened or closed in response to the time and action needed.

**Transformer**     An electrical device that transforms electrical energy from one circuit to another by electrical induction. For example, the voltage supplied to the control system of a residential air-conditioning unit is reduced from line voltage (120 or 240 VAC) to 24 VAC by the transformer.

# U

**Underwriter's Laboratories**    A testing agency whose primary function is to ensure that products are manufactured to meet specific safety standards. A listing of a product by the Underwriter's Laboratories indicates that the product was tested and met the recognized safety requirements. For example, a furnace nameplate will have a statement saying that the equipment has passed the Underwriter's Laboratories specifications for that type of furnace or heating equipment.

# V

**Voltage relay**    *See* Potential relay.

**Voltmeter**    An instrument used to measure electrical voltage. For example, a voltmeter is used to measure the voltage across two terminals, such as the contacts in a contactor. It may be used to measure the voltage across any two points in the electrical circuit.

# W

**Walk-in cooler**    A large commercial refrigerated cabinet often found in supermarkets or places for wholesale meat distribution. For example, a walk-in cooler is used to store products until they are needed. It is large enough to allow the personnel to walk inside to get the products needed. These coolers are available in many different sizes, from small, one person operated to larger ones that could hold several forklifts which are driven inside to get the products.

**Warp switch**    A thermal relay switch caused to operate by the heating effect of an electric current. For example, a warp switch is used to delay the ON cycle of heating elements in an electric furnace. Some air-conditioning units have warp switches to keep the compressor from restarting too soon after it has stopped. This is to keep from overloading the compressor motor.

**Water-cooled condenser**    A condenser cooled by water rather than air. For example, the larger air-conditioning and refrigeration equipment is usually water cooled because they use less electricity than air-cooled condensers. The water is pumped through the condenser to condense the refrigerant. It is then cooled in a cooling tower and recirculated through the condenser. Water cooling also keeps the cost of operation as low as possible.

**Water-regulating valve**    A valve used to automatically control the amount of water flowing through a water-cooled condenser to keep the needed head pressure. For example, a water-regulating valve may be operated by a modulating motor or it may be controlled directly by the head pressure. In either case it works to keep the head pressure high enough to allow the pressure drop needed by the flow control device. It may also operate to keep the head pressure low enough to cause economical and efficient operation. When the head pressure is too high, the efficiency will drop in response to the pressure rise.

**Wet-bulb temperature**    The temperature indicated on a wet-bulb thermometer; it indicates the total amount of heat in a mixture of air and water vapor. For example, the wet-bulb temperature is taken with a thermometer having its bulb covered by a wet cloth, commonly called a *sock*. Evaporation of the water from the sock causes the reading on the thermometer to be lower than the dry-bulb temperature. The wet-bulb temperature is used in calculating the relative humidity in a space.

# Index

## A

Air-conditioning thermostats, 165–83
    Anticipators, 168
    Bellows-operated, 172
    Bimetal, 165
    Circuit wiring diagram, 174
    Discharge air averaging, 180
    Heating anticipator, 169
    Fan-coil, 178
    Fan switches, 174
    Forced warm-air system—anticipated, 172
    Forced warm-air system—nonanticipated, 171
    Location, 176
    Modulating, 174
    Outdoor, 177
    Outdoor heat pump, 178
    Outdoor reset, 179
    Setting an adjustable heat anticipator, 170
    Snap action versus mercury switch, 166
    Staging, 173
    Types of room, 165
    Voltage, 177
Air-cooled condenser discharge pressure controls, 240–247
    Condenser damper modulation, 243
    Low-ambient kit, 243
    Two-speed condenser fan controls, 241
Airstats and enthalpy controllers, 186–88
Automatic control systems and basic functions, 22

## B

Balancing relay, 22, 26
Bimetal, 8
Brownout protectors and low-voltage monitors, 336–40
    Applications, 338
    Operation, 336
    Uses, 338
Bypass timers, 333–36
    Applications, 334
    Operation, 334
    Uses, 334

## C

Classification of electrical control circuits, 28–34
Combination fan and limit controls, 209–12
Combustion chamber, 4
Commercial defrost systems, 271–78
    Hot gas defrost system, 271
    Types of hot gas defrost systems, 272
Compressor motor overloads, 123
    Bimetal type, 124
    Externally mounted, 123
    Hydraulic type, 125
    Internal, 125
Contactors and starters, 87
    Coil, 89
    Contacts, 90

Definitions, 88
Operation, 88
Overload relays, 91
Pole configuration, 90
Two-speed compressor, 92–98
Control equipment, 20
Control point, 21
Control system components, 6–11
Controls process, the, 6
Controlled medium, 8
Controlled system characteristics and element, 19–24
Controlled systems, 19
Controlled variable, 7, 11, 20
Controller, 8
Corrective action, 21
Current-sensing relay, 346–47
Operation, 346
Uses, 346
Cycling, 21

**D**

Delay-on-break adjustable solid-state timers, 327–30
Applications, 327
Operation, 327
Uses, 327
Delay-on-make adjustable solid-state timer, 323–27
Applications, 324
Operation, 323
Uses, 324
Desirable value, 21
Deviation, 21
Differential gap, 21
Discharge gas temperature protector, 320–23
Applications, 320
Discharge pressure control (cooling tower), 235–40
Cooling tower fan control, 235
Motor-actuated three-way water-regulating valve, 238
Pressure-actuated water-regulating valve, 237
Temperature-actuated modulating water valves, 236
Three-way water-regulating valve, 238
Disturbance sensing element, 6
Domestic refrigeration controls, 279–281
Defrost limiter, 280

**E**

Economizer control package, 364
Economizer use, 364
Electromagnetic relays, 100
Electronic thermostats, 291–93
Saverstat, 291
Enthalpy controllers, 187

**F**

Fan controls, 201–05
Electrically operated, 202
Temperature-actuated, 201
Fan relays and fan centers, 101–102
Final control element, 8, 22
Floating controls, 45–50
Floating control combinations, 49
Floating control equipment, 46
Floating control operation, 47
Fundamentals of electric control circuits, 24–28

**G**

Gas valves, 188–96
Automatic spark-ignition, 192
Definitions, 189
Heat motor, 190
High-low-off fire, 192
Magnetic diaphragm, 190
Redundant, 190

**H**

Heat pump defrost controls, 281–88
Air-pressure differential (demand) defrost systems, 281
Outdoor coil temperature, 282
Time, 284
Time and temperature, 285
Heating relay, 103
Electric, 104
Humidistats, 184–86
Controls and sensors, 184
Hydronic cooling controls, 231–35
Chiller protector, 231
Fan center, 233
Flow switch, 232
Freezestat/low-limit control, 232

Hydronic heating controls, 218–31
    Automatic changeover control, 228
    Boiler pressure control, 223
    Boiler temperature control (aquastat), 221
    Combination feeder and low water cutoff, 219
    Fan center, 228
    High-limit control, 224
    High-low-off fire valves, 225
    Liquid-level float switch, 227
    Water level control, 218

# I

In-contacts and out-contacts,
    22, 26
Integrated furnace control modules, 365
    Incorrectly wired modules, 366
    Operation, 365

# L

Laboratory 1, Control system components, 379
    2, Fundamentals of electric control circuits, 381
    3, Starters and contactors, 383
    4, Electromagnetic relays, 385
    5, Thermal relays, 387
    6, Potential starting relays, 389
    7, Sizing potentials starting relays, 392
    8, Pressure controls, 394
    9, Temperature controls, 396
    10, Fan controls, 399
    11, Limit controls, 401
    12, Discharge gas temperature protector, 403
    13, Delay-on-make adjustable solid-state timers,
      405
    14, Current-sensing relay, 408
    15, Solid-state crankcase heaters, 410
Lag, 21
Limit controls, 205–09
    Adjustment, 205
    Air switches, 207
    High-limit controller, 207
Line monitors, 340–42
    Applications, 340
    Uses, 340
Line-voltage, 25
Line-voltage two-position control applications,
    61–63

Line-voltage two-position control combinations,
    62
Line-voltage two-position equipment, 62
    Controllers, 62
    Motor units, 62
Line-voltage two-position operation, 62
Lockout relay, 102, 288–90
    Circuit, 288

# M

Manipulated variable, 8, 20
Modulating motors, 247–52
    Modulating spring return motor, 247
    Reversing two-position and proportional motors,
      248
    Two-position spring-return motors, 248
Motor starting devices, 112
    Amperage (current), 114
    Hard-start kit, 116
    Kickstart TO–5®, 117–122
Multiposition (multistage) control, 14

# O

Off-delay-on-delay timers, 330–33
    Applications, 330
    Operation, 330
    Uses, 331
Offset, 21
Oil burner controls, 212–18
    Burner-mounted combustion thermostat, 214
    Burner safety controls, 215
    Ignition timing, 213
    Kwik-sensor combination oil burner-hydronic
      control, 215
    Magnetic valves, 215
    Protectorelay (stack mounted) control, 212
Oil-failure controls, 145–57
    Operation, 155
ON-OFF (line voltage) control operations, 30
ON-OFF (two-position) control, 13
ON-OFF two-position (line-voltage) applications,
    29
ON-OFF two-position (line-voltage) control
    combinations, 31
    High-limit control, 31
    Low-limit control, 32

Two-stage control, 32
Unit heater control, 31
ON-OFF two-position (line-voltage) equipment, 29
Controllers, 30
Motor units, 30
Relays, 30
Solenoid valves, 30
ON-OFF two-position (low-voltage) controls, 34–38
ON-OFF (low-voltage) control applications, 34
ON-OFF (low-voltage) control combinations, 36
Solenoid valve control, 36
ON-OFF (low-voltage) control equipment, 35
Relays and motors, 35
Valves, 35
ON-OFF (low-voltage) control operation, 35
Heat anticipation, 36
Overload relays, 91
Overshoot, 21

## P

Pilot safety devices, 196–200
Sail switch, 197
thermocouple, 196
Plug-in time delay relays, 342–45
Applications, 344
Operation, 343
Uses, 343
Potentiometer, 25
Pressure controls, 140–54
Definitions, 141
High-pressure sensing for condenser fan control, 151
High-pressure sensing for high-limit control, 149
Operation, 141
Proportional pressure controls, 152
Suction-pressure sensing for alarm control, 147
Suction-pressure sensing for capacity control, 145
Suction-pressure sensing for low-pressure limit control, 148
Suction-pressure sensing for pumpdown control, 145
Primary element, 22
Proportioning band, 21

Proportioning (modulating) control, 15
Complex variations of the, 15
Proportioning (modulating) control applications, 64–80
Proportioning control combinations, 68
Low-limit control, 68–72
High-limit control, 72
Manual and automatic switching, 74
Manual minimum positioning, 78
Reversing for heating and cooling control, 74
Recycling step controllers, 78
Sequence control, 76
Transfer of motor from one thermostat to another, 74
Transfer of thermostat from one motor to another, 76
Two-position limit controls, 72–74
Unison control, 77
Proportioning control equipment, 64
Controllers, 64
Motors, 65
Proportioning control operation, 66
Balancing relay operation, 66
Motor balancing potentiometer, 66–68

## R

Refrigeration temperature controls, 158–65
Evaporator sensing for temperature control, 162
Ice thickness sensing for temperature control, 163
Methods of, 159–65
Product-temperature sensing for temperature control, 161
Space and return-air sensing for temperature control, 159
Reversing (four-way) valve, 133–40
Operation, 133
Touch test chart, 136–39

## S

Self-contained control system, 4–6
Sensing element, 4
Set point, 7, 21
Six-step troubleshooting procedure, 367–78
Logical approach, 368
maintenance (upkeep), 367

Six-step procedure, 368
Step 1: Symptom recognition, 369
Step 2: Symptom clarification, 370
Step 3: Listing of probable faulty functions, 371
Step 4: Localizing the faulty function, 372
Step 5: Localizing trouble to the circuit, 374
Step 6: Failure analysis, 375
Troubleshooting, 367
Solenoid valves, 128–132
Solid-state crankcase heaters, 362–64
Checking the heater, 363
Operation, 362
Solid-state fan controls (condensers), 293–96
Discharge pressure controls, 293
Solid-state gas burner ignition controls, 296–315
Application Guidelines, 302
Automatic gas burner ignition system, 297
Direct electronic ignitors, 299
Flame rectification, 300
Flame rectification sensor, 312
Hot surface ignitors (silicon carbide HSI), 303
Intermittent ignition systems, 297
Pilot relight control, 312
Principles of operation, 299
Retrofit, intermittent pilot, gas burner ignition systems, 312
Spark plug and sensor, 305
Sequence of operation, 301
Sensing methods, 299
Spark ignitor, 310
The 100-percent lockout module, 310
Solid-state heat pump controls, 358–59
Operation, 359
Solid-state heat pump defrost controls, 353–55
Operation, 353
Solid-state humidistats, 360
Operation, 360
Solid-state type controllers, 360
Solid-state motor protectors, 355–58
Operation, 355
Troubleshooting, 356
Solid-state motor starting devices, 347–52
Operation, 348
Positive-temperature-coefficient starting device, 350
Solid-state hard-start kit, 349

Solid-state oil burner controls, 315–19
Burner safety controls, 316
Cadmium sulphide flame detector, 316
Kwik-sensor cad-cell (protectorelay) burner controls, 315
Kwik-sensor combination oil burner-hydronic control, 315
Step controllers, 253–55
Auxiliary (end) switches, 254

**T**

Temperature controls, 157–84
Thermal relays, 107
Fan, 107
Heating, 108
Time-delay sequencing, 108
Timers and time clocks, 256–70
Definitions, 256
Defrost time clocks, 261
Multicircuit defrost controls, 264
Seven-day four-pole switch with carryover mechanism, 263
The 24-hour time switch, skip-a-day series, 264
Time-initiated pressure-terminated control, 263
Time-initiated temperature-terminated control, 261
Time-initiated time-terminated control, 261
Wiring diagrams, 265–69
Transducers, 7
Transformer, 2
Transformers, 81
Phasing, 83–86
Purpose, 81
Rating, 82
Sizing, 82
Two-position, spring return control applications, 38–45
Two-position, spring control combinations, 42
High- and low-limit two-position, spring control circuits, 42
Two controllers and one relay, 43
Two-position, spring return control equipment, 38
Controllers, 38
Relays, 38
Two-position, spring control operation, 39

## U

Unidirection control applications, 51
Unidirection control equipment, 51
    Controllers, 51
    Motor units, 51
    Relays, 51

Unidirection control operation, 52–56
Unidirectional control combinations, 56
    Controlling two motors with one controller, 57
    Low-limit control, 57
    Unidirectional controller and universal relay,
       58–60